Forgotten Calculus

A Refresher Course

With Applications to Economics and Business
(and the optional use of the graphing calculator)

Third Edition

Barbara Lee Bleau, Ph.D.

Formerly, Jefferson-Pilot Professor
of Management Science
Division Head
Business and Economics
Pfeiffer University
Misenheimer, North Carolina

BARRON'S

To MAЯY LOU
Again, your concern, support, and encouragement made my writing hours more enjoyable.

All inquiries should be addressed to:
Barron's Educational Series, Inc.
250 Wireless Boulevard
Hauppauge, New York 11788
http://www.barronseduc.com

Library of Congress Catalog Card No. 2001052751

ISBN-13: 978-0-7641-1998-9
ISBN-10: 0-7641-1998-2

Library of Congress Cataloging-in-Publication Data

Bleau, Barbara Lee.
    Forgotten calculus : a refresher course with applications to economics and business and
the optional use of the graphing calculator / Barbara Lee Bleau.—3rd ed.
        p.   cm.
    Includes bibliographical references and index.
    ISBN 0-7641-1998-2
    1. Calculus.   I. Title.
QA303.2 .B47   2002
515—dc21   2001052751

PRINTED IN THE UNITED STATES OF AMERICA

19 18 17 16 15 14 13 12 11 10 9 8 7

Contents

OTHER BOOKS CITED IN THE TEXT

Barnett, Raymond A., Michael R. Ziegler, and Karl E. Byleen, *Applied Calculus for Business, Economics, Life Sciences, and Social Sciences*, Seventh Edition, Prentice Hall Publishing Company.

Bleau, Barbara Lee, *Forgotten Algebra*: *A Refresher Course*, Second Edition, Barron's Educational Series, Inc.

Harshbarger, Ronald J., and James J. Reynolds, *Mathematical Applications for the Management, Life, and Social Sciences*, Sixth Edition, Houghton Mifflin Company.

Hoffmann, Laurence D., and Gerald L. Bradley, *Calculus for Business, Economics, and the Social and Life Sciences*, Seventh Edition, McGraw Hill Book Company.

Waner, Stefan, and Steven R. Costenoble, *Applied Calculus*, Second Edition, Brooks/Cole Company.

Preface

Welcome to the wonderful world of calculus! If you want to develop an intuitive understanding of calculus as well as an appreciation of its usefulness in solving managerial, business, economic, and social science problems, then this third edition was written for you.

This edition differs from the usual calculus text in one major way. There is no attempt to develop, derive, or prove any of the formulas, principles, or operational techniques of the calculus. These are simply presented, explained in detail, and then applied in examples. In other words, this text is aimed at the uses of calculus, not its development as a subject in mathematics.

Although the title, *Forgotten Calculus*, implies that this is not the reader's first experience with the subject, such need not be the case. The book is suitable as a self-teaching guide for anyone without the slightest previous knowledge of what calculus is all about. Or the book can be used by persons wishing to "brush up" their calculus skills, regardless of where the need originated.

Calculus is a significant part of the mathematical hierarchy. As such, it is essential that you learn to do the math manually, that is, with paper and pencil. Notwithstanding, graphing calculators have become a way of life for many people and I want to encourage the use of any new technology available to us. All the material in this text is designed to be done manually. But at the end of most units, there is an optional section entitled "Calculus and the Calculator." Explanations are provided for using the TI-83 graphing calculator. I bought mine, used, over the Internet, and for a modest price. It does everything I need. I use it mostly when the numerical calculations are too "messy" to be done by hand and for checking my work. Optional exercises and solutions are included as well. Remember, though, your time and energy is best spent learning calculus, rather than learning how to use a graphing calculator.

Forgotten Calculus was designed as a self-pacing workbook. Each unit provides explanations and includes numerous examples, problems, and exercises with detailed solutions to facilitate self-study. Special emphasis has been placed on the techniques you will need to solve practical problems in the social and managerial sciences.

January 2002   Barbara Lee Bleau McKinley, Ph.D.

UNIT 1

Functions

The early units of this book contain a review of topics from algebra. This is helpful if you have "forgotten" algebra. However, if you do know algebra, you might prefer to jump ahead and start with either Unit 6 or Unit 9.

In this first unit of your workbook, you will learn the meaning of function, a term that we will be using throughout the book. When you have completed the unit, you will be able to read and to use functional notation and identify the domain.

We are going to start with a picture.

This is a picture of a machine. We can define our machine to do most anything. Since this is a workbook about calculus, our machine is going to do calculations with numbers. Suppose that the numbers 1, 2, and 3 are located as shown and that the machine takes a number, squares it, and adds 1.

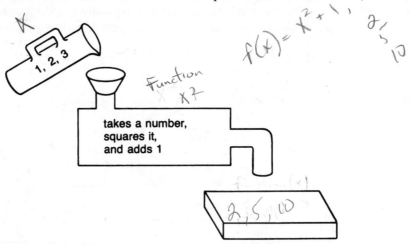

It should be clear what happens as each number is put into the machine.

If 1 is entered, the machine takes the number, squares it, adds 1, and out comes 2.

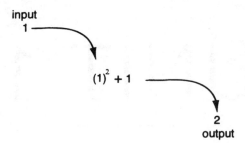

In a similar fashion, if 2 had been put in, the output would have been 5. If 3 had been put in, the output would have been 10. In the machine drawing, write the numbers 2, 5, and 10 in the box.

The values that can be put into the machine are referred to as the domain. The **domain** is the set of all possible values that can be used as input. In this example, the domain is the set of numbers 1, 2, and 3, or $D = \{1, 2, 3\}$. Only those three numbers can be put into this particular machine. Why? Because that is the way we created it on page 1.

The values that come out are referred to as the range. The **range** is the set of all possible values that are output. In the picture, the range is the set of numbers 2, 5, and 10, or $R = \{2, 5, 10\}$.

The description "takes the number, squares it, adds 1" is referred to as the **rule**.

Notice our picture has three parts: the domain, the range, and the rule.

The domain, often denoted by x, is the **independent variable**, whereas the range, often denoted as y, is the **dependent variable**. A way to remember the distinction between the two is that we may select any value from the domain to put into the machine; but once a specific value of x is selected, such as 3, the output, or the y-value, is dependent upon it.

By using x to denote the independent variable, it is now possible to rewrite the rule algebraically, $x^2 + 1$. At the same time, we will give the machine the name f. Typically f is the most common name used, although g and h are popular, followed by some of the Greek letters, such as α, β, and γ.

The picture has been drawn again to include all the terminology.

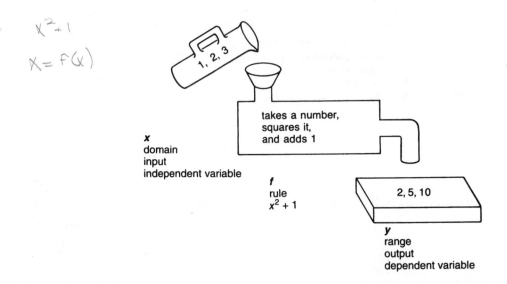

Definition: A **function** is a rule that assigns to each element in the domain one and only one element in the range.

Like the picture, the definition refers to three parts: domain, range, and rule. The key concept, though, is the *one and only one* phrase. For each value put in, only a single value may come out if the expression is to be a function. Otherwise the expression is classified as a **relation**.

The machine named *f* is a function because for each value put in only a single value came out.

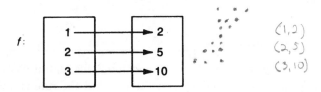

Rather than having to draw pictures each time, we can combine all of the above information into a single statement using functional notation.

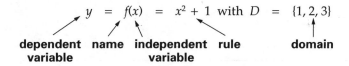

$$y = f(x) = x^2 + 1 \text{ with } D = \{1, 2, 3\}$$

dependent   name   independent   rule   domain
variable   variable

The above statement is read "*y* equals *f* of *x* equals *x* squared plus 1 with domain, *D*, equal to the set of numbers 1, 2, and 3."

Caution: $f(x)$ is read "*f* of *x*" and does not mean multiplication. It is used to indicate that the name of the function is *f* and that the variable inside the parentheses, in this example *x*, is to be the independent variable.

A variation on the above notation is:

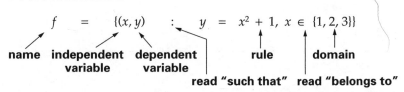

$$f = \{(x, y) : y = x^2 + 1, x \in \{1, 2, 3\}\}$$

name   independent   dependent   rule   domain
            variable   variable

read "such that"   read "belongs to"

The above statement is read "*f* equals the set of all ordered pairs *x* and *y*, such that *y* equals *x* squared plus 1 and *x* belongs to the set of numbers 1, 2, and 3." With this notation, the name of the function is first and the dependent variable is always the last one of the grouped variables listed inside the parentheses, in this case *y*.

It is time to stop and try some questions. Using *f* as defined by the picture, answer the following questions.

1. If $x = 2$, what is y?   Answer: 5

2. If $x = 3$, what is y?   Answer: 10

3. If $y = 2$, what is x?   Answer: 1

The confusing part occurs when the same questions are asked using functional notation. Here are some examples. Continue to use *f* as defined earlier.

EXAMPLE 1

Find $f(2)$.

Solution: $f(2)$, read "f of 2," is asking, If $x = 2$, what is y? It is exactly the same as question 1 above, and the answer is 5, or $f(2) = 5$.

EXAMPLE 2

Find $f(3)$.

Solution: $f(3)$ is asking the same thing as question 2 above, but using function notation. The question is asking, If the independent variable is 3 and the rule is f, what is the value of the dependent variable? As in question 2 above, the answer is 10, or $f(3) = 10$.

EXAMPLE 3

Find $f^{-1}(2)$.

Solution: $f^{-1}(2)$ is read "f inverse of 2." The question is asking, If the dependent variable is 2, what is x? As in question 3 above, the answer is 1, or $f^{-1}(2) = 1$.

Here are a few problems for you to try. Use f as defined in the picture on page 1 to answer each of the following.

Problem 1

Find $f(1)$, $f^{-1}(10)$, and $f^{-1}(5)$.

Solution:

Answers: 2; 3; 2

Problem 2

Find $f(4)$. Be careful; the answer is not 17.

Solution:

Answer: $f(4)$ is undefined, because 4 is not an element of the domain. f is defined only for the numbers 1, 2, and 3, which was an arbitrary decision on my part by defining the machine for only the numbers 1, 2, and 3.

Problem 3

Imagine that you are in a class with other students named Smith, Jones, and Howard. The machine is defined to assign grades at the conclusion of the course. If the choice were yours, what would the range be?

Solution:

Answer: Many answers are possible. The most popular answer is that the range would contain only the letter grade A. This would mean that everyone in the class would receive an A.

Problem 4

Is the example used in Problem 3 a function?

Solution:

Answer: Yes. For each name entered, only one grade is given. It is a function even if everyone might be assigned the same grade.

Problem 5

Suppose that the example in Problem 3 was reversed. In other words, when A was entered, the output would be the names of the students receiving A grades. Would the example then be a function?

Solution:

✓ Answer: No. For the letter grade A, four names would be printed. That violates the concept
of "for each x there is a single y value."

Unless specified, the **domain** of a function is all values for which the dependent variable is defined and
real. This is usually the set of real numbers, with two exceptions that will be discussed later. Remember the
real numbers are all the counting numbers, zero, whole numbers, positive and negative integers, fractions,
rationals, and irrational numbers.

EXAMPLE 4

$3(2) + 1 = 7$
$3(-5) + 1 = -14$
$3(3x) + 1 = 9x + 1$
$3(a+b) + 1 = 3a + 3b + 1$

Given: $y = g(x) = 3x + 1$.
Identify each part of the expression and
find: $g(2)$, $g(-5)$, $g(a)$, $g(3x)$, $g(a + b)$, and $g^{-1}(7)$.

Solution: y is the dependent variable.

x is the independent variable.

g is the name of the function.

The rule is: 3 times the number plus 1.

The domain is assumed to be the set of all real numbers since nothing to the contrary is
specified.

$g(x)$ $= 3x$ $+ 1$

$g(2)$ $= 3(2)$ $+ 1 =$ $6 + 1 =$ 7

$g(-5)$ $= 3(-5)$ $+ 1 = -15 + 1 = -14$

$g(a)$ $= 3(a)$ $+ 1 = 3a + 1$

$g(3x)$ $= 3(3x)$ $+ 1 = 9x + 1$

$g(a + b) = 3(a + b) + 1 = 3a + 3b + 1$

$g^{-1}(7)$ $= 2$ because, as shown above, when $y = 7$, the x value is 2.

Try the following problem before continuing.

Problem 6

Given: $y = f(x) = 5 + 2x$.
Find: $f(0)$, $f(2)$, $f(1/2)$, $f(-9)$, $f(x - 1)$, and $f^{-1}(9)$.

Solution: $f(x) = 5 + 2x$

$f(0) =$

$f(2) =$

$f(1/2) =$

$f(-9) =$

$f(x - 1) =$

$f^{-1}(9) =$

Answers: 5; 9; 6; −13; 3 + 2x; 2

The next example is a function defined **piecewise** by three different function rules. Each function rule is associated with a different nonoverlapping domain. The choice of which one to use is dependent upon the value of x.

EXAMPLE 5

Given: $h(x) = \begin{cases} x^2 + 1 & \text{if } x > 0 \\ 10 & \text{if } x = 0 \\ 2x & \text{if } x < 0 \end{cases}$

Find: $h(2), h(3), h(0),$ and $h(-1)$.

Solution: $h(2) = (2)^2 + 1 = 4 + 1 = 5$ The top rule was used because 2, the x value, is greater than 0.

$h(3) = (3)^2 + 1 = 9 + 1 = 10$

$h(0) = 10$ The middle rule was used because $x = 0$.

$h(-1) = 2(-1) = -2$ The bottom rule was used because −1, the x-value, is less than 0.

Although functions commonly are expressed in algebraic form, sets and graphs also can be used.

EXAMPLE 6

Given: $F = \{(1, 2), (1, 3), (2, 4)\}$.
Find: a.   The domain.
            b.   The range.
            c. $F(2)$.
            d.   Is F a function?
            e. $F^{-1}(2)$.

Solution: F is the name. The ordered pair (1, 2) means that, when $x = 1, y = 2$. Similarly, (1, 3) means that, when $x = 1, y = 3$, and (2, 4) means when $x = 2, y = 4$.

a.   The domain is the set of all x values: $D = \{1, 2\}$. Note: there is no need to write the 1 twice.

b   The range is the set of y values: $R = \{2, 3, 4\}$.

c. $F(2) = 4$. The notation is asking, When x is 2, what is y?

d.   No, F is not a function. When x is 1, there are two different values for y, 2 and 3.

e. $F^{-1}(2) = 1$. The notation is asking, When the y value is 2, what is x?

Problem 7

Does the graph at the right specify a function?

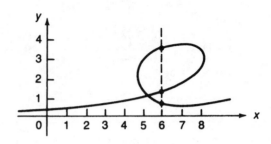

Solution:

Answer: No; for example, when $x = 6$, there are three values for y.

Notice that, if you are able to draw a vertical line that intersects the graph at more than one point, then the graph does not represent a function. This is referred to as the **vertical line test**.

FUNCTIONS WITH MORE THAN ONE VARIABLE

Thus far we have looked at functions with a single independent variable. A **bivariate function** has two independent variables. Think of a bivariate function as having two separate sets of input values rather than just one.

EXAMPLE 7

Given: $z = h(x, y) = 3x + y^2$.
Identify each part of the expression and
find: $h(2, 3)$ and $h(5, -1)$.

Solution: $z \quad = \quad h(x, y) = 3x + y^2$

        dependent   **two**   └─ **rule: 3 times the first number plus the second number squared**
        variable   **independent**
             name **variables**

$$z = h(2, 3) \ = 3(2) + (3)^2 \ = \ 6 + 9 = 15$$

$$z = h(5, -1) = 3(5) + (-1)^2 = 15 + 1 = 16$$

A **multivariate function** has more than two independent variables. The following example has four independent variables. Subscripts can be used rather than different letters; x_1 is read "x sub 1." $x_1, x_2, x_3,$ and x_4 denote different variables.

EXAMPLE 8

Given: $y = f(x_1, x_2, x_3, x_4) = 2x_1 + x_2x_3x_4$.
Find: $y = f(23, -1, 10, 0)$.

Solution: $y = f(x_1, x_2, x_3, x_4) = 2x_1 + x_2x_3x_4$

$y = f(23, -1, 10, 0) = 2(23) + (-1)(10)(0) = 46 + 0 = 46$

LIMITATIONS ON THE DOMAIN

When a function is expressed in algebraic form, the domain is seldom mentioned. The domain, remember, is the set of values of the independent variable for which a function can be evaluated. In most cases, this is the set of all real numbers. There are two exceptions.   *or is defined*

First Exception

If the function contains a quotient, all values of the independent variable that would result in the denominator equaling a value of 0 must be excluded from the domain. The reason is that division by 0 is undefined.

EXAMPLE 9

Given: $f(x) = \dfrac{3}{x-2}$, find the domain.

Solution: The domain is the set of all real numbers except 2 because 2 would result in the denominator equaling 0.

EXAMPLE 10

Given: $g(x) = \dfrac{1}{x}$, find the domain.

Solution: The domain is the set of all real numbers except 0.

EXAMPLE 11

Given: $G(y) = \dfrac{y-7}{3}$, find the domain.

Solution: The domain is the set of all real numbers. There is no number that must be excluded because the denominator will never be 0.

EXAMPLE 12

Given: $H(x) = \dfrac{3}{(x-2)(x-1)(x+6)}$, find the domain.

Solution: The domain is the set of all real numbers except 2, 1, and –6 because each of those would result in the denominator assuming a value of 0.

Second Exception

If an even root is being taken, the independent variable must be defined so that the quantity under the radical sign is 0 or positive, that is, nonnegative.

EXAMPLE 13

Given: $f(x) = \sqrt{x}$, find the domain.

Solution: The quantity under the radical sign must be 0 or positive; therefore the domain is the set of all x such that $x \geq 0$.

EXAMPLE 14

Given: $g(x) = \sqrt{x-5}$, find the domain.

Solution: To say that the quantity under the radical sign must be 0 or positive is the same as saying $x - 5 \geq 0$
             or $x \geq 5$; therefore the domain is the set of all x such that $x \geq 5$ or, using set notation: $D = \{x \mid x \geq 5\}$.

EXAMPLE 15

Given: $h(x) = \sqrt{x^2 + 1}$, find the domain.

Solution: The rule states that the quantity under the radical sign must be 0 or positive (nonnegative). Since $x^2 + 1$ is always positive, no numbers need be excluded. The domain is the set of reals.

EXAMPLE 16

Given: $t(x) = \sqrt[3]{x}$, find the domain.

Solution: This example requires finding the cube root, which is an odd root. The second exception does not apply. The domain is the set of reals.

Calculus is an essential part of the mathematical hierarchy. As such, it is essential that you learn to do the math manually, that is, with paper and pencil. Notwithstanding, graphing calculators have become a way of life for many people, and I want to suggest that we use the technology available to us. All the material in this text is designed to be done manually. But at the end of most units, there is an optional section entitled "Calculus and the Calculator." Explanations are provided for using the TI-83 graphing calculator. I bought mine, used, over the Internet, and for a modest price. It does everything I need. I use it mostly when the numerical calculations are too "messy" to be done by hand and for checking my work. Optional exercises for practice and solutions are included as well. Remember, though, I want you to spend your time and energy learning calculus rather than learning how to use a graphing calculator.

CALCULUS AND THE CALCULATOR (Optional)

So that we start at the same place, after pressing the (**ON**) key,

   press (**CLEAR**) (**MODE**) (**CLEAR**) and

   press (**2nd**) (**TblSet**) to view the table setup menu.

   Now assign   TblStart=0   by pressing (**0**) (**ENTER**).

   ΔTbl=1   by pressing (**1**) (**ENTER**).

   Indpnt: Ask   by highlighting Ask and pressing (**ENTER**).

   Depend: Auto by highlighting Auto and pressing (**ENTER**).

You should see the following.

To return to the home screen,

press (**2nd**) (**QUIT**) or (**CLEAR**).

```
TABLE SETUP
 TblStart=■
 ΔTbl=1
Indpnt: Auto ASK
Depend: AUTO Ask
```

Creating a Table of Values for a Function

EXAMPLE 17

Given: $y = f(x) = 1.3x^3 - 4x^2 + 5$.

Find: $f(14)$, $f(7.5)$, $f(1.2)$, $f(0)$, $f(-0.023)$, $f(-5)$, and $f(-8.1)$.

Solution: This example is the same as what you have been doing except that the function has more terms and the x-values include decimals; thus manual calculations would be tedious and time-consuming. We will do it on the calculator.

To enter the function, press the following keys.

(**Y=**) (**CLEAR**) (**1**) (**·**) (**3**) (**X,T,θ,n**) (**∧**) (**3**) (**−**) (**4**) (**X,T,θ,n**) (**∧**) (**2**) (**+**) (**5**) (**ENTER**)

You should have the following.

To find the desired *y*-values,
press (2nd) (TABLE).

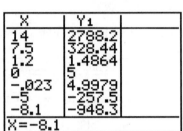

To find $f(14)$, press (1) (4) (ENTER).

To find $f(7.5)$, press (7) (•) (5) (ENTER).

Repeat by entering 1.2 and 0.

To find $f(-0.023)$, press ((–)) (0) (•) (0) (2) (3) (ENTER).

Note: The ((–)) key is used to denote a negative number. An error message will appear if the subtraction key (–) is used by mistake. Finish the example by entering –5 and –8.1.

You should see the following.

X	Y₁
14	2788.2
7.5	328.44
1.2	1.4864
0	5
-.023	4.9979
-5	-257.5
-8.1	-948.3

X=-8.1

===

EXAMPLE 18

Given: $y = g(x) = \sqrt{(2x - 3)} + \pi x$.
Find: $g(10)$, $g(3.4)$, and $g(-8)$.

Solution: To enter the function, press the following keys.

(Y=) (CLEAR) (2nd) (√) (2) (X,T,θ,n) (–) (3) ()) (+) (2nd) (π) (X,T,θ,n) (ENTER)

To find the desired *y*-values, press (2nd) (TABLE).

To find $g(10)$, press (1) (0) (ENTER).

Note: If an ERROR message appears for the *y*-value, go back and check your function to be sure the right parenthesis was keyed after the 3. If the) was omitted, the $x\pi$ term was included under the square root symbol rather than as a separate term.

Finish by entering 3.4 and –8.

Answers: 35.539; 12.631; ERROR

Comment: The ERROR message associated with –3 occurs because –3 is not in the domain for $g(x)$. Substituting –3 for *x* causes the quantity under the square root to be negative, which is undefined in the set of real numbers.

You should now understand the basic concept of a function. Given a function, you should be able to identify the domain, range, rule, name, and independent and dependent variables. These basic terms will be used continually throughout the following units.

You should also be able to read and evaluate questions written in functional notation, including those for bivariate and multivariate functions as well as functions defined piecewise.

Remember also that, if the domain for a function f is not specified, it is assumed to be the set of real numbers with two exceptions:

1. If f contains a quotient, the domain must exclude any numbers that result in the denominator being 0.

2. If f contains a radical sign for an even root, the domain must be only those values that result in the quantity under the radical sign being 0 or positive (nonnegative).

Before beginning the next unit you should do all the following exercises. When you have completed them, check your answers against those at the back of the book. Those marked with a C are optional calculator problems.

EXERCISES

Given: $f(x) = x^2 - 2x + 3$, find each of the following:

1. $f(0)$

2. $f(10)$

3. $f(-2)$

4. $f(a)$

5. $f(x + 1)$

6. $f(-x)$

7.   The domain for f

Given: $g(x) = 1 / (x - 3)$, find each of the following:

8. $g(5)$

9. $g(c)$

10. $g(a + 3)$

11.   The domain for g

Given: $q = h(p) = \sqrt{13 - p}$, find each of the following:

12.   The letter that denotes the independent variable

13.   The domain for h

14. $h(12)$

15. $h(13)$

16. $h(-3)$

17. $h(2)$

18.   Given $\alpha(x, y, z) = x - 3z + xy$, find $\alpha(5, 3, -2)$.

Given: $H(x) = \begin{cases} x + 1 & \text{if } x \le 3, \\ x - 1 & \text{if } x > 3, \end{cases}$ find:

19. $H(7)$

20. $H(0)$

21. $H(-4)$

22. $H(3)$

Given: $G = \{(5, 6), (5, 7), (8, 5)\}$

23.   State the domain for G.

24.   State the range for G.

25.   Is G a function?

26.   Given the function $F = \{(3, 4), (2, 1)\}$, find the range.

Given the function $w = z(m) = 3m^2 - 10$, find each of the following:

27.   The letter that denotes the independent variable

28. $z(-2)$

29. $z(x - 1)$. Be sure to simplify your answer.

30.   The domain for the function

C31. Given the function $y = f(x) = 2x^2 + 3x - 7$, find $f(24)$, $f(3.01)$, and $f(-0.25)$.

C32. Given the function $y = f(x) = -0.6x^5 - 4x^3 + 2x$, find $f(5)$, $f(1.02)$, $f(0)$, $f(-2.3)$, and $f(4.1)$.

If additional practice is needed:

  Barnett, Ziegler, and Byleen, pages 18–19, problems 1–80
  Harshbarger and Reynolds, pages 85–87, problems 1–22
  Hoffmann and Bradley, page 10, problems 1–21
  Waner and Costenoble, pages 11–12, problems 1–20

UNIT 2

Functions: Composite and Applications with Restricted Domains

In Unit 2 several more topics related to functions and notation will be explained. You will learn what is meant by a composite function. When you have completed the unit, you will be able to evaluate composite functions. You also will be able to interpret functional notation and to specify restricted domains for application problems.

Cost functions with both fixed and variable costs, revenue functions, and profit functions will be introduced in the application problems at the end of the unit.

COMPOSITE FUNCTIONS

Operations on functions occur when we take existing functions and start combining them to form new functions. The new function sometimes is referred to as a **composite** function.

We will look at some easy examples first.

EXAMPLE 1

Given: $f(x) = x^2 + 1$ and $g(x) = \dfrac{x}{2}$.

Find: $f(7) + g(10)$.

Solution: Now the reason for naming functions should be apparent. The name serves to indicate which rule is to be used. Obviously the functions must have different names.

The rule for f: takes a number, squares it, and adds 1.

The rule for g: takes a number, divides it by 2.

15

$f(7)$ indicates that 7 is to be put into the function f, and $g(10)$ indicates that 10 is to be put into the function g.

$$f(7) + g(10) = \quad f(7) \quad + g(10)$$
$$= [(7)^2 + 1] + [10/2]$$
$$= \quad 50 \quad + 5$$
$$= \quad 55$$

EXAMPLE 2

Given: $f(x) = x^2 + x - 5$ and $g(x) = \sqrt{x}$.

Find: $\dfrac{g(0)}{f(3)}$.

Solution: $g(0) = \sqrt{0} = 0$ and $f(3) = (3)^2 + 3 - 5 = 7$

Therefore $\dfrac{g(0)}{f(3)} = \dfrac{0}{7} = 0$

EXAMPLE 3

Given: $f(x) = x^2 + 2x - 1$ and $g(x) = 2x^2 + 5x$.
Find: $h(x) = g(x) - 2f(x)$.

Solution: $h(x) = g(x) - 2f(x)$ means 2 times the function f

$$= (2x^2 + 5x) - 2(x^2 + 2x - 1)$$

$$= 2x^2 + 5x - 2x^2 - 4x + 2$$

$$= x + 2$$

Note that h is a new function, created by combining g and f.

The rule for h: takes the number and adds 2.

Before we look at more complicated composite functions, Problem 1 introduces additional notation used when combining various functions.

Problem 1

Given: $f(x) = 3x + 1$ and $g(x) = 4x^2$.
Find: a. $(f + g)(5)$.
           b. $(f - g)(5)$.
           c. $(fg)(5)$.
           d. $\left(\dfrac{f}{g}\right)(5)$.

Solution: The function notation being used here is nothing more than a condensed version from that used in earlier examples. I will rewrite each part to illustrate the meaning of the notation and then let you finish the problem for practice.

a. $(f + g)(5) = f(5) + g(5) = ?$

b. $(f - g)(5) = f(5) - g(5) = ?$

c. $(fg)(5) = f(5)g(5) = ?$

d. $\left(\dfrac{f}{g}\right)(5) = \dfrac{f(5)}{g(5)} = ?$

Answers: a. 116.   b. –84.   c. 1600.   d. $\dfrac{16}{100} = \dfrac{4}{25}$.

Thus far our new functions have been formed by adding, subtracting, multiplying, or dividing functions. A *composite* function exists when one function is considered a function of another function. To illustrate this idea, we will draw another picture, this time with two function machines.

The first function, g, takes the number and doubles it.

The second function, f, takes the number and adds 5.

Find the value of y if first 10 is put into the function g, and then that output is put into the function f.

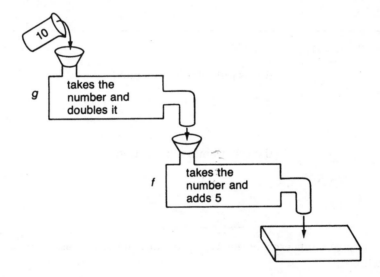

With the help of the drawing, the answer should be straightforward to calculate. The answer is 25. When the same question is asked using functional notation, which can be written in two different ways, the issue becomes confused:

Given $y = f(x) = x + 5$ and $y = g(x) = 2x$, find $y = (f \circ g)(10) = f(g(10))$.

It will help if you recall that the rule for removing groups of parentheses is to work from the inside out. The same rule applies here.

To find $f(g(10))$:

First put 10 into the function g.
Then put the result of that into the function f.

In case you wondered, both $(f \circ g)(10)$ and $f(g(10))$ are read "f of g of 10" and *not* "fog of 10."

EXAMPLE 4

Given: $y = f(x) = x + 5$ and $y = g(x) = 2x$.
Find: $(g \circ f)(10)$.

Solution: $y = (g \circ f)(10) = g(f(10))$

10 first is put into the function f.

$$f(10) = 10 + 5 = 15$$

$= g(15)$

15 then is put into the function g.

$= 2(15)$

$= 30$

EXAMPLE 5

Given: $y = f(x) = x + 5$.
Find: $y = (f \circ f)(3)$.

Solution: $y = (f \circ f)(3) = f(f(3))$.

3 first is put into the function f.

$$f(3) = 3 + 5 = 8$$

$= f(8)$

8 then is put into the function f.

$= 8 + 5$

$= 13$

As you can see, composite functions of this type are really not as bad as at first glance. Try Problem 2 with some new functions.

Problem 2

Given: $f(x) = x^2 + 1$ and $g(x) = \dfrac{x}{2}$.

Find: a. $(g \circ f)(3)$.
          b. $(g \circ f)(5)$.
          c. $(f \circ g)(8)$.

Solution:

Answers: a. 5.   b. 13.   c. 17

Now that you got all of those correct, here are a few more that use the same functions. Problem 3 is for fun.

Problem 3

Given: $f(x) = x^2 + 1$ and $g(x) = \dfrac{x}{2}$.

Find: a. $f(f(f(0)))$.
           b. $g(g(g(16)))$.
           c. $f(g(f(1)))$.
           d. $g(f(g(g(20))))$.

Solution:

Answers: a. 5.   b. 2.   c. 2.   d. 13

So far, each of the composite functions that was a function of a function has been numerical; that is to say, we have been finding such values as $f(g(\text{some number}))$. Examples 6–8 explain how to find a new function, say h, that is a function of a function of a variable.

EXAMPLE 6

Given: $g(x) = 2x$ and $f(x) = x + 5$.
Find: $h(x) = (f \circ g)(x) = f(g(x))$.

Solution: $h(x) = f(g(x))$

x first is put into the function g.

$$g(x) = 2x$$

$= f(2x)$

$2x$ then is put into the function f.

$= 2x + 5$

The functions g and f are the same functions used in the drawing on page 17. $h(x)$ should be thought of as a new function that combines f and g into a single machine with the rule "takes a number, doubles it, and adds 5."

EXAMPLE 7

Using f, g, and h as defined in Example 6, find $(f \circ g)(7)$.

Solution: There are two approaches to solving this problem.

1. $(f \circ g)(7) = f(g(7))$

    $= f(14)$ because $g(7) = 2(7) = 14$

    $= 14 + 5$

    $= 19$

or, using the function computed from Example 6,

2. $f(g(x)) = 2x + 5$

    $f(g(7)) = 2(7) + 5$

    $= 19$

Use whichever method you are more comfortable with; obviously the answer should be the same either way.

EXAMPLE 8

Given: $y = H(x) = x^2 - 5$ and $y = G(x) = x + 3$.

Find: $y = (H \circ G)(x)$ and simplify.

Solution: $y = (H \circ G)(x) = H(G(x))$

$= H(x + 3)$ Working from the inside out, we find that $G(x) = x + 3$.

$= (x + 3)^2 - 5$ H takes the number, squares it, and subtracts 5.

$= (x^2 + 6x + 9) - 5$ Reminder: $(x + 3)^2 = (x + 3)(x + 3)$

$= x^2 + 6x + 4$ $= x(x + 3) + 3(x + 3)$

$= x^2 + 3x + 3x + 9$

$= x^2 + 6x + 9$

RESTRICTED DOMAINS

In Unit 1, domains and ranges were discussed from a purely mathematical point of view. When considering application problems (maybe you call them word problems), however, the domain and range often need to be restricted so that the values of the variables "make sense."

Problem 4

Given: $y = f(x) = 5x$.
Find the domain for f.

Solution:

Answer: The domain is the set of reals, because there is no quotient or radical to an even power requiring that any numbers be excluded.

EXAMPLE 9

Given: $y = f(x) = 5x$

where x = the number of hours worked per day

and y = the wages earned per day, in dollars.

Find the restricted domain for f.

Solution: This is the same function as given in Problem 4, but now x has been defined as the number of hours worked per day. We should reason that, since there are 24 hours in a day and negative numbers would be meaningless in this situation, the only values that "make sense" for x are 0 to 24, inclusive.

The restricted domain is $0 \leq x \leq 24$.

Problem 5

Given: $y = f(x) = 5x$ with restricted domain of $0 \leq x \leq 24$

where x = the number of hours worked per day

and y = the wages earned per day, in dollars.

Calculate and interpret: $f(3), f(0)$, and $f(24)$.

Solution:

Answers: $f(3) = 15$; $15 are the wages earned per day for working 3 hours.

$f(0) = 0$; $0 are the wages earned per day for working 0 hour.

$f(24) = 120$; $120 are the wages earned per day for working 24 hours.

EXAMPLE 10

Find the restricted range for Problem 5 and interpret.

Solution: Since the restricted domain is $0 \le x \le 24$

and $f(0) = 0$

and $f(24) = 120$,

the restricted range is $0 \le y \le 120$.

In other words, the minimum amount of wages that can be earned per day is $0 for working no hours. The maximum amount of wages that can be earned per day is $120 for working 24 hours. The actual wages earned will be anywhere from $0 to $120, inclusive.

In summary, you should be able to read, interpret, and evaluate composite functions, both numerical and algebraic. When two or more functions are combined as a function of a function, such as $y = f(g(x))$, the composite function is evaluated from the inside out.

Application problems were introduced to illustrate the necessity for sometimes specifying a restricted domain to ensure that the value of the variable "makes sense."

Now try to work the following exercises. Some of the application problems contain formulas that will be used again in later units. Reminder, answers are provided at the back of the book.

EXERCISES

Given: $f(x) = x + 1$

$g(x) = x - 2$

$h(x) = 2x + 3$

$H(x) = x^2 + x$

$F(x) = x^2 - 3x$

$G(x) = 3x - 5$

Find and simplify:

1. $(f + g)(5)$

2. $(H - f)(5)$

3. $(GF)(2)$

4. $g(F(G(2)))$

5. $F(w + 1)$ Be careful with this one.

6. $f(h(x))$

7. $2f(x) + g(x) - h(x)$

8. $(h \circ f)(x)$

9. $H(2x) - h(x)$

10. $(G \circ F)(x)$

11. It is estimated that t years from now the population of an urban community will be $N(t) = -t^2 + 400t + 50{,}000$.

   a. What is the current population of the community?

   b. What will the population be 10 years from now?

12. The Alexander Company produces hand-crocheted rugs. The overhead cost for such items as insurance and rent is $6,452. In addition, each rug costs $72 to make, including all labor and materials. Let $C(x)$ represent the total cost of producing x hand-crocheted rugs; then

$$C(x) = 6{,}452 + 72x \quad x \geq 0.$$

   The function C is called a **total cost function**.

   a. Determine the total cost of producing 10 rugs.

   b. Calculate and interpret $C(100)$.

   c. Find $C(0)$.
      $C(0)$ is called the **fixed cost**. It is the cost to the company when no rugs are produced. The cost of producing each additional rug, $72, is called the **variable cost per unit**.

Cost Function

Total cost = fixed cost + (variable cost/unit)(number of units)
   or, if the *total* variable cost is given,
Total cost = fixed cost + total variable cost

---

13. At another location, the Alexander Company has a fixed cost of $13,200 and a variable cost per unit of $43 on braided rugs. Let x represent the number of braided rugs produced and $C(x)$ equal the total cost of producing x braided rugs.

   a. Write the equation that defines the cost function.

   b. Calculate and interpret $C(0)$, $C(10)$, and $C(100)$.

   c. State the restricted domain for C.

14. Crocheted rugs are popular at the present time, and the Alexander Company is able to sell each rug for $153. If $R(x)$ represents the total sales revenue gained from selling x crocheted rugs, then

$$R(x) = 153x \quad x \geq 0.$$

   The function R is called a **total revenue function**.

   a. Find the total revenue gained from selling 50 rugs.
   b. Calculate and interpret $R(100)$ and $R(0)$.

Revenue Function

Total revenue = (selling price per unit)(number of units sold),
   which is often shortened to
Total revenue = (price)(quantity).

---

15. The Alexander Company's braided rugs sell for only \$78, but the company is able to sell quite a few. Let $R(x)$ represent the total revenue gained from selling x braided rugs.

   a. Write the total revenue equation relating $R(x)$ and x.

   b. Calculate and interpret $R(1,000)$.

   c. State the restricted domain for R.

16. Profit for an organization is the difference between total revenue and total cost. If $P(x)$ represents the total profit from the production and sale of x crocheted rugs, then

$$P(x) = R(x) - C(x)$$

$$= 153x - (6,452 + 72x)$$

$$= 81x - 6,452 \quad x \geq 0$$

   a. Find the profit (or loss) gained from the production and sale of 200 rugs.
   b. Calculate and interpret $P(50)$.

Profit Function

Profit = total revenue − total cost

---

17. $C(x_1, x_2, x_3)$ is the cost function, in dollars, for Mary Lou's Toy Company with

$$C(x_1, x_2, x_3) = 1,750 + 7.5x_1 + 6x_2 + 10.6x_3$$

where $x_1 =$ the number of toy soldiers,

$x_2 =$ the number of teddy bears,

and $x_3 =$ the number of dolls manufactured.

   a. Calculate and interpret (10, 25, 100)

   b. Interpret each of the numbers in the cost function C.

18. H. Jones is an excellent, but persnickety, gardener who enjoys warm weather. In fact, the warmer the weather, the more hours he is willing to work. Each morning at 8 A.M., he checks the temperature and makes a decision as to the number of hours he will work that day, which in turn determines his salary for the day. His decision rules are:

$$y = f(x) = 4x$$

$$x = g(t) = \frac{t - 30}{2}$$

where $x =$ the number of hours worked per day,

$y =$ the salary earned per day by H. Jones,

and $t =$ the temperature, in F°, at 8 A.M.

   a. Calculate and interpret $y = f(5)$.

   b. Calculate and interpret $x = g(32)$.

   c. Calculate and interpret $y = f(g(32))$.

    d.   Find and interpret $y = h(t) = f(g(t))$.

    e.   Find and interpret $y = h(50)$, where h is the function from part d.

19.   Use the information provided in Exercise 18 to:

    a.   Name the independent variable for function f.

    b.   Find the restricted domain for f.

    c.   Name the independent variable for function g.

    d.   Find the restricted domain for g.

    e.   Find the restricted range for f.

C20. Carlos, a high school senior who operates the Dirty Dog Car Wash after school, determines that his profit, measured in dollars, from washing x cars per week is given by

$$P(x) = 0.0002x^3 + 10x \quad x \geq 0.$$

    Calculate and interpret $P(51)$.

If additional practice is needed:

   Barnett, Ziegler, and Byleen, pages 20–22, problems 85–96
   Harshbarger and Reynolds, pages 87–90, problems 23–58
   Hoffmann and Bradley, page 10, problems 22–66
   Waner and Costenoble, pages 12–14, problems 21–34

UNIT 3

Graphing Linear Functions

The purpose of this unit is to provide you with an understanding of how to graph a linear function of one variable. When you have finished the unit, you will be able to recognize a linear function and to graph it quickly and accurately.

Definition: The **graph of a function** f is the set of all points whose coordinates (x, y) satisfy the function $y = f(x)$.

Traditionally the horizontal axis is labeled as the independent variable.

The vertical axis is labeled as the dependent variable.

LINEAR FUNCTIONS

Definition: $y = f(x) = mx + b$, with m and b being real numbers, m and b not both zero, is called a **linear function** involving one independent variable x and a dependent variable y. Sometimes this is shortened to read "a linear function of one variable," meaning there is only one independent variable.

In other words, in a linear function of one variable:

1. There are one independent variable and one dependent variable.

2. Neither the independent nor the dependent variable is raised to any power other than 1.

3. Neither the independent nor the dependent variable appears in any denominator.

4. No term contains a product of the independent and dependent variables.

Here are some examples of linear functions of one variable:

$$y = f(x) = 3x - 1$$

$$y = g(x) = 1 + 5x$$

$$y = h(x) = -4x$$

$$q = F(p) = 1,743 + 0.03p$$

$$y = G(x) = 25$$

Here are some examples that are *not* linear functions of one variable:

$$y = f(x) = x^{-4} + 3x$$

$$z = g(x, y) = 5xy$$

$$y = h(x) = x^2 - x + 1$$

$$y = N(x) = 1/(x - 2)$$

$$w = G(x, y, z) = 3x + 2y - z$$

Before proceeding, determine why each of the above is *not* a linear function of one variable.

To avoid confusion in future courses it should be noted that the standard form of linear functions used in business and economics is $y = a + bx$. In this text, however, the mathematical form, $y = f(x) = mx + b$, will be used throughout. Note the difference between the two forms.

In the mathematical form: $y = mx + b$

slope   y-intercept

In the business form: $y = a + bx$

y-intercept   slope

Thus, a major source of confusion can occur if you fail to remember that b is used to denote the slope in business and economics texts, while b, in this book and in all mathematical texts, denotes the y-intercept.

GRAPHING

The graph of a linear function is a straight line.

There are three basic procedures for graphing linear functions: plotting points, using intercepts, or graphing with slope-intercept.

Plotting Points

To graph a linear function by plotting points, locate three points whose coordinates satisfy the equation and connect them with a straight line. Actually only two points are needed; the third point serves as a check.

To locate each point:

1. Select some convenient value for x. Any value will do because the domain is the set of all real numbers.

2. Substitute this value into the function.

3. Solve for y.

EXAMPLE 1

Graph: $y = f(x) = -\dfrac{1}{2}x + 6$.

Solution: $y = f(x) = -\dfrac{1}{2}x + 6$ is a linear function; therefore its graph will be a straight line.

Locate the first point:

1. Select a convenient value for x.   Let $x = 2$.

2. Substitute into the function. $y = f(2)$

3. Solve for y.
$$= -\frac{1}{2}(2) + 6$$
$$= -1 + 6$$
$$= 5$$

Thus $(2, 5)$ is a point on the graph of the function because it satisfies the equation:

$$5 \overset{?}{=} -\frac{1}{2}(2) + 6$$

$$5 = 5$$

Repeat the procedure to find a second point.

1. Select a value for x.   Let $x = 0$.

2. Substitute in the function. $y = f(0)$

3. Solve for y.
$$= -\frac{1}{2}(0) + 6$$
$$= 6$$

Thus $(0, 6)$ is a second point on the graph because $(0, 6)$ satisfies the equation:

$$6 \overset{?}{=} -\frac{1}{2}(0) + 6$$

$$6 = 6$$

Repeat the procedure to find a third point.

1. Select a value for x.   Let $x = 4$.

2. Substitute into the function. $y = f(4)$

3. Solve for y.
$$= -\frac{1}{2}(4) + 6$$
$$= 4$$

Thus (4, 4) is a third point on the graph.

Plot the three points and connect with a straight line.

Answer:

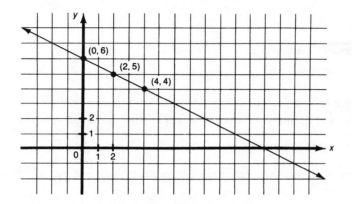

Note that there was nothing special about the three values selected for x. The domain of the <u>function</u> was the set of all reals so any values could have been selected. It often happens that $x = 0$ is a convenient value to use. The other two were selected so that y would result in an <u>integral value</u>.

The point where the line crosses the y-axis is called the **y-intercept**. In other words, the y-intercept is the value of y when $x = 0$. In Example 1 the y-intercept is 6.

The point where the line crosses the x-axis is called the **x-intercept**. The x-intercept is the value of x when $y = 0$.

EXAMPLE 2

Find the x-intercept for $y = f(x) = -\dfrac{1}{2}x + 6$.

Solution: The x-intercept is the value of x when $y = 0$.

The equation is $\qquad y = -\dfrac{1}{2}x + 6$

By substitution: $\qquad 0 = -\dfrac{1}{2}x + 6$

Clear of fractions: $\quad 0 = -x + 12$

$\qquad\qquad\qquad\quad x = 12$

The x-intercept is 12.

Refer to the graph of Example 1 to verify that the line crosses the x-axis at 12.

Do not let the term *linear function* confuse you.

Example 1 was the explicit function $y = f(x) = -\frac{1}{2}x + 6$.

$$y = -\frac{1}{2}x + 6$$

If it is rewritten as an implicit equation $2y = -x + 12$

we have $x + 2y = 12,$

which is a linear equation in two variables. A linear equation in two variables is the equation of a straight line, so all along we have been talking about graphing a line by plotting three points.

Try graphing the next linear function yourself.

Problem 1

Graph: $y = f(x) = -4x + 5.$

Solution: It is a linear function; the graph will be a straight line.
Locate three points whose coordinates satisfy the equation.

For example, let $x = 0$; then $y = f(0) =$ _____

let $x = 2$; then $y =$ _____

let $x =$ _____

Plot the three points on the graph provided, and connect them with a straight line.

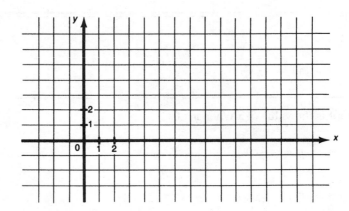

Reminder: Be sure to put arrows at each end to indicate that the line continues on in both directions.

Answer:

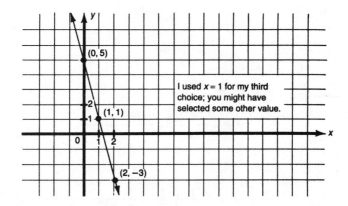

If you need additional practice or explanation on plotting points to graph a line, refer to Unit 24, "Graphing Linear Equations in Two Variables," in *Forgotten Algebra*.

Using Intercepts

To graph a linear function using intercepts, locate the x- and y-intercepts and connect them with a straight line.

To locate each point:

1. Rewrite the function as an equation. Whenever possible, an equation should be simplified before locating the points used to graph it.

2. Find the x-intercept, which is the value of x when $y = 0$.

3. Find the y-intercept, which is the value of y when $x = 0$.

EXAMPLE 3

Graph: $y = f(x) = \dfrac{3}{5}x + 3$ using intercepts.

Solution: 1.   Rewrite the function as an equation.

$$y - \frac{3}{5}x = 3$$

Simplify the equation by clearing of fractions first.

$$5y - 3x = 15$$

2.   Find the x-intercept, which is the value of x when $y = 0$.

By substitution: $5(0) - 3x = 15$

$$-3x = 15$$

$$x = -5$$

The x-intercept is −5 or point (−5, 0).

3. Find the y-intercept, which is the value of y when $x = 0$.

    By substitution: $5y - 3(0) = 15$

                                                                 $5y = 15$

                                                                 $y = 3$

The y-intercept is 3 or point $(0, 3)$.

Plot the two points and connect with a straight line.

Answer:

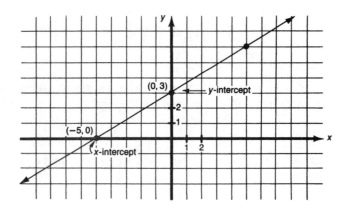

The danger with using intercepts to graph a line is that you are locating only two points. Since there is no third point to serve as a check, it is easy to make a mistake. I recommend finding a third point just to be safe.

SLOPE

Before continuing on with the third procedure for graphing linear functions, we need to introduce the word *slope*. One of the most important characteristics of a line is its slope.

> Definition: **Slope of a line** $= m = \dfrac{\text{change in } y}{\text{change in } x} = \dfrac{\Delta y}{\Delta x}$

Δ is the Greek letter delta. $\dfrac{\Delta y}{\Delta x}$ is read "delta y over delta x." It means the change in y over the change in x.

Think of the slope of a line as a rate of change. It is the rate at which y changes with respect to x.

A fundamental property of a straight line is that the slope is constant.

It can be shown that, if y is written as a function of x, the coefficient of x is the slope of the line and the constant term is the y-intercept.

> Definition: $y = f(x) = mx + b$ is said to be in **slope-intercept form**.
> m is the slope of the line and b is the y-intercept.

EXAMPLE 4

Identify the slope and y-intercept for $y = f(x) = 2x + 3$.

Solution: $y = f(x) = mx + b$ is in slope-intercept form.

$y = f(x) = 2x + 3$ is written in slope-intercept form. Therefore the coefficient of x is the slope and the constant term is the y-intercept. $y = f(x) = 2x + 3$

$$\underset{\textbf{slope}}{\uparrow} \quad \underset{\textbf{\textit{y}-intercept}}{\uparrow}$$

The slope of the line is 2.

The y-intercept is 3.

You are probably wondering, So the slope is 2, what does that mean?

It might be clearer if we rewrite 2 as $\frac{2}{1}$ and think of the slope as a rate of change, $m = \frac{\Delta y}{\Delta x} = \frac{2}{1}$. Usually the independent variable is read first. The interpretation of this slope would be: if x increases by 1, the y value will increase by 2.

Graphically, an interpretation of the slope $m = \frac{\Delta y}{\Delta x} = \frac{2}{1}$ is that you can start at any point on the line; and, as x increases by 1 (moves one unit to the right), and y increases by 2 (moves two units up), the new point also is located on the line. This ratio remains constant all along the line. It makes no difference where you start on the line. Each time you let x increase by 1 and y increase by 2, the new point also will be located on the line. Several examples are provided in the solution to Example 5.

EXAMPLE 5

Graph: $y = f(x) = 2x + 3$.

Answer:

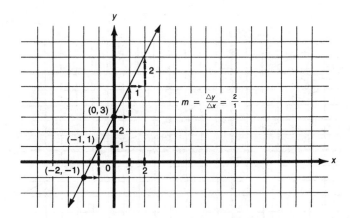

In this example the slope was a positive number. A **positive** slope means that, as x increases, y increases. In other words, as you move from left to right, the line is rising.

Before proceeding, we need to comment on the sign of a fraction.

$$\frac{-a}{b} = \frac{a}{-b} = -\frac{a}{b}$$

These three are equivalent fractions. The negative sign may be placed either in the numerator, in the denominator, or in front of the entire fraction. My personal choice is to always put the negative sign in the numerator.

EXAMPLE 6

Identify the slope and interpret for $y = f(x) = -\frac{1}{2}x + 4$.

Solution: $y = f(x) - \frac{1}{2}x + 4$ is written in slope-intercept form.

Thus the slope is the coefficient of x.

The slope of the line is $\frac{-1}{2}$ or $\frac{1}{-2}$ or $-\frac{1}{2}$, whichever you prefer.

There are three interpretations depending on which form you used. Each one is correct.

$\dfrac{\Delta y}{\Delta x} = \dfrac{-1}{2}$ If x increases by 2, the y value will *decrease* by 1.

$\dfrac{\Delta y}{\Delta x} = \dfrac{1}{-2}$ If x *decreases* by 2, the y value will increase by 1.

$\dfrac{\Delta y}{\Delta x} = -\dfrac{1}{2}$ If x increases by 1, the y value will *decrease* by $\dfrac{1}{2}$.

For the rest of the book, my answers will be written with the negative sign in the numerator.

A graph will illustrate that all three of the interpretations are correct. Each one will result in points located on the line.

Answer:

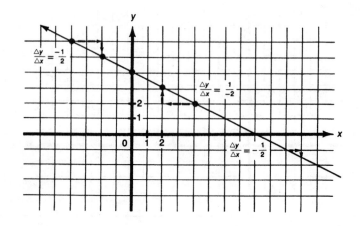

In Example 6 the slope was a negative number. A **negative** slope means that, as x increases, y decreases. In other words, as you move from left to right, the line is falling.

Problem 2

Identify the slope and y-intercept for $g(x) = 0.53x - 2.17$.

Solution:

Answer: The slope is 0.53, and the y-intercept is –2.17.

Now we are ready to use the slope and y-intercept for graphing linear functions.

Graphing with Slope-Intercept

To graph a linear function using slope-intercept, plot the y-intercept and locate two additional points using the slope. Connect the points with a straight line.

To locate the points:

1. Rewrite the equation in slope-intercept form, if necessary.

2. Identify the slope and y-intercept.

3. Plot the y-intercept.

4. **Start at the y-intercept**; use the slope to locate a second point.

5. Continue from the second point; use the slope to locate a third point.

EXAMPLE 7

Graph $y = f(x) = 3x + 1$ using the slope and y-intercept.

Solution: 1. $y = f(x) = 3x + 1$ is in slope-intercept form.

slope   y-intercept

2.   The y-intercept is 1.

The slope is 3.

$$m = \frac{\Delta y}{\Delta x} = 3 = \frac{3}{1}$$

If x increases by 1, the y value increases by 3.

3.   Plot the y-intercept at 1, the first point.

4.   **Start at the y-intercept of 1**. Let x increase by 1 (move one unit to the right), then y increases by 3 (move three units up); that is your second point. Put a dot at the second point.

5. Continue from the second point. Let x increase by 1 (move one unit to the right), then y increases by 3 (move three units up); that is your third point. Put a dot at the third point.

Connect the three points with a straight line.

Answer:

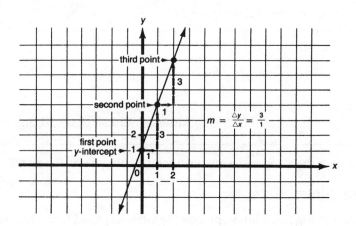

EXAMPLE 8

Graph $y = f(x) = -\dfrac{2}{3}x + 5$ using the slope and y-intercept.

Solution: 1. $y = f(x) = -\dfrac{2}{3}x + 5$ is in slope-intercept form.

slope   y-intercept

2.   The y-intercept is 5.

The slope is $\dfrac{-2}{3}$

$$m = \frac{\Delta x}{\Delta y} = \frac{-2}{3}$$

If x increases by 3, the y value will *decrease* by 2.

3.   Plot the y-intercept of 5, the first point.

4.   **Start at the y-intercept of 5.** Let x increase by 3 (move three units to the right), then y decreases by 2 (move two units down). Put a dot at the second point.

5.   Continue from the second point. Let x increase by 3 (move three units to the right), then y decreases by 2 (move two units down). Put a dot at the third point.

Connect the three points with a straight line.

Answer:

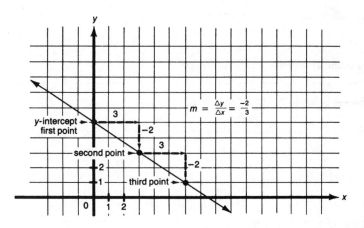

EXAMPLE 9

Graph $y = f(x) = \dfrac{x}{5} - 2$ using the slope and y-intercept.

Solution: Rewrite $y = f(x) = \dfrac{x}{5} - 2$ in slope-intercept form.

$$y = f(x) = \frac{1}{5}x - 2$$

slope   y-intercept

The y-intercept is -2 and the slope is $\dfrac{1}{5}$.

$$m = \frac{\Delta y}{\Delta x} = \frac{1}{5}$$

If x increases by 5, the y value will increase by 1.

Plot the y-intercept.

Start at the y-intercept. Use the slope to locate a second point. Continue from the second point. Use the slope to locate a third point.

Answer:

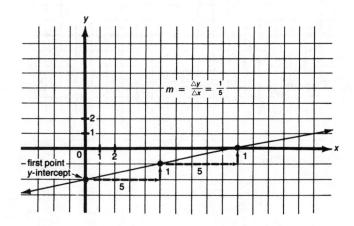

Now try to graph a linear function yourself using this procedure.

Problem 3

Graph $y = f(x) = 5x - 4$ using the slope and y-intercept.

Solution: $y = f(x) = 5x - 4$ is in slope-intercept form.

The y-intercept is _____ .

The slope is _____ .

$$m = \frac{\Delta y}{\Delta x} = \text{_____} \ .$$

If x increases by 1, the y value will _____ .

Plot the y-intercept on the graph provided, and find two additional points using the slope. Connect the points with a straight line.

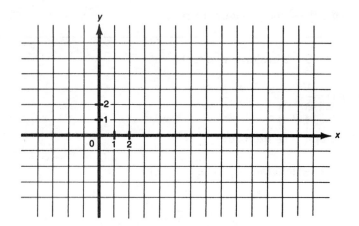

Answer:

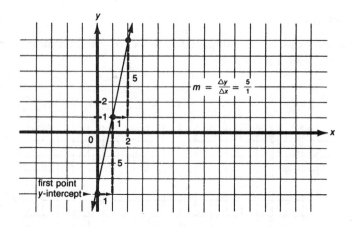

Try the next graph without any clues.

Problem 4

Graph $y = f(x) = -\dfrac{4}{3}x + 3$ using the slope and y-intercept.

Solution:

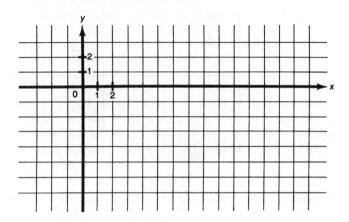

Answer:

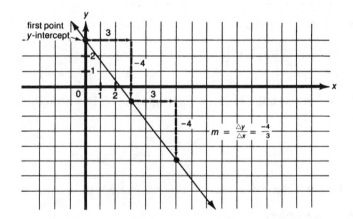

Examples 10 and 11 illustrate what happens when either m or b is equal to zero.

EXAMPLE 10

Graph $y = f(x) = 2x$ using the slope and y-intercept.

Solution: Recall that a linear function is of the form $y = f(x) = mx + b$ with m and b not both zero.

Compare: $y = f(x) = \boxed{m}\ x + \boxed{b}$

$\qquad\quad y = f(x) = \boxed{2}\ x\ \ \boxed{}$ so $m = 2$ and $b = 0$.

Therefore $y = f(x) = 2x$ is classified as a linear function; its graph will be a straight line with slope of 2 and y-intercept of 0.

$$m = \frac{\Delta y}{\Delta x} = \frac{2}{1}$$

Plot the y-intercept at 0. Starting at 0 (the y-intercept), locate two additional points using the slope of 2.

Answer:

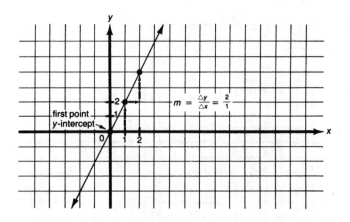

EXAMPLE 11

Graph $y = f(x) = 7$ using the slope and y-intercept.

Solution: By a similar line of reasoning,

compare: $y = f(x) = \boxed{m}\ \ x + \boxed{b}$

$y = f(x) = \boxed{}\boxed{7}$ so $m = 0$ and $b = 7$.

Thus, $y = f(x) = 7$ is also a linear function; its graph will be a straight line with slope equal to 0 and y-intercept of 7.

$$m = \frac{\Delta y}{\Delta x} = 0 = \frac{0}{1}$$

Think of a slope of 0 as meaning that, if x increases by 1, the y value does *not* change.

The y-intercept is plotted at 7.

Start at the y-intercept of 7. If x increases by 1, the y value does *not* change and remains at 7. There is your second point.

Locate the third point yourself.

Answer:

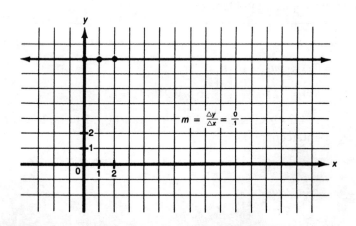

Observe: The graph of $y = f(x) = 7$ is a horizontal line.

Each point on the line has a y-coordinate of 7.

The y-intercept is 7.

There is no x-intercept.

In this example the slope was zero. A **zero** slope means that, as x increases, y remains the same. In other words, the graph of the line with a slope of 0 is horizontal.

> The graph of a linear function $y = f(x) = p$, where p is a real number, is a **horizontal line** with y-intercept of p. Its slope is 0.

EXAMPLE 12

Graph $y = f(x) = -2$.

Solution: The graph of the linear function $y = f(x) = -2$ is a horizontal line with y-intercept of -2.

Answer:

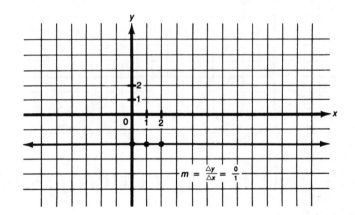

The advantage of graphing linear functions using the slope and y-intercept should be obvious. Most linear functions are written in slope-intercept form to start with, thus making the slope and y-intercept easily identifiable. Starting with the y-intercept, we can then quickly plot three points with a minimum amount of work and draw the line.

Example 13 is a function in terms of p and q.

EXAMPLE 13

Graph $q = D(p) = -2p + 8$.
Identify the slope and interpret.

Solution: Recall if $q = D(p) = -2p + 8$, then p is the independent variable.

dependent   independent
variable   variable

Since the horizontal axis is always the independent variable, the horizontal axis will be labeled p, with the vertical axis being labeled q.

Compare: $y = f(x) = \boxed{m}\ x + \boxed{b}$

$q = D(p) = \boxed{-2}\ p + \boxed{8}$ so $m = -2$ and $b = 8$.

Thus $q = D(p) = -2p + 8$ is a linear function; its graph will be a straight line.

The y-intercept is 8. It is easier to refer to 8 as the y-intercept even though the vertical axis is labeled q in this situation.

The slope is -2, or $m = \dfrac{\Delta y}{\Delta x} = \dfrac{-2}{1} = \dfrac{\Delta q}{\Delta p}$.

To interpret a slope of -2, remember that the independent variable is read first; if p increases by 1, the q value will *decrease* by 2.

Answer:

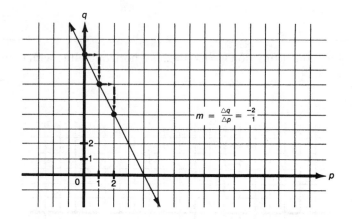

CALCULUS AND THE CALCULATOR (Optional)

Graphing Linear Functions

So that we start at the same place, for now we will use the standard viewing window for our graphs. Its dimensions are [–10, 10] by [–10, 10]. That is, the x-axis will be shown from –10 to 10 and the y-axis will be shown from –10 to 10 as pictured here.

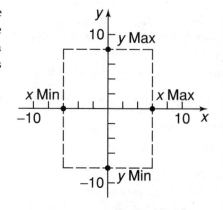

Press $\boxed{\text{Y=}}$ $\boxed{\text{CLEAR}}$ $\boxed{\text{ZOOM}}$ $\boxed{6}$ to use the standard viewing window, and a blank standard viewing window should appear on the screen.

Press **WINDOW** to confirm that the viewing window's dimensions agree with the screen shown here. If they don't, reset them to match.

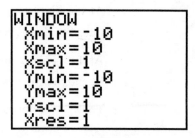

Press **2nd** **FORMAT** and select RectGC to use rectangular coordinates.

Return to the home screen by pressing **CLEAR** or **2nd** **QUIT**.

EXAMPLE 14

Graph $y = f(x) = -\dfrac{5}{2}x + 4$ using the graphing calculator.

Find: a. $f(3.74)$ using the **CALC** key.

b.   the y-intercept using the **TRACE** key.

c.   the x-intercept using the **CALC** key.

Solution: Create a table of values for the function using the following keystrokes:

Y= **CLEAR** **(** **(−)** **5** **÷** **2** **)** **X,T,θ,n** **+** **4** **ENTER**

To see the various x- and y-values, press **2nd** **TABLE**.

To see the graph, press **GRAPH**. Surprisingly easy, wasn't it?

a.   To find $f(3.74)$ using the **CALC** key, press **2nd** **CALC** **1** **3** **·** **7** **4** **ENTER**.

The function appears in the upper right, the x- and y-values of 3.74 and −5.35 are at the bottom, and a star appears on the graph at the point with coordinates (3.74, −5.35).

b.   To find the y-intercept using the **TRACE** key, press **TRACE**. Using only the ▶ and ◀ arrows, move the blinking star (cursor) along the graph until the x- value is 0. The y-intercept, in this example, $y = 4$, appears at the bottom of the screen.

c.   To find the x-intercept, also called a zero of the function, using the **CALC** key requires a bit more work.

Press **2nd** **CALC** **2**.

At the <u>Left Bound?</u> prompt, using only the ▶ and ◀ arrows, move the blinking star (cursor) to a point just to the left of the x-intercept and press **ENTER**. If you find it difficult to decide where to place the cursor, keep in mind that both the x- and y-values will be positive.

At the <u>Right Bound?</u> prompt, move the cursor to a point just to the right of the x-intercept. Here x will be positive and y will be negative; then press (**ENTER**).

At the <u>Guess?</u> prompt, move the cursor between the boundaries shown on the screen by the two arrows, ▶ ◀, and press (**ENTER**).

The answer appears at the bottom of the screen. Zero $x = 1.6$, $y = 0$.

Before continuing, I suggest you redo some of the earlier examples for practice with the calculator. Also, when keying functions, experiment with the (**DEL**) and (**INS**) keys.

EXAMPLE 15

Graph $2x + 5y = 60$ using the graphing calculator.

Solution: Before the equation can be entered into the calculator, it first must be rewritten in slope-intercept form, $y = f(x)$, as follows:

$$2x + 5y = 60$$
$$5y = -2x + 60$$
$$y = -\frac{2}{5}x + 12$$

Press (**Y=**) (**(**) (**(−)**) (**2**) (**÷**) (**5**) (**)**) (**X,T,θ,n**) (**+**) (**1**) (**2**) (**ENTER**).

This time, however, when we press (**GRAPH**), only a small portion of the line is visible in our standard viewing window. Obviously the viewing window needs to be enlarged. Several methods are available.

One method is to press (**TRACE**) (**ZOOM**) (**3**) (**ENTER**).

If you are curious about the new dimensions, press (**WINDOW**).

A second method is to manually reset the dimensions of the window. For example, you might find the x- and y-intercepts, which are 30 and 12, respectively, and reset the window's dimensions accordingly.

To reset the dimensions of the window,

press (**WINDOW**) and reset Xmax to 35 to accommodate the x-intercept of 30 and reset Ymax to 15 to accommodate the y-intercept of 12. Press (**GRAPH**) to view the graph in the new enlarged viewing window.

Now I encourage you to do a bit of experimenting with the settings and the various keys. For instance, press (**2nd**) (**FORMAT**), select various options, and observe the results on a graph. Or, change the viewing window's dimensions again. (**ZOOM**) (**6**) will always take you back to the standard viewing window. Or, use (**TRACE**) to move along the line and observe what happens when you reach the edge of the viewing window.

You should now be able to recognize a linear function of one variable and graph it using one of three methods: plotting points, locating the x- and y-intercepts, or using the slope and y-intercept.

To plot points:

select three convenient values for x;

substitute these values into the function and solve for y;

connect the three points with a straight line.

To graph using the x- and y-intercepts:

locate the y-intercept (the value of y when $x = 0$);

locate the x-intercept (the value of x when $y = 0$);

connect the two points with a straight line.

To graph using the slope and y-intercept:

locate the y-intercept;

start at the y-intercept, use the slope to locate two additional points;

connect the three points with a straight line.

Also, given any linear function of one variable, you should be able to identify the slope and both the x- and y-intercepts, if they exist.

Before beginning the next unit you should graph the functions in the following exercises.

EXERCISES

Identify the slope and y-intercept and graph each function.

1. $y = f(x) = 2x + 5$

2. $y = f(x) = \dfrac{7}{3}x - 1$

3. $y = \dfrac{2}{5}x + 3$

4. $f(x) = x - 4$

5. $g(x) = -\dfrac{3}{2}x + 7$

6. $h(x) = -x + 2$

7. $y = f(x) = \dfrac{x}{3} + 1$

8. $y = f(x) = 0.25x + 1.50$ Hint: $0.25 = \dfrac{25}{100} = \dfrac{1}{4}$

9. $q = D(p) = -3p + 11$

10. Given $x - 3y = -12$:

    a. Rewrite, in slope-intercept form, $y = f(x)$.

   b.   Identify and interpret the slope.

   c.   Find both the x- and y-intercepts.

   d.   Graph.

C11.   Identify the slope and y-intercept and graph $y = f(x) = 2.53x + 1.75$.

C12.   Given $y = f(x) = -3.08x - 7.15$:

   a.   Graph.

   b.   Identify the slope.

   c.   Find both the x- and y-intercepts.

   d.   Find y so that the point $(2.45, y)$ is on the line.

13.   Given $2x + 3y = 21$:

   a.   Rewrite, in slope-intercept form, $y = f(x)$.

   b.   Identify the slope.

   c.   Find the x-intercept.

   d.   Find the y-intercept.

   e.   Find w so that point $(w, 15)$ is on the line.

14.   Given $5p - 4q = 20$:

   a.   Rewrite in slope-intercept form, $q = D(p)$.

   b.   Identify and interpret the slope.

15.   Graph a line through $P(3, 7)$ with slope of $-\dfrac{2}{5}$.

16.   Graph a line through $Q(-4, -3)$ with slope of 0.

17.   Graph a line through $S(2, -3)$ with slope of -4.

If additional practice is needed:

Barnett, Ziegler, and Byleen, pages 48–49, problems 1–28
Harshbarger and Reynolds, page 99, problems 1–22
Hoffmann and Bradley, page 41, problems 7–18
Waner and Costenoble, pages 24–25, problems 1–18; page 37, problems 15–28

UNIT 4

Writing Linear Functions

The purpose of this unit is simply the reverse of Unit 3. Given the graph of a linear function, we want to be able to write its equation. When you have finished the unit, you will be able to write a linear function that describes the relationship between the independent and dependent variables.

There are several approaches for writing linear functions. We will use only one, the point-slope formula.

> Point-slope Formula
>
> $y - y_1 = m(x - x_1)$ is the **point-slope formula** for a line that has slope m and contains point (x_1, y_1).

The point-slope formula can be used to write the equation of a line. In order to use the formula, two items of information are required: the slope of the line and a given point (x_1, y_1) on the line.

EXAMPLE 1

Write the linear function whose graph contains point (1, 2) with slope 3.

$$(y - y_1) = m(x - x_1)$$
$$(y - 2) \quad 3(x - 1)$$
$$y = -1 + 3x$$
$$= 3x - 1$$

Solution: The graph of a linear function is a straight line.

To write the equation of a line, use the point-slope formula.

In order to use the formula, we need two items, the slope and a given point on the line. The slope was given as 3, so $m = 3$. The line is to contain point (1, 2), so $(x_1, y_1) = (1, 2)$.

Use point-slope formula.	$y - y_1 = m(x - x_1)$
Substitute.	$y - 2 = 3(x - 1)$
Remove the parentheses.	$y - 2 = 3x - 3$
Solve for y.	$y = 3x - 1$
So the answer is:	$y = f(x) = 3x - 1$

Before continuing, let's make sure you are able to write a linear function yourself.

Problem 1

Write the linear function whose graph contains point (–10, 2) with slope $\frac{1}{2}$.

Solution: Use the point-slope formula. $y - y_1 = m(x - x_1)$

Answer: $y = f(x) = \frac{1}{2}x + 7$

To summarize what has been said so far:

$y - y_1 = m(x - x_1)$ is the point-slope formula. It can be used to write the equation of a line, given the slope and a point (x_1, y_1) on the line.

The point-slope formula is not always the fastest method for writing linear functions, but I like it because it works for all situations. For that reason you need to remember only the one formula. It is the only method we will use in this book.

CALCULATING THE SLOPE

Thus far we always have been given the slope. What happens when that is not the case? First, the definition of the slope of a line needs to be expanded.

Given two points on a line such as $P(x_1, y_1)$ and $Q(x_2, y_2)$, the slope is found by taking the difference of the y-coordinates divided by the difference of the x-coordinates. In symbols the definition looks like this:

Definition: **Slope of a line** $= m = \dfrac{\Delta y}{\Delta x} = \dfrac{y_2 - y_1}{x_2 - x_1}$

EXAMPLE 2

Use the definition to find the slope of the line connecting points (0, 1) and (2, 5).

Solution: Let (0, 1) be the first point; therefore $x_1 = 0$ and $y_1 = 1$.

Then (2, 5) is the second point and $x_2 = 2$ and $y_2 = 5$.

Substituting into the definition gives

$$m = \frac{\Delta y}{\Delta x} = \frac{y_2 - y_1}{x_2 - x_1} = \frac{5 - 1}{2 - 0} = \frac{4}{2} = 2$$

The slope of the line is 2.

Points (0, 1) and (2, 5) have been plotted on the grid below and the line drawn connecting them. From Unit 3, remember that, if the slope is 2, $m = \dfrac{\Delta y}{\Delta x} = 2 = \dfrac{2}{1}$. If we start at any point on the line, as x increases by 1 (moves one unit to the right), and y increases by 2 (moves two units up), a new point is located on the line. This ratio remains constant all along the line, as illustrated below.

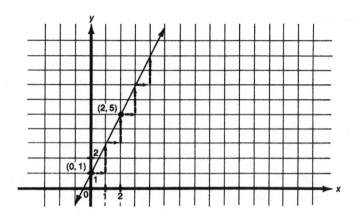

One of the troublesome aspects about finding the slope of a line is remembering to subtract the coefficients in the same order. It makes no difference which of the two points is specified as the first and which is the second, but the same designation must be used for calculating both Δx and Δy.

EXAMPLE 3

Find the slope of the line connecting points (–3, 5) and (7, –2).

Solution: Let (–3, 5) be the first point and (7, –2) be the second point.

Then

$$m = \frac{\Delta y}{\Delta x} = \frac{y_2 - y_1}{x_2 - x_1} = \frac{(-2) - 5}{7 - (-3)} = \frac{-2 - 5}{7 + 3} = \frac{-7}{10}$$

The slope of the line is $\dfrac{-7}{10}$.

Had we let (7, –2) be the first point and (–3, 5) be the second point, the result would have been the same.

$$m = \frac{\Delta y}{\Delta x} = \frac{y_2 - y_1}{x_2 - x_1} = \frac{5 - (-2)}{(-3) - 7} = \frac{5 + 2}{-3 - 7} = \frac{7}{-10}$$

This way the slope is calculated to be $\dfrac{7}{-10}$, but that equals $\dfrac{-7}{10}$.

Unfortunately, students often want to simply take the larger number minus the smaller number. Using the numbers from Example 3, we would then take $5 - (-2) = 7$ and $7 - (-3) = 10$ and conclude that the slope is a positive $\dfrac{7}{10}$. Doing so disregards which point is considered the first and all too frequently yields the wrong answer, as it would in this case.

An alternative method for finding the slope is to actually subtract the coordinates of the points as they are written as ordered pairs. This procedure eliminates the necessity for remembering which is to be the first point and which is to be the second point for substituting into the definition.

Example 4 will be worked using this alternative method. I think you will find it much easier.

EXAMPLE 4

Find the slope of the line connecting points $(0, -1)$ and $(-5, 3)$.

Solution: Subtract the coordinates of the points as written. Remember that the rule for subtracting integers is to change the sign of the bottom number and then treat the problem as addition.

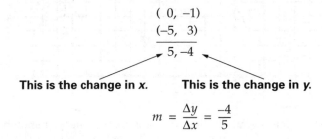

$$m = \frac{\Delta y}{\Delta x} = \frac{-4}{5}$$

Isn't that a neat way to find the slope?

It is time to return to our original purpose, that of writing linear functions.

EXAMPLE 5

Write the linear function containing points $(0, -1)$ and $(-5, 3)$.

Solution: The graph of a linear function is a straight line.

To write the equation of a line, we use the point-slope formula.

In order to use the formula, we need to know the slope and a given point on the line.

From Example 4, the slope was found to be $\frac{-4}{5}$.

Two points on the line are given; select one to use in the formula. I chose to use $(0, -1)$, but it would not make any difference which one was selected.

Use formula. $y - y_1 = m(x - x_1)$

Substitute. $y - (-1) = \frac{-4}{5}(x - 0)$

Remove parentheses. $y + 1 = \frac{-4}{5}x$

Solve for y. $y = \frac{-4}{5}x - 1$

So the answer is: $y = f(x) = \frac{-4}{5}x - 1$

The procedure then for writing a linear function when given two points on the line is as follows:

1. Use the two points to first calculate the slope.

2. Select one of the points to be used as the given point on the line. Either of the two points may be used; select the simpler.

3. Use the point-slope formula to write the equation of the line.

Now try a few problems yourself.

Problem 2

Write the linear function whose graph contains points (5, 8) and (4, 3).

Solution:

$$\frac{8-3}{5-4} \quad \frac{5}{1} = 5 \qquad\qquad 5\checkmark$$

$$y - 3 = 5(x-4)$$
$$y = 5(x-4) + 3$$
$$= 5x - 20 + 3$$
$$= 5x - 17$$

Answer: $y = f(x) = 5x - 17$

Problem 3

Find the slope of the line connecting points (5, 1) and (–3, –4).

Solution:

$$\frac{1-(-4)}{5-(-3)} = \frac{5}{8} = m \qquad \frac{y-(-4)}{x-(-3)} = \frac{5}{8}$$

$$y + 4 = \frac{5}{8}(x+3)$$
$$y + 4 = \frac{5}{8}x + \frac{15}{8}$$
$$y = \frac{5}{8}x + \frac{15}{8} - \frac{32}{8}$$
$$\frac{5}{8}x - \frac{17}{8}$$

Answer: $m = \dfrac{5}{8}$

Problem 4

Write the linear function whose graph contains points (5, 1) and (–3, –4).

Solution:

$$\frac{1-(-4)}{5-(-3)}$$

$$y = \frac{5}{8}(x) - \frac{17}{8}$$

Answer: $y = f(x) = \dfrac{5}{8}x - \dfrac{17}{8}$

EXAMPLE 6

The lines connecting the points from Problems 2 and 4 are shown on the graph below. Verify, by starting at any point and moving to a new point, that the slope of the first line is 5 and the slope of the second line is $\frac{5}{8}$.

Solution:

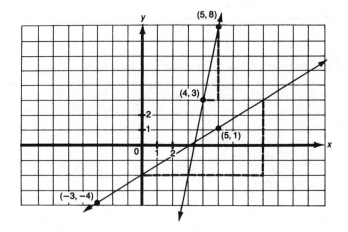

Observe that both lines have positive slopes. Graphically a positive slope means that, as you move left to right, the line is rising. Stated in symbols, if $m > 0$, the line is rising.

Also notice that the first line, with a slope of 5, is rising much faster than the second line with a slope of $\frac{5}{8}$. Think of the comparison this way: if x increases by 1, the y value of the first line will increase by 5, whereas the y value of the second line will increase by only $\frac{5}{8}$.

The larger the value of m, the faster the line will rise. Visualize what two lines would look like if one goes through the origin with a slope of 3 and the second also goes through the origin but with a slope of 100. Which line will be rising faster, or in other words, which is the steeper line?

Let's continue with writing linear functions.

Problem 5

Write the linear function whose graph contains points $(3, -2)$ and $(1, 1)$.

Solution:

Answer: $y = f(x) = \dfrac{-3}{2}x + \dfrac{5}{2}$

Problem 6

Write the linear function whose graph contains points (–4, 5) and (–1, 2).

Solution:

$$\frac{+5-2}{-4-(-1)} \qquad \frac{3}{-3}-1=m$$

$$\frac{y-2}{x-(-1)} = -1(x+1) \quad , \quad y-2=-1(x+1) \quad -x+1$$

$$y=-1(x+1)+2$$

$$= -x-1+2 \qquad \text{Answer:} \quad y=f(x)=-x+1$$

The lines connecting the points from Problems 5 and 6 are shown on the graph below. We can verify, by starting at any point and moving to the next point, that the slope of the line from Problem 5 is $\frac{-3}{2}$ and the slope of the line from Problem 6 is –1.

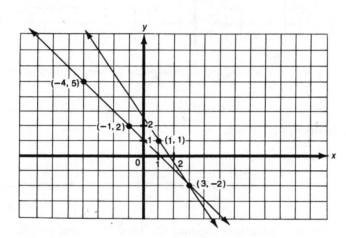

This time both lines have a negative slope or, as you move from left to right, the lines are falling. Stated in symbols, if $m < 0$, the line is falling.

Think what a line would look like with a slope of –50. The more negative the slope, the faster the line falls.

The next problem is easy, but be careful.

Problem 7

Write the linear function whose graph contains points (3, 4) and (–2, 4).

Solution:

Answer: $y = f(x) = 4$

EXAMPLE 7

Verify, by graphing, that the slope of the line connecting points (3, 4) and (–2, 4) is 0.

Solution:

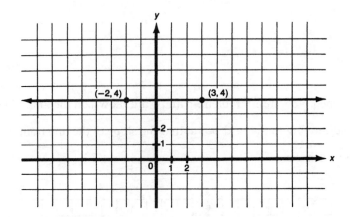

Starting at any point on the line, as x increases by 1, the y value does not change, but remains at 4.

In symbols, if $m = 0$, the line is horizontal.

Example 8 is a special situation.

EXAMPLE 8

Write the linear function whose graph contains points (3, 7) and (3, 5), and graph.

Solution: First use the two points to calculate the slope.   (3, 7)
                                                                     (3, 5)
                                                                     ‾‾‾‾‾‾
                                                                     (0, 2)

Then $m = \dfrac{\Delta y}{\Delta x} = \dfrac{2}{0}$ = undefined

As I am sure you remember, division by zero is undefined. Therefore, since the slope, m, in this example is undefined, the point-slope formula cannot be used.

Maybe the graph will help.

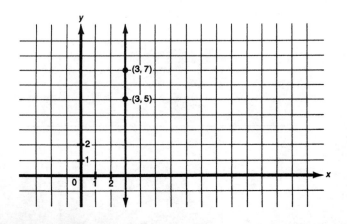

Observe: The graph is a vertical line.

The x-intercept is 3.

There is no y-intercept.

Each point on the line has an x-coordinate of 3.

No linear function can be written whose graph contains points (3, 7) and (3, 5).

In symbols, if m is undefined, the line is vertical.

However, even though the line graphed in Example 8 is not a function, it is possible to write an equation for the line.

> Equation of a Vertical Line
> The **equation of a vertical line** through a given point, (r, p), is $x = r$.

The graph of a linear equation $x = r$, where r is a real number, is a vertical line with x-intercept of r. Its slope is undefined.

EXAMPLE 9

Write the equation of a vertical line through point (3, 5).

Solution: The equation of a vertical line through the given point (3, 5), is $x = 3$.

You should now be able to write a linear function to describe the relationship between variables if you are given either the slope and one point on the line or two points on the line. If you are given two points, the slope must be calculated first. In either situation, the point-slope formula could be used to write the linear function.

We can generalize the relationship between the slope and the graph of a line, always moving from left to right, as follows:

if $m > 0$, the line is rising,

if $m < 0$, the line is falling,

if $m = 0$, the line is horizontal,

if m is undefined, the line is vertical.

A vertical line is a special case. It is not a function, but the equation of the line can be written as $x = r$, where the line contains point (r, p).

Before beginning the next unit you should write the linear functions called for in the exercises, given the information in each problem.

EXERCISES

1.   Find the slope of the line connecting points (2, 2) and (0, 3).

For each of the following, find, if possible, a linear function, relating x and y with the given properties. Write the final answer in slope-intercept form, $y = mx + b$.

2. Line passes through $(4, -3)$ with slope 5.

3. Line passes through $(-5, 3)$ with slope $\dfrac{-4}{5}$.

4. Line passes through $(6, 7)$ with slope $\dfrac{-1}{6}$.

5. Line passes through $(2, 7)$ with slope $\dfrac{-1}{5}$.

6. Line passes through $(7, 5)$ and $(-3, 5)$.

7. Horizontal line passes through $(-2, -8)$.

8. Slope equals 0.15 and point $(2.35, 100)$ lies on the line.

9. Line passes through $(-5, -2)$ and $(4, -2)$.

10. Points $(-5, -2)$ and $(-3, -8)$ lie on the line.

11. Line contains point $(2, -10)$ and the origin.

12. Line contains points $(400, 200)$ and $(3,600, 1,000)$.

13. Vertical line passes through $(-9, 5)$.

14. Line passes through $(2.75, 300)$ and $(3.95, 435)$.

15. Vertical line passes through $(15, 37)$.

16. Write the equation of the line connecting points $(-2, 5)$ and $(-2, 13)$.

17. Find the slope of the line connecting points $(5, -1)$ and $(5, 7)$.

18. Write the linear function whose graph is a line passing through $(6, 4)$ and parallel to $y = 5x + 2$. Hint: Recall that, if two nonvertical lines are *parallel*, then they have the same slope. Or, if m is the slope of one line, any line parallel to it must also have a slope of m.

19. Write the linear function whose graph is parallel to $y = -3x + 2$ and passes through the origin.

20. Write the linear function whose graph is the line through $(-7, 2)$ and perpendicular to $y = 3x - 1$. Hint: Recall that, if two nonvertical lines are *perpendicular*, then their slopes are negative reciprocals of each other. In other words, if m is the slope of one line, any line perpendicular to it will have a slope of $\dfrac{-1}{m}$.

21. Determine the linear function whose graph is a line passing through $(4, 7)$ and parallel to $y = 6$.

22. Determine the equation of the line that passes through $(4, 7)$ and is parallel to $x = 9$.

23. Write the linear function whose graph is a line passing through $(-7, -4)$ and perpendicular to $x = 8$.

If additional practice is needed:

Barnett, Ziegler, and Byleen, page 49, problems 29–62
Harshbarger and Reynolds, pages 99–100, problems 23–42
Hoffmann and Bradley, pages 41–42, problems 1–6, 19–34
Waner and Costenoble, pages 37–38, problems 31–60

UNIT 5

Linear Applications

In this unit we will consider several applications of linear functions. The emphasis will be not only on the mathematical answer, but also on the interpretation of the answer in terms of the problem. Demand functions, supply functions, and equilibrium will be introduced. Total cost, total revenue, and total profit functions will be revisited with break-even analysis introduced.

DEMAND FUNCTIONS

Demand functions describe the relationship between the demand for a product and its price. Suppose that p denotes the price of a certain product. Let D denote the demand function. Then $q = D(p)$ is the number of units of the product that consumers are willing to buy when the market price is p, or quantity demanded = D(market price).

EXAMPLE 1

In a small farm town, the demand function for Margaret's homemade cakes is defined by

$$q = D(p) = -2p + 10 \quad 0 \le p \le 5$$

where p = the price, in dollars, per cake

and q = the number of cakes sold at price p.

a. Calculate the demand for Margaret's cakes if the price is $3 per cake.

b. Calculate and interpret $D(1)$.

Solution: a. $q = D(p) = -2p + 10$

$$= D(3) = -2(3) + 10$$

$$= -6 + 10$$

$$= 4$$

If the price of a cake is $3, Margaret will sell 4 cakes.

b. $q = D(p) = -2p + 10$

$\quad q = D(1) = -2(1) + 10$

$\qquad\qquad = -2 + 10$

$\qquad\qquad = 8$

If the price of a cake is $1, Margaret will sell 8 cakes.

EXAMPLE 2

Graph: $q = D(p) = -2p + 10$ with $0 \le p \le 5$.

Solution: $q = D(p) = -2p + 10$ is a linear function and its graph is a straight line. But this function has a restricted domain; so, instead of being the entire line, its graph will be only a line segment. The line segment will start at the lower limit of the restricted domain, $p = 0$, and end at the upper limit of the restricted domain, $p = 5$.

Lower limit: Let $p = 0$; then $q = D(0) = -2(0) + 10 = 10$.

The starting point of the line segment is $(0, 10)$.

Upper limit: Let $p = 5$; then $q = D(5) = -2(5) + 10 = 0$.

The endpoint of the line segment is $(5, 0)$.

The notation $D(p)$ indicates that p is to be the independent variable; therefore, the horizontal axis is labeled p and q is on the vertical axis.

So as not to confuse you in other courses, I need to point out that mathematicians often view the relationship between price and quantity differently than economists. In economics texts, the convention is that price, p, is placed on the vertical axis and quantity, q, on the horizontal axis.

Answer:

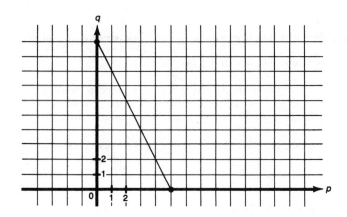

EXAMPLE 3

Interpret the ordered pair $(0, 10)$ from Example 1.

Solution: (0, 10) corresponds to $p = 0$ and $q = 10$.

If the price of a cake is $0, Margaret will sell 10 cakes; or, in more common terms, if the cakes are free, Margaret can give away only 10. Remember that this is a small town.

EXAMPLE 4

Identify and interpret the slope from Example 1.

Solution: The slope is –2, or $m = \dfrac{\Delta y}{\Delta x} = \dfrac{\Delta p}{\Delta q} = -2 = \dfrac{-2}{1}$.

Remember that the independent variable is read first; if p increases by 1, the q value will decrease by 2. Although that is the mathematical interpretation of the slope, since this is an application problem, the definitions of p and q as stated in the problem should be used.

If the price in dollars per cake increases by $1, the number of cakes sold at that price will decrease by 2. Or, in still other words, for each additional dollar increase in price, the number of cakes sold will decrease by 2.

A typical demand function will have a negative slope because, as the price increases for a product, the demand for that product usually decreases, all else being equal. Of course there are exceptions.

Here is a problem for you to do.

Problem 1

Use $D(p)$ from Example 1 to calculate and interpret $D(4)$, $D(5)$, and $D(15)$.

Solution:

Answers: $D(4) = 2$. If the price per cake is $4, Margaret will sell 2 cakes.

$D(5) = 0$. If the price per cake is $5, Margaret will sell 0 cake.

From this information we can conclude that, if the price per cake is $5 or more, Margaret won't sell any cakes; at this level she has priced herself out of the market.

At a price of $15 per cake, Margaret won't be able to sell any cakes.

SUPPLY FUNCTIONS

Supply functions relate the market price to the quantities that suppliers are willing to produce and sell. Suppose that p is the price of a certain product and S denotes the supply function. Then $q = S(p)$ is the number of units of this product that suppliers are willing to produce and sell when the market price is p, or quantity supplied = S(market price).

In most cases, the supply function has a positive slope because, as the market price increases, suppliers usually are willing to produce and sell more.

EXAMPLE 5

On the basis of her costs and other interests, Margaret has determined that the supply function for her cakes is given by

$$q = S(p) = 2p - 2 \quad p \geq 1$$

where p = the price, in dollars, per cake

and q = the number of cakes she is willing to bake and sell at price p.

How many cakes is Margaret willing to bake if she sells them for $4 each?

Solution: The supply function is $q = S(p) = 2p - 2$.

If $p = \$4$, then $q = S(4) = 2(4) - 2$

$$= 6$$

If the market price of cakes is $4, Margaret is willing to bake and sell 6 cakes.

Problem 2

Calculate and interpret $S(15)$.

Solution:

Answer: If the market price of cakes is $15 per cake, Margaret is willing to bake and sell 28 cakes.

EQUILIBRIUM

Equilibrium occurs if there is a price at which the supply and demand are equal. In symbols, equilibrium occurs when $S(p) = D(p)$.

EXAMPLE 6

Find the equilibrium price for Margaret's cakes.

Solution: Equilibrium occurs when $S(p) = D(p)$.

By substitution $2p - 2 = -2p + 10$

$$4p = 12$$

$$p = 3$$

At a price of $3 per cake, the number of cakes supplied and demanded will be equal.

Although this was not asked for, either of the two functions can be used to find q:

$$q = S(p) = 2p - 2$$

$$S(3) = 2(3) - 2$$

$$= 4$$

At a price of $3 per cake, the number of cakes supplied *and* sold will be 4.

___

The supply and demand functions are graphed on the same set of axes. Observe where the equilibrium point occurs.

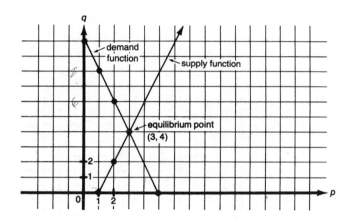

EXAMPLE 7

If cellular telephones are priced at $200 each, the Wygand Manufacturing Company is willing to produce 400 telephones. If the telephone's price drops to $150, the company is willing to produce only 200 telephones.

If p is the price per telephone and q corresponds to supply, use the concept of slope to find the rate of change of supply with respect to price. Assume the relationship is linear.

___

Solution: At price of $p = \$200$, quantity supplied is $q = 400$ or $(200, 400)$

At price of $p = \$150$, quantity supplied is $q = 200$ or $\dfrac{(150, 200)}{50, 200}$

The slope $= m = \dfrac{\Delta y}{\Delta x} = \dfrac{\Delta q}{\Delta p} = \dfrac{200}{50} = 4$

The rate of change of supply with respect to price is 4.

Problem 3

If the price of a cellular telephone increases by $1, the supply will increase by how many telephones?

___

Solution:

Answer: 4

EXAMPLE 8

Determine the supply function for cellular telephones with $q = S(p)$.

Solution: What do we know about this supply function?

It is assumed to be linear; therefore its graph is a line.

When $p = 200$, $q = 400$, or (200, 400) is a point on the line.

When $p = 150$, $q = 200$, or (150, 200) is also on the line.

The slope has been calculated to be 4.

I draw pictures whenever possible. A picture usually makes the problem easier for me to comprehend. The graph below is a quick sketch of the information provided in the problem.

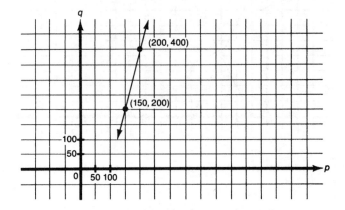

To determine the supply function, we are actually being asked to write the equation of the line. To write the equation of a line, use the point-slope formula.

Point-slope formula: $y - y_1 = m(x - x_1)$

Use one point and slope. $y - 400 = 4(x - 200)$

$$= 4x - 800$$

$$y = 4x - 400$$

Be careful; this is not the answer. The problem asks for $q = S(p)$. In other words, the answer is to be written in terms of p and q rather than x and y. But that is easy. $S(p)$ indicates that p is to be the independent variable. By exchanging variables, we find that the correct answer is:

$$q = S(p) = 4p - 400$$

EXAMPLE 9

a.   Where does the supply function from Example 8 cross the horizontal axis?

b.   What is the economic significance of this point?

Solution: a.   The point where the supply function crosses the horizontal axis is the x-intercept. The x-intercept is where $y = 0$. Actually it is the p-intercept, but it is easier to think of the x-intercept.

$$y = 4x - 400$$

$$0 = 4x - 400$$

$$4x = 400$$

$$x = 100$$

The point where the supply function crosses the horizontal axis is 100. Mathematically, this means that the restricted domain for the supply function is $p \geq 100$ because, if p were less than 100, q would be negative and meaningless. The company cannot produce a negative number of telephones.

b. The economic significance of this point is that, at a price of $100 or less, the Wygand Manufacturing Company would be unwilling to produce and sell any cellular telephones.

Ready to try one yourself?

Problem 4

In an attempt to forecast the demand for a particular style of running shoes, Bill noted that, if the price per pair was $50, he sold only 30 pairs. But when he had a sale and lowered the price to $45, 45 pairs were sold. Bill assumes there is a linear relationship between the price and the number of pairs sold.

a. Determine $q = D(p)$, where p is the price per pair of running shoes and q is the number of pairs of running shoes demanded at price p.

A quick sketch of the information might help.

b. Use your function to predict how many pairs of running shoes will be sold if Bill has a half-price sale and the price of the shoes is lowered to $25.

c. Graph the function.

d. What is the restricted domain?

e. What is the restricted range.

Solution:

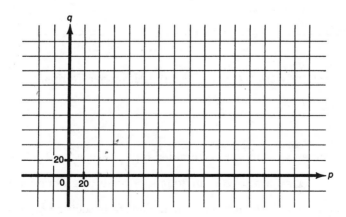

Answers: a. $q = D(p) = -3p + 180$

b. If the price of shoes is $25, it is estimated that Bill will sell 105 pairs.

c.

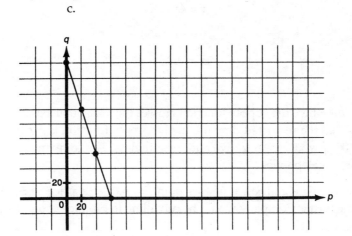

d. $0 \le p \le 60$

e. $0 \le q \le 180$

Let's leave supply and demand functions for the time being and look at a few other types of applications. The objective here will be to interpret the meaning of points, slope, and intercepts in the context of the application.

The following functions were defined in the exercises in Unit 2 and are restated here for convenience.

Total revenue = (selling price/unit)(number of units sold)

Total cost = fixed cost + (variable cost/unit)(number of units)
or, if the total variable cost is given,
= fixed cost + total variable cost

Profit = total revenue − total cost

EXAMPLE 10

In Unit 2, the following functions were determined for the Alexander Company:

$C(x) = 13{,}200 + 43x$

$R(x) = 78x$

where x = the number of braided rugs produced and sold by the company,

$C(x)$ = the total cost, in dollars, of producing x braided rugs,

and $R(x)$ = the total revenue, in dollars, from selling x braided rugs.

a.   Identify and interpret the slope of the total cost function.

b.   Identify and interpret the slope of the total revenue function.

Solution: a.   The slope of $C(x)$ is 43, or $m = \dfrac{\Delta y}{\Delta x}$ or $\dfrac{\Delta C}{\Delta x} = 43 = \dfrac{43}{1}$.

Notice that the variable cost per unit is the slope.

For each additional braided rug produced, total cost will increase by $43.

b.   The slope of $R(x)$ is 78, or $m = \dfrac{\Delta y}{\Delta x}$ or $\dfrac{\Delta R}{\Delta x} = 78 = \dfrac{78}{1}$.

For each additional braided rug sold, total revenue will increase by $78.

BREAK-EVEN ANALYSIS

Break-even analysis identifies the level of operation or level of output at which a manufacturer is just able to break even on operations, neither incurring a loss nor earning a profit. The **break-even point** represents the level of output at which total revenue equals total cost, $R(x) = C(x)$.

EXAMPLE 11

Determine how many braided rugs the Alexander Company must sell in order to break even.

Solution: Break even is where total revenue equals total cost,

$$R(x) = C(x)$$

By substitution: $78x = 13{,}200 + 43x$

$$35x = 13{,}200$$

$$x = 377.14$$

The Alexander Company must make and sell approximately 377 braided rugs in order to break even.

As previously stated, the break-even point is the level of output at which the producer is just able to break even, neither incurring a loss nor earning a profit. Any changes from the break-even level of operation will result in either a profit if the output is increased, or a loss if the output is decreased.

EXAMPLE 12

Verify that, if the Alexander Company produced and sold 450 braided rugs, it would make a profit.

Solution: First determine the profit function for braided rugs.

$$\text{Profit} = \text{total revenue} - \text{total cost}$$

$$P(x) = R(x) - C(x)$$

$$= 78x - (13{,}200 + 43x)$$

$$= 78x - 13{,}200 - 43x$$

$$= 35x - 13{,}200$$

Use the profit function to determine the profit if 450 rugs are sold.

$$P(450) = 35(450) - 13{,}200$$

$$= 15{,}750 - 13{,}200$$

$$= 2{,}550$$

If 450 braided rugs are produced and sold, the Alexander Company will make $2,550 profit.

Problem 5

Verify, by selecting some number less than 377, that, if the number of braided rugs produced and sold is less than the break-even level, the Alexander Company will experience a loss.

Solution:

$$\frac{y_2 - y_1}{x_2 - x_1} \qquad \frac{1225 - 625}{150 - 50} \qquad \frac{600}{100} = 6.0$$

$(50, 625)$

$(150, 1225)$ $\qquad m = 6$

Answers will vary depending upon the number selected.

EXAMPLE 13

Mary Lou's Toy Company can produce 50 teddy bears at a total cost (fixed plus variable) of $625, while 150 such teddy bears cost $1,225 to produce. Mary Lou assumes there is a linear relationship between cost and the number of teddy bears produced.

Determine total cost = f(number of teddy bears produced).

Solution: What do we know about this function?

Neither fixed costs nor variable costs are given, so we cannot simply substitute into the total cost function.

The way the question has been phrased—determine total cost = f(number of teddy bears produced)—indicates that the number of teddy bears is to be the independent variable, while total cost is to be the dependent variable.

Therefore, let x = the number of teddy bears produced.

and y = the total cost.

The function is assumed to be linear, the graph is a line.

When $x = 50$, $y = 625$, or (50, 625) must be a point on the line.

When $x = 150$, $y = 1{,}225$, or (150, 1,225) must also be on the line.

Does all of this sound familiar?

Again a quick sketch of the information may be helpful.

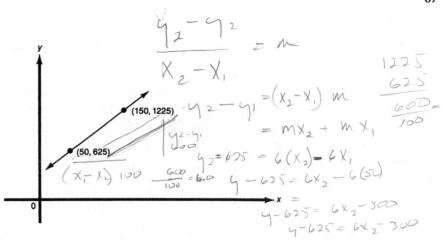

$$\frac{y_1 - y_2}{x_2 - x_1} = m$$

$$y_2 - y_1 = (x_2 - x_1)\, m$$

$$= m x_2 + m x_1$$

$$y_2 = 625 = 6(x_2) - 6 x_1$$

$$y - 625 = 6 x_2 - 6(50)$$

$$y - 625 = 6 x_2 - 300$$

$$y - 625 = 6 x_2 - 300$$

$$\frac{1225}{625}$$
$$\frac{600}{100} =$$

$$y = 6 x_2 + 325$$

Actually we are being asked to do nothing more than write the equation of a line, given two points on it.

Calculate the slope.   (150, 1,225)

(50,   625)

—————————

100,   600 $m = \dfrac{\Delta y}{\Delta x} = \dfrac{600}{100} = 6$

Point-slope formula: $y - y_1 = m(x - x_1)$ $y - y_1 = m(x_0 - y_0)$

Use one point and slope. $y - 625 = 6(x - 50)$ $y - 625 = 6(x - 50)$

$$y - 625 = 6x - 300$$ $y - 625 = 6x - 300$

$$y = 6x + 325$$ $y = 6x + 325$

$$y = 325 + 6x$$

Answer: $y = f(x) = 6x + 325$ $x \geq 0$

$$y = 6(300) + 325$$

where $x =$ the number of teddy bears produced

$$1800 + 325$$

and $y =$ the total cost, in dollars, of producing x teddy bears.

$$2125$$

Problem 6

Use the information from Example 13 to answer the following questions.

a.   What will the cost be if 300 teddy bears are produced?

b.   Find the variable cost per unit. — the slope — the amt of increase in y (TC) per unit increase

c.   Find the fixed cost.

Solution:

Answers: a. $2,125.   b. $6.   c. $325

Example 14 looks at an application problem defined piecewise.

EXAMPLE 14

My friend Rosalie is a freelance writer and is quite good at her job. She specializes in writing annual reports for corporations. Annual reports average between 12 and 16 pages, and her fee, which is a function of the number of pages she writes, includes the cost of research, on-site visits, and all necessary rewrites. Rosalie's schedule of fees is given by

$$F(x) = \begin{cases} 7{,}000 & \text{if } 1 \le x \le 5 \\ 5{,}000 + 500x & \text{if } 5 < x \le 10 \\ 1{,}000x & \text{if } 10 < x \end{cases}$$

where $F(x)$ = Rosalie's fee, in dollars, for writing x pages.

a.   Graph $F(x)$.

b.   Interpret $F(x)$ in this application.

c.   What would Rosalie's fee be for a 15-page annual report?

d.   Determine her fee for writing an 8-page annual report.

Solution: a.   The function is defined piecewise with three parts.

One part is:

$$F(x) = 7{,}000 \quad \text{if } 1 \le x \le 5$$

This part of the function is linear. The graph will be a line segment.

Lower limit: Let $x = 1$ and find $F(1)$.

$$F(x) = 7{,}000$$

$$F(1) = 7{,}000$$

One endpoint of the line segment is $(1, 7{,}000)$.

Upper limit: Let $x = 5$ and find $F(5)$.

$$F(x) = 7{,}000$$

$$F(5) = 7{,}000$$

The other endpoint of the line segment is $(5, 7{,}000)$.

The second part is:

$$F(x) = 5{,}000 + 500x \quad \text{if } 5 < x \le 10$$

This part of the function is linear. The graph will be a line segment too.

Lower limit: Let $x = 5$ and find $F(5)$.

$$F(x) = 5{,}000 + 500x$$

$$F(5) = 5{,}000 + 500(5)$$

$$= 5{,}000 + 2{,}500$$

$$= 7{,}500$$

One endpoint of the line segment is $(5, 7{,}500)$.

This endpoint is graphed as an open circle because 5 is not included in the interval $5 < x \le 10$.

Upper limit: Let $x = 10$ and find $F(10)$.

$$F(x) = 5{,}000 + 500x$$

$$F(10) = 5{,}000 + 500\,(10)$$

$$= 5{,}000 + 5{,}000$$

$$= 10{,}000$$

The other endpoint of the line segment is (10, 10,000).

The third part is:

$$F(x) = 1{,}000x \quad \text{if } 10 < x$$

This part of the function is also linear but the graph will be what is called a **half line.** The half line has one endpoint at the lower limit but continues on since it has no upper limit.

Lower limit: Let $x = 10$ and find $F(10)$.

$$F(x) = 1{,}000x$$

$$F(10) = 1{,}000(10)$$

$$= 10{,}000$$

The one endpoint of the half line is (10, 10,000).

This endpoint also would be graphed as an open circle because 10 is not included in the interval $10 < x$. But notice that the open circle will be "filled in" by one of the endpoints of the second part.

Since there is no upper limit, we need to either find the coordinates of a second point or use the slope to locate a second point. In this case it is just as easy to pick another value for x, but remember that it must be greater than 10.

Suppose we let $x = 11$. Find $F(11)$.

$$F(x) = 1{,}000x$$

$$F(11) = 1{,}000(11)$$

$$= 11{,}000$$

(11, 11,000) is a second point on the half line.

The graph of $F(x)$ is shown on the grid. Do you see why the third part is referred to as a half line?

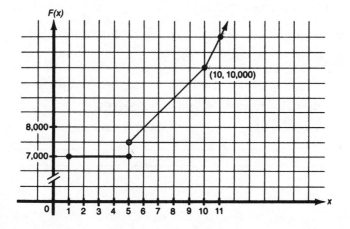

b. Try doing the interpretation of $F(x)$ yourself before reading the answer. It might help to look at the graph.

Rosalie charges a flat fee of $7,000 for writing an annual report anywhere from one page to five pages in length. Anything over five pages, even a fraction of a page, up to and including ten pages is charged at the rate of $5,000 plus $500 per page. Any report in excess of ten pages is charged at the rate of $1,000 per page.

c. and d.   I will leave the solutions of the last two questions for you.

Here are the answers: c. $15,000;   d. $9,000.

CALCULUS AND THE CALCULATOR (Optional)

Finding Points of Intersection

We will revisit Example 7, but this time we will find the solution using the graphing calculator.

EXAMPLE 15

Given: $y = D(x) = -2x + 10 \quad 0 \le x \le 5$

$y = S(x) = 2x - 2 \quad x \ge 1.$

Graph and find their point of intersection.

Solution: Use the standard viewing window (**ZOOM**) (**6**).

Key in *both* functions as follows:

$\backslash Y_1$ (**Y=**)(**(-)**)(**2**)(**X,T,θ,n**)(**+**)(**1**)(**0**)(**ENTER**)

$\backslash Y_2$ (**Y=**)(**2**)(**X,T,θ,n**)(**-**)(**2**)(**ENTER**)

and view the graph. The part of the graph in Quadrant I should match the graph shown in Example 6.

To find the point of intersection, press (**2nd**) (**CALC**) (**5**).

At the <u>First curve?</u> prompt, move the cursor along the first line to where the intersection appears to be and press (**ENTER**).

At the <u>Second curve?</u> prompt, move the cursor along the second line to where the intersection appears to be and press (**ENTER**).

At the <u>Guess?</u> prompt, move the blinking star between the boundaries shown on the screen by the two arrows, ▶ ◀, and press (**ENTER**).

Answer: Intersection X = 3 Y = 4 appears at the bottom of the screen.

You should now be able to interpret points, slopes, and intercepts for various application problems, especially those involving cost and revenue functions and break-even points. In some instances, you should be able to write the linear function that relates the variables as described by the given information. Often a quick sketch proves helpful.

In addition you should have a reasonably good sense of what is meant by supply and demand functions and of how the equilibrium price can be found.

Before beginning the next unit you should work the following exercises. These are probably some of the more difficult ones that you will encounter in this book, but try them. If you need help, remember that the answers are given in the back.

EXERCISES

Be sure to *interpret* each answer in terms of the application problem.

1. A street vendor in Harrisburg has noticed that the amount of lemonade he sells during the lunch hour is closely related to the temperature. He estimates that

$$L(t) = 1.5t - 60$$

where t = the temperature, in F°,

and $L(t)$ = the number of lemonades sold at lunch.

   a. Find $L(86)$.

   b. Determine the significance of $L(40)$.

   c. Identify the slope.

2. Marguerite is attending school and earning extra money by providing a cleaning service. For $10 an hour she will come and clean your house. Marguerite brings her own cleaning supplies and often has a helper, Christine. She estimates that her weekly costs and revenue are given by

$$C(x) = 38.50 + 4.5x$$

$$R(x) = 10x$$

where x = the number of hours worked per week,

$C(x)$ = Marguerite's weekly costs, in dollars,

and $R(x)$ = Marguerite's total revenue, in dollars.

   a. How many hours must Marguerite work to break even?

   b. Marguerite would like to attend an upcoming summer concert. Tickets are priced at $22.00 and go on sale in a week. How many hours must she work to have sufficient money for a ticket?

3. Graph: $f(x) = \begin{cases} 2x + 1 & \text{if } 0 \le x \le 4 \\ 3 & \text{if } 4 < x < 9. \\ \dfrac{-2}{3}x + 12 & \text{if } 9 \le x \end{cases}$

4. The police in a major city have released crime figures indicating that reported crime was up 11.4% in May from April of this year. One police consultant has estimated that the number of reported crimes can be predicted by the function

$$n = f(t) = 1,850 + 322t$$

> where n = the number of reported crimes per month

> and t = the time, in months,

since January ($t = 0$ corresponds to January of this year).

a. Which is the independent variable? Do a quick *sketch* of information.

b. Identify the y-intercept (vertical intercept) and interpret the meaning in this application.

c. Identify the slope and interpret the meaning in this application.

d. Predict the number of reported crimes that will occur this coming October.

5. The demand function for cellular telephones at the Wygand Manufacturing Company is defined by

$$q = D(p) = 15,000 - 30p \quad 0 \le p \le 500$$

> where p = the price, in dollars, per telephone

> and q = the quantity demanded at price p.

a. Calculate and interpret $D(100)$.

b. Identify and interpret the slope.

c. If the price of a telephone *decreases* by \$1, demand will increase by how many telephones?

6. Consider the relation

$$p + 2q - 5,000 = 0$$

> where p = the price, in dollars, of a product

> and q = the quantity demanded, in units.

a. Determine the corresponding demand function $q = D(p)$.

b. Determine $D(2,000)$.

c. Identify the y-intercept (vertical intercept).

d. Comment about $D(5,000)$.

e. Identify and interpret the slope.

7. The ABC Dairy Company uses the following demand function:

$$q = D(p) = -3,000p + 4,200$$

> where p = the price, in dollars, for a quart of milk

> and q = the number of quarts demanded at price p

to predict the weekly demand for milk at various prices. All answers should be in sentence form. You may wish to do a quick sketch.

a. How many quarts of milk will be demanded if the price is \$0.70 per quart?

b. Determine the y-intercept (vertical intercept) and interpret its meaning.

c. Determine the x-intercept (horizontal intercept) and interpret its meaning.

d. If the ABC Dairy Company wants to sell 1,000 quarts per week, what price should be charged?

8. If city maps are priced at $5 each, suppliers are willing to produce 46 maps. If the price of maps increases to $9 each, suppliers are willing to produce 86. It is assumed that the relationship is constant, that is, linear.

 a. Determine quantity supplied = S(price).

 b. Calculate $S(15)$.

 c. Find the x-intercept.

 d. What is the restricted domain for S?

9. The Spaniel Company manufactures cordless toy telephones and nothing else. The company has limited facilities and can manufacture no more than 300 telephones. The company also has determined that its cost function (the total cost of manufacturing x telephones) is given by

$$y = C(x) = 8.35x + 3,400$$

where $C(x)$ = the total cost, in dollars,

and x = the number of telephones manufactured.

 a. What are the company's fixed costs, that is, the costs to the company when no telephones are manufactured?

 b. Find and interpret $C(200)$.

 c. State the restricted domain for C. Be careful.

10. The Spaniel Company has found that the toy telephones are extremely popular, and the company is able to sell all it produces when the telephones are priced at $17.95. Let $R(x)$ represent the total revenue gained from selling x toys telephones.

 a. Write the total revenue function for $R(x)$.

 b. Find and interpret $R(200)$.

11. Let $P(x)$ represent the total profit, in dollars, for the Spaniel Company from the production and sale of x toy telephones.

 a. Write the total profit function for $P(x)$, and simplify.

 b. Find and interpret $P(200)$.

12. Continuing with the information on city maps as given in Exercise 8, if city maps are priced at $5 each, there will be a demand for 65 maps. If they are priced at $10 each, the demand drops to 40. Assume that the relationship is linear.

 a. Determine quantity demanded = D(price).

 b. Find both intercepts.

 c. What is the restricted domain for D?

 d. Determine the equilibrium price and corresponding quantity for city maps.

13. Must there be an equilibrium price?

C14. In marketing a certain item, a business estimates that its daily costs and revenue are given by

$$y = C(x) = 0.5x + 500$$

$$y = R(x) = 0.75x$$

where x = the number of items produced per day,

$C(x)$ = the daily costs, in dollars,

and $R(x)$ = the daily revenue, in dollars.

Graph both functions and use the INTERSECT function to determine the number of items that should be produced and sold daily to break even. Hint: Use a viewing window reset to the values shown here.

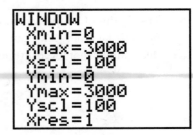

If additional practice is needed:

Barnett, Ziegler, and Byleen, pages 50–51, problems 67–77
Harshbarger and Reynolds, pages 100–101, problems 43–57; pages 131–135, problems 1–56
Hoffmann and Bradley, pages 42–43, problems 35–44
Waner and Costenoble, pages 49–54, problems 1–48

UNIT 6

Quadratic Functions

$$y = ax^2 + bx + c, \quad a \neq 0$$

In this unit you will learn to recognize quadratic functions of one variable and to graph them using a minimum number of well-chosen points.

Definition: $y = f(x) = ax^2 + bx + c$, with a, b, and c being real numbers, $a \neq 0$, is called a **quadratic** or **second-degree function of one variable.**

In other words, a second-degree function of one variable must contain a squared term, x^2, and no higher powered term.

In this unit we will deal only with second-degree or quadratic functions of one variable.

There are basically two ways to graph quadratic functions:

1.   Find and plot a large number of points—this can be time consuming, and the key points missed.

2.   Plot a few well-chosen points based on knowledge of quadratic functions—this is the approach we will use.

$$y = f(x) = ax^2 + bx + c, \quad a \neq 0$$

Definition: $y = f(x) = ax^2 + bx + c$ is called the **standard form** of a quadratic function of one variable.

Note: *All* terms are on the right side of the equal sign with *only* y or $f(x)$ on the left.

The graph of a quadratic function, like that of a linear function, is very predictable. The following are some basic facts that can be used when graphing a quadratic function of one variable.

BASIC FACTS ABOUT THE GRAPH OF A QUADRATIC FUNCTION IN STANDARD FORM: $y = f(x) = ax^2 + bx + c$

1. The graph of a quadratic function is a smooth, ⌣-shaped curve called a **parabola**.

2. If a, the coefficient of the squared term, is positive, the curve opens up: ⌣.

   If a is negative, the curve opens down: ⌢.

3. The y-intercept is c, the constant. Remember that the y-intercept is the value of y when $x = 0$; thus $y = a(0)^2 + b(0) + c = c$.

4. There are **at most two x-intercepts**, "at most" meaning there can be two, one, or none. The x-intercepts are the values of x when $y = 0$; thus the solution to $0 = ax^2 + bx + c$ yields the x-intercepts.

5. The low point on the curve (or the high point if the curve opens down) is called the **vertex.**

6. The vertex is located at the point

$$\left(\frac{-b}{2a}, \frac{4ac - b^2}{4a}\right) \quad \text{or} \quad \left(\frac{-b}{2a}, f\left(\frac{-b}{2a}\right)\right).$$

7. The curve is symmetric to a vertical line through the vertex. The vertical line is called the **axis of symmetry**. Its equation is $x = -\dfrac{b}{2a}$.

On the basis of the above information about quadratic functions of one variable, it is now possible to graph such functions using only a few well-chosen points. By "few" I mean at least three, but no more than five.

SUGGESTED APPROACH FOR GRAPHING QUADRATIC FUNCTIONS OF ONE VARIABLE

1. Write the quadratic function in standard form.

2. Determine whether the parabola opens up or down.

3. Find the y-intercept at c.

4. Solve $0 = ax^2 + bx + c$ to find the x-intercepts, if the equation is easily factored.

   Note: For a review of factoring and solving second-degree equations of this type, see Units 18–22 in *Forgotten Algebra*.

5. Find the coordinates of the vertex:

$$\left(\frac{-b}{2a}, \frac{4ac - b^2}{4a}\right) \quad \text{or} \quad \left(\frac{-b}{2a}, f\left(\frac{-b^2}{2a}\right)\right)$$

6. Draw the axis of symmetry.

x = intercept value of x when y = 0

7. Locate one point on either side of the vertex.

8. Plot the above points, and connect them with a smooth, $\smile$-shaped curve.

The following examples will illustrate this approach.

EXAMPLE 1

Graph: $y = f(x) = x^2 - 8x + 7.$

$y = x^2 - 8x + 7$

Solution: 1.   Write in standard form and compare:

$0 = (x-7)(x-1)$ $x = 7, x = 2$

$$y = f(x) = \boxed{a}\, x^2 \boxed{+b}\, x \boxed{+c}$$
$$y = f(x) = \boxed{}\, x^2 \boxed{-8}\, x \boxed{+7}$$

$b = -8, c = 7$

$a = 1$

Thus $a = 1$, $b = -8$, $c = 7$.

Graph will be a parabola.

2.   Since a is positive, curve opens up: $\smile$.

3.   The y-intercept is $c = 7$.

4.   Find the x-intercept by solving:

$$0 = x^2 - 8x + 7$$
$$0 = (x - 1)(x - 7)$$
$$x = 1, \qquad x = 7$$

The x-intercepts are 1 and 7.

5.   Find the coordinates of the vertex.

$$x = \frac{-b}{2a} = \frac{-(-8)}{2(1)} = \frac{8}{2} = 4$$

$$y = \frac{4ac - b^2}{4a} = \frac{4(1)(7) - (-8)^2}{4(1)} = \frac{28 - 64}{4} = \frac{-36}{4} = -9$$

The vertex is located at (4, –9).

6.   Draw the axis of symmetry with a dashed line. It is a vertical line through the vertex, (4, –9).

7.   Locate one point on either side of the vertex. Since the vertex is at $x = 4$:

$$\text{let } x = 3; \text{ then } y = f(3) = (3)^2 - 8(3) + 7 = 9 - 24 + 7 = -8$$

$$\text{let } x = 5; \text{ then } y = f(5) = (5)^2 - 8(5) + 7 = 25 - 40 + 7 = -8$$

8.   Plot the above points, and connect them with a smooth curve.

Answer:

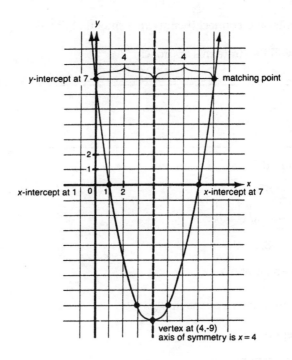

Before going on to another example, let's mention a few points about the above graph. Be sure to notice that every ordered pair whose coordinates satisfy the function, $f(x) = x^2 - 8x + 7$, must correspond to some point on the curve. And every point on the curve must have coordinates that satisfy the function.

The axis of symmetry is indicated with a dashed line. Because the curve is symmetric about this line, the y-intercept at 7 will have a matching point the same distance from, but on the other side of, the axis of symmetry. As shown above, the y-intercept is four spaces to the left of the axis of symmetry. Since the curve is symmetric about this line, there must be a matching point four spaces to the right of the axis of symmetry, which corresponds to the y-intercept of 7. The actual coordinates of the point are not important—only the point's location, which is shown, need concern us.

The arrows on each end of the parabola indicate that the curve continues upward.

EXAMPLE 2

Graph: $y = f(x) = 3 + x^2$.

Solution: 1.   Write in standard form and compare:

$$y = f(x) = \boxed{}\, x^2\, \boxed{}\, \boxed{+3}$$
$$y = f(x) = \boxed{a}\, x^2\, \boxed{+b}\, x\, \boxed{+c}$$

Thus $a = 1$, $b = 0$, $c = 3$.

Graph will be a parabolic curve.

2.   Since a is a positive, curve will open up: ⌣.

3.   The y-intercept is $c = 3$.

4. Find the x-intercepts, at most two, by solving:

$$0 = x^2 + 3$$

$$-3 = x^2 \text{ has no solution}$$

Therefore there are no x-intercepts.

5. Find the coordinates of the vertex:

$$x = \frac{-b}{2a} = \frac{0}{2(+1)} = 0$$

To find the y-coordinate for the vertex, we can, of course, use the formula $(4ac - b^2)/4a$, but there is an alternative method.

Once the x-coordinate is determined, the y-coordinate can be calculated by substituting into the original function:

if $x = 0$, then $f(0) = 3 - (0)^2 = 3$.

The vertex is located at $(0, 3)$.

For this particular example, using the original function was an easier way to determine y than using the formula for the vertex, but the choice is yours.

6. In this example the axis of symmetry is the y-axis. Draw a dashed line right next to the y-axis as shown.

7. Locate one point on either side of the vertex:

$$\text{let } x = 1; \text{ then } y = f(1) = 3 + (1)^2 = 3 + 1 = 4$$

$$\text{let } x = -1; \text{ then } y = f(-1) = 3 + (-1)^2 = 3 + 1 = 4$$

8. Plot the above points, and connect them with a smooth curve.

Answer:

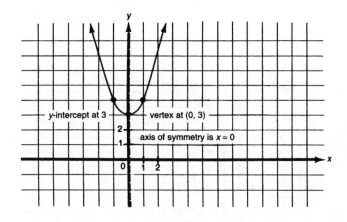

If more detail is needed, plot additional points.

Now it is time for you to try a few problems.

Problem 1

Graph: $y = f(x) = x^2 - 2x - 8$.

Solution: 1.   Write in standard form and compare.

2.   Curve will open ___*up*___ .

3.   The y-intercept is _____ .

4.   The x-intercepts, at most two, are ___(-2,0)___ and ___(4,0)___ .

5.   The vertex is located at ___(1,-9)___ .

6.   Draw the axis of symmetry.

7.   Locate one point on either side of the vertex.

Step 7 may be omitted because we already have points on either side of the vertex, the y-intercept, its matching point, and the two x-intercepts.

8.   Plot the above points on the grid provided and connect them with a smooth curve.

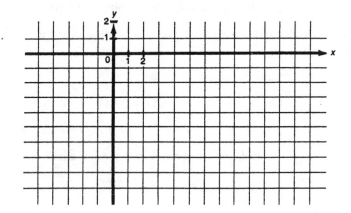

Answer:

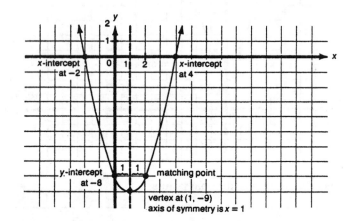

Problem 2

Graph: $y = f(x) = 2x^2$.

Solution:

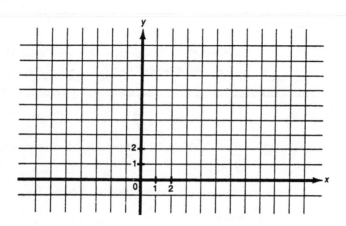

Answer:

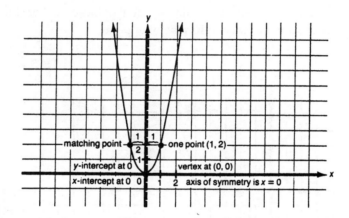

Example 3 illustrates still another method for locating the vertex.

EXAMPLE 3

Graph: $y = f(x) = x^2 - 6x + 5$.

Solution: 1.   This is a quadratic function with $a = 1$, $b = -6$, and $c = 5$.

2.   Parabolic curve opens up.

3.   The y-intercept is 5.

4.   The x-intercepts are:

$$0 = x^2 - 6x + 5$$

$$0 = (x - 1)(x - 5)$$

$$x = 1, \quad x = 5$$

5. As before, we could locate the vertex by using the formula, but for this example there is an easier method. Recall that the axis of symmetry is a vertical line through the vertex that cuts the curve in the middle; the two sides of the curve must match. Since in this example the x-intercepts are 1 and 5, where must the axis of symmetry be? It must be midway between the two points. Thus the x-coordinate of the vertex is 3, the value in the middle of 1 and 5.

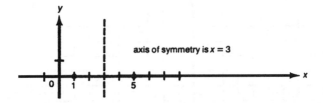

The y-coordinate may be calculated by either method discussed earlier. In this case I will substitute into the original function:

$$\text{if } x = 3, f(3) = (3)^2 - 6(3) + 5 = 9 - 18 + 5 = -4$$

The vertex is located at $(3, -4)$.

6. The axis of symmetry was found earlier in the problem.

Step 7 may be omitted because we already have points on either side of the vertex, the two x-intercepts, the y-intercept, and its matching point.

Answer:

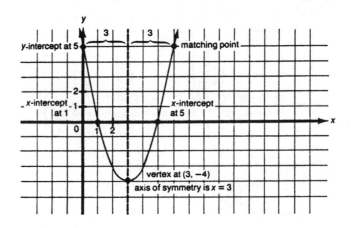

As you probably suspected, all quadratic functions are not as easy to work with as the preceding ones. Consider the next example.

EXAMPLE 4

Graph: $y = f(x) = x^2 + 2x - 6$.

Solution: 1.   This is a quadratic function with $a = 1$, $b = 2$, and $c = -6$.

2.   Parabolic curve opens up.

3.   The y-intercept is -6.

4. Find the x-intercepts, at most two, by solving:

$$0 = x^2 + 2x - 6$$

Since the right-hand side is not easily factored, we will go on to step 5.

Note: If the x-intercepts were required, the quadratic formula could be used to find them since we are unable to factor.

$$x = \frac{-b \pm \sqrt{b^2 - 4ac}}{2a}$$

$$= \frac{-2 \pm \sqrt{(2)^2 - 4(1)(-6)}}{2(1)}$$

$$= \frac{-2 \pm \sqrt{4 + 24}}{2}$$

$$= \frac{-2 \pm \sqrt{28}}{2}$$

If you have a calculator you will find that $\sqrt{28} \approx 5.29$.

$$x = \frac{-2 + 5.29}{2} \quad \text{and} \quad x = \frac{-2 - 5.29}{2}$$

$$= 1.6 \qquad\qquad\qquad\qquad = -3.6$$

The x-intercepts are approximately 1.6 and –3.6.

5. The vertex is located at (–1, –7) because:

$$\frac{-b}{2a} = \frac{-2}{2(1)} = -1 \quad \text{and} \quad f(-1) = (-1)^2 + 2(-1) - 6 = 1 - 2 - 6 = -7$$

6. Draw the axis of symmetry.

Step 7 may be omitted because again we have a point on either side of the vertex, the two x-intercepts.

Answer:

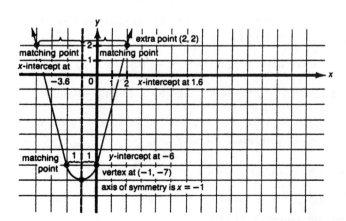

Comments: To get a better sense of where the curve actually crosses the x-axis, I plotted a convenient extra point in quadrant I, (2, 2). There would be a corresponding point in quadrant II. When the points are connected with a smooth curve, the locations of the x-intercepts are well determined.

CALCULUS AND THE CALCULATOR (Optional)

Graphing Quadratic Functions

Graphing a quadratic function and finding x-intercepts is basically the same as working with a linear function. Finding the vertex, though, can be done several different ways. In this unit, I will explain my preference, which uses the steps and keystrokes introduced in Units 1 and 3. Another method will be presented later.

EXAMPLE 5

Graph $y = f(x) = -2x^2 + 5x + 3.$

    a.   Find the coordinates of the vertex.

    b.   Find the x-intercepts.

Solution: Use the standard viewing window $\boxed{\text{ZOOM}}$ $\boxed{6}$.

Create a table of values for the function with the following keystrokes:

$\boxed{\text{Y=}}$ $\boxed{\text{(−)}}$ $\boxed{2}$ $\boxed{\text{X,T,θ,n}}$ $\boxed{\wedge}$ $\boxed{2}$ $\boxed{+}$ $\boxed{5}$ $\boxed{\text{X,T,θ,n}}$ $\boxed{+}$ $\boxed{3}$ $\boxed{\text{ENTER}}$

To see the y-values, press $\boxed{\text{2nd}}$ $\boxed{\text{TABLE}}$.

To see the graph, press $\boxed{\text{GRAPH}}$.

    a.   Return to the table of y-values, press $\boxed{\text{2nd}}$ $\boxed{\text{TABLE}}$.

         This is a quadratic function with $a = -2$, $b = 5$, and $c = 3$.

         The x-coordinate of the vertex is given by the formula $y = \dfrac{-b}{2a}$.

         Using the formula, key in the x-coordinate of the vertex as follows:

        $\boxed{\text{(−)}}$ $\boxed{5}$ $\boxed{÷}$ $\boxed{(}$ $\boxed{2}$ $\boxed{\times}$ $\boxed{\text{(−)}}$ $\boxed{2}$ $\boxed{)}$ $\boxed{\text{ENTER}}$

         Note: The entire denominator must be enclosed in parentheses so that −5 is divided by the product of 2 and −2, not just the first 2.

Answer: The vertex is located at the point (−1.25, −6.375).

        As a check, use $\boxed{\text{TRACE}}$ to verify that the answer is reasonable.

    b.   To find the positive x-intercept, also called a zero of the function, the steps are the same as those introduced in Unit 3.

        Press $\boxed{\text{2nd}}$ $\boxed{\text{CALC}}$ $\boxed{2}$.

        At the <u>Left Bound?</u> prompt, move the cursor to a point just to the left of the x-intercept and press $\boxed{\text{ENTER}}$.

At the <u>Right Bound?</u> prompt, move the cursor to a point just to the right of the x-intercept and press (**ENTER**).

At the <u>Guess?</u> prompt, move the cursor between the boundaries and press (**ENTER**). The answer appears at the bottom of the screen.

Answer: The positive x-intercept is 3.

For practice, find the negative x-intercept on your own.

Answer: The negative x-intercept is –1.5

From here on, I expect you always to verify the reasonableness of your analytical answers by reworking the problem on the graphing calculator.

You should now be able to recognize a quadratic function of one variable, of the type $y = f(x) = ax^2 + bx + c$ with $a \neq 0$. Also you should be able to graph it using a minimum number of well-chosen points based on your knowledge of quadratic functions.

Briefly, the graphing approach is as follows:

1.   Write the function in standard form and compare.

2.   Determine whether the parabola opens up or down.

3.   Find the y-intercept at c.

4.   Solve $0 = ax^2 + bx + c$ to find the x-intercepts, if the equation is easily factored.

5.   Find the coordinates of the vertex:

$$\left(\frac{-b}{2a}, \frac{4ac - b^2}{4a}\right) \quad \text{or} \quad \left(\frac{-b}{2a}, f\left(\frac{-b}{2a}\right)\right)$$

6.   Draw the axis of symmetry with a dashed line.

7.   Locate a point on either side of the vertex, if needed.

8.   Plot the points and connect them with a smooth, $\smile$-shaped curve.

Now try the following exercises.

EXERCISES

Graph each of the following by using your knowledge of quadratic functions of one variable to plot a few well-chosen points:

1. $y = f(x) = x^2 + 6x + 9$

2. $y = f(x) = x^2 + 2x + 1$

3. $y = f(x) = -x^2 + 4x$

4. $y = f(x) = -x^2 + 2x + 3$

5. $y = f(x) = 5 - x^2 - 4x$

6. $y = f(x) = -2x^2 + 4x$

7. $y = f(x) = 2x^2 - x - 10$

8. $y = f(x) = -x^2 - 7$

9. $y = f(x) = x^2 - 4x + 2$

10. $y = f(x) = 3x^2 - 9x$

C11.   Find the x-intercepts for the quadratic function given in Exercise 9.

If additional practice is needed:

Barnett, Ziegler, and Byleen, pages 63–64, problems 1–12
Bleau, page 185, problems 1–10
Harshbarger and Reynolds, page 164, problems 1–20
Hoffmann and Bradley, page 26, problems 9–16
Waner and Costenoble, page 77, problems 1–14

UNIT 7

Quadratic Applications

Now that you have mastered the mechanics of graphing quadratic functions, we will look at some applications. These will be similar in nature to those discussed in Unit 5, except that the functions will be quadratics.

In the examples in this unit the emphasis will be on making sketches rather than detailed graphs, in part because of the large numbers that frequently occur in application problems. However, even with a sketch, the key points—the vertex and intercepts—often must be shown.

Recall that the suggested approach for graphing functions is as follows:

1. Write the quadratic functions in standard form.

2. Determine whether the parabola opens up or down.

3. Find the y-intercept at c.

4. Find the x-intercepts, if any, by solving $0 = ax^2 + bx + c$.

    Note: Often this step may be omitted unless the x-intercepts are required for the problem.

5. Find the coordinates of the vertex:

$$\left(\frac{-b}{2a}, \frac{4ac - b^2}{4a}\right) \quad \text{or} \quad \left(\frac{-b^2}{2a}, f\left(\frac{-b}{2a}\right)\right).$$

6. Draw the axis of symmetry.

7. If a detailed graph is required, locate one additional point on either side of the vertex.

8. Plot the above points and connect them with a smooth, ⌣-shaped curve.

Notice in the following examples that the scales on the x- and y-axes have had to be adjusted to accommodate the larger numbers.

EXAMPLE 1

Graph: $y = f(x) = 2.5x^2 - 6{,}250.$

Solution: 1.   This is a quadratic function with $a = 2.5$, $b = 0$, and $c = -6{,}250$.

2.   Parabolic curve opens up.

3.   The y-intercept is $-6{,}250$.

4.   Find the x-intercepts, at most two, by solving:

$$0 = 2.5x^2 - 6{,}250$$

$$-2.5x^2 = -6{,}250$$

$$x^2 = 2{,}500$$

$$x = \pm 50$$

The x-intercepts are 50 and -50.

5.   Recall that the axis of symmetry is a vertical line through the vertex that cuts the curve in the middle. Since in this example the x-intercepts are -50 and 50, the axis of symmetry must be midway between the two points. Thus the x-coordinate of the vertex is 0, the value in the middle of -50 and 50.

The y-coordinate will be calculated by using the original function:

$$\text{if } x = 0, f(0) = 2.5(0)^2 - 6{,}250 = -6{,}250.$$

The vertex is located at $(0, -6{,}250)$.

Answer:

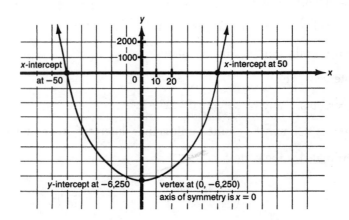

You may be wondering why we are spending so much time graphing. I believe that all this practice will make life easier as you progress through the book.

Remember that supply functions relate the market price to the quantities that suppliers are willing to produce and sell.

EXAMPLE 2

The supply function for ceiling-hugger fans is defined by

$$S(p) = 2.5p^2 - 6{,}250$$

where $p =$ the price, in dollars, per fan

and $S(p) =$ the number of fans the manufacturer is willing to supply at price p.

    a. Sketch the function.

    b. Calculate and interpret $S(60)$.

    c. Is there any restriction on the domain?

Solution: a. $S(p) = 2.5p^2 - 6{,}250$ is the same function as in Example 1 except that the independent variable is changed to p. The sketch is repeated here for convenience.

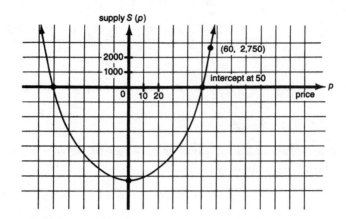

    b. $S(p) = 2.5p^2 - 6{,}250$

$$S(60) = 2.5(60)^2 - 6{,}250$$

$$= 9{,}000 - 6{,}250$$

$$= 2{,}750$$

At a price of \$60 each, the manufacturer is willing to supply 2,750 ceiling-hugger fans.

    c. Since p represents price, p must be positive or zero, that is, nonnegative. From the graph, it should be apparent that, if $p < 50$, then q will be negative and meaningless. Remember that q is quantity and therefore cannot be negative.

Restricted domain: $50 \le p$

EXAMPLE 3

Suppose the total cost of manufacturing x items is given by the function

$$y = C(x) = 3x^2 + x + 50$$

where $C(x)$ = the total cost, in dollars,

and x = the number of items manufactured.

Graph the function.

Solution: 1. This is a quadratic function with $a = 3$, $b = 1$, and $c = 50$.

    2. The parabola opens up.

    3. The y-intercept is 50.

4. Find the x-intercepts, at most two, by solving:

$$0 = 3x^2 + x + 50$$

Since the right-hand side is not easily factored, go on to step 5.

5. Find the coordinates of the vertex.

$$x = \frac{-b}{2a} = \frac{-1}{2(3)} = \frac{-1}{6}$$

$$y = \frac{4ac - b^2}{4a} = \frac{4(3)(50) - (1)^2}{4(3)} = \frac{600 - 1}{12} = \frac{599}{12}$$

The vertex is located at $\left(\frac{-1}{6}, \frac{599}{12}\right)$ or, if you prefer decimals, at $(-0.17, 49.92)$.

6. Draw the axis of symmetry.

7. I choose to locate an additional point on each side of the vertex because of the difficulty in graphing the vertex accurately.

$$\text{let } x = 1; \text{ then } f(x) = 3(1)^2 + 1 + 50 = 54$$

$$\text{let } x = -1; \text{ then } f(x) = 3(-1)^2 + (-1) + 50 = 52$$

Answer:

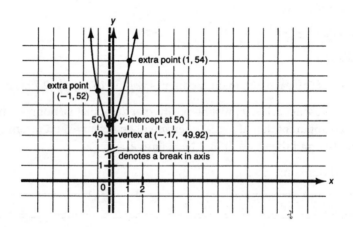

EXAMPLE 4

Use the information from Example 3 to answer the following questions.

a. What is the cost of manufacturing 20 items?

b. Interpret the y-intercept.

c. Calculate and interpret $C(100)$.

Solution: a. $y = C(x) = 3x^2 + x + 50$

$y = C(20) = 3(20)^2 + 20 + 50$

$= 3(400) + 20 + 50$

$= 1{,}270$

The cost of manufacturing 20 items will be $1,270.

b.   The y-intercept is 50.

If no units are manufactured, the cost will be $50. Or, in other words, the fixed costs are $50.

c. $y = C(x) = 3x^2 + x + 50$

$y = C(100) = 3(100)^2 + 100 + 50$

$= 3(10{,}000) + 100 + 50$

$= 30{,}150$

The cost of manufacturing 100 units will be $30,150.

Recall that demand functions describe the relationship between the demand for a product and its price.

====

EXAMPLE 5

Betsy owns a boutique on an exclusive resort island. Over the past few years she has determined that the demand function for T-shirts with the resort's logo is defined by

$$q = D(p) = -p^2 - 10p + 5{,}600$$

where p = the price, in dollars, for the T-shirts

and $D(p)$ = the quantity sold (the demand) at price p.

Use the graph of the function to determine the restricted domain and the range.

Solution: 1.   This is a quadratic function with $a = -1$, $b = -10$, $c = 5{,}600$.

2.   The parabolic curve opens down.

3.   The y-intercept is 5,600.

4.   The x-intercepts are -80 and 70 because

$$0 = -p^2 - 10p + 5{,}600$$

$$0 = p^2 + 10p - 5{,}600$$

$$0 = (p + 80)(p - 70)$$

$$p = -80 \quad p = 70$$

5.   The vertex is located at $(-5, 5{,}625)$ because

$$\frac{-b}{2a} = \frac{-(-10)}{2(-1)} = \frac{10}{-2} = -5$$

and $D(-5) = -(-5)^2 - 10(-5) + 5{,}600 = -25 + 50 + 5{,}600 = 5{,}625$

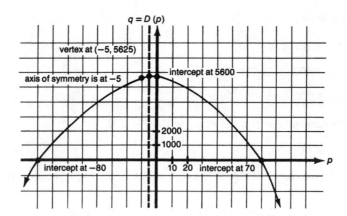

Reminder: In economics texts, the convention is that price (p) is placed on the vertical axis and quantity (q) on the horizontal axis.

To determine the restricted domain and the range, we should reason as follows. Since p is price, any negative number will be meaningless. At the same time, q is quantity and again any negative number will be meaningless. In other words, *both p and q must be positive or zero*, that is, nonnegative.

What restrictions are necessary to ensure that both p and q are nonnegative? The graph above should be of help. Both p and q are nonnegative only in quadrant I.

The restricted domain is the set of all possible values that can be used as input and make sense with reference to the problem. The domain always describes the independent variable, which in this application is p. The only relevant portion of the graph applicable to this problem is the part in quadrant I.

The smallest value p takes on in quadrant I is 0.

The largest value p takes on in quadrant I is 70.

Therefore the restricted domain is $0 \leq p \leq 70$.

The restricted range describes the dependent variable, in this case q, and is the set of all possible values that are output and make sense in terms of the problem. In quadrant I, the smallest value q takes on is 0 and the largest value q takes on is 5,600.

Therefore the restricted range is $0 \leq q \leq 5,600$.

The graph of $D(p)$ with the restricted domain is shown below. Verify for yourself that, if the price of T-shirts is greater than \$70, Betsy will not sell any.

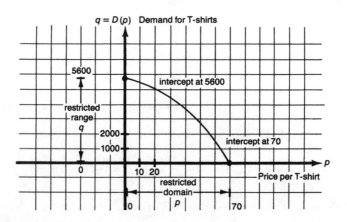

Problem 1

It is near the start of the season and Betsy has received a shipment of 4,400 T-shirts. She decides to price them at $50 each. How many T-shirts should she expect to sell at that price?

Solution:

Answer: According to the demand function, at a price of $50 per shirt Betsy will sell only 2,600 shirts, leaving her with 1,800 unsold at the end of the season.

Problem 2

Suppose Betsy does not want to carry any T-shirts over to the following season, and a sale at the end of the season is impractical. What is the highest price she can put on the T-shirts and still be reasonably sure of selling all of them? I will get you started, but you are to finish the problem yourself.

Solution: $q = -p^2 - 10p + 5{,}600$

$$4{,}400 = -p^2 - 10p + 5{,}600$$

$$p^2 + 10p - 1{,}200 = 0$$

Answer: To sell all 4,400 T-shirts, the highest price Betsy should put on them is $30 each.

EXAMPLE 6

The local country club, founded in 1990, has a membership estimated by the following function:

$$M(t) = -t^2 + 22t + 240$$

where $M(t)$ = the number of members in the country club at time t

and t = the time, in years, since 1990.

Graph the function.

Solution: 1.   This is a quadratic function with $a = -1$, $b = 22$, and $c = 240$.

         2.   Parabolic curve opens down.

         3.   The y-intercept is 240.

         4.   Find the x-intercepts, at most two, by solving:

$$0 = -t^2 + 22t + 240$$

I am better at factoring if the squared term is positive, so I always multiply the entire equation by –1. Be careful when you do this. The entire equation must be multiplied by –1; that means *both* sides. You should not attempt to do this with a function, as it usually complicates the expression rather than simplifies it.

$$0 = t^2 - 22t - 240$$

$$0 = (t + 8)(t - 30)$$

$$t = -8 \qquad t = 30$$

The *x*-intercepts are –8 and 30.

5. The vertex is located at (11, 361) because

$$\frac{-b}{2a} = \frac{-22}{2(-1)} = 11$$

and $f(11) = -(11)^2 + 22(11) + 240 = -121 + 242 + 240 = 361$

Answer:

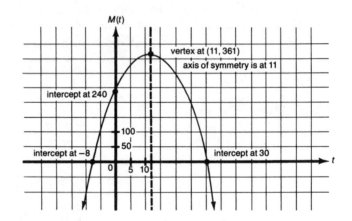

Try answering the questions in Example 7 before looking at the answers.

EXAMPLE 7

Use the information from Example 6 to do the following:

a. Calculate and interpret $M(5)$.

b. Interpret the vertex.

c. Interpret the intercepts.

Solution: a. $M(5)$ denotes $t = 5$ and $t = 5$ means 5 years since 1990.

$$M(t) = -t^2 + 22t + 240$$

$$M(5) = -(5)^2 + 22(5) + 240$$

$$= -25 + 110 + 240$$

$$= 325$$

In the year 1995, the country club had 325 members.

b.   The vertex is located at (11,361).

In 2001 (1990 + 11), the membership of the country club was 361.

c.   The y-intercept is 240 or point (0, 240).

In 1990 (1990 + 0), the year the country club was founded, membership started at 240.

The x-intercept is 30 or point (30, 0).

In the year 2020 (1990 + 30), it is estimated that the membership of the country club will be 0.

EXAMPLE 8

Use the information, especially the graph, from Example 6 to determine the restricted domain and range for $M(t)$.

Solution: Again, *both* t and $M(t)$ must be positive or zero because t is time measured in years since the club's founding and $M(t)$ is the number of members. Neither could be a negative number.

Both t and $M(t)$ are nonnegative only in quadrant I.

The domain always refers to the independent variable; in this application it is t. Using the graph from Example 6, we find that the smallest value t takes on in quadrant I is 0, and the largest value t takes on in quadrant I is 30.

Therefore the restricted domain is $0 \le t \le 30$.

To analyze the problem yet another way, if t is less than 0, it would be before the founding of the club, and hence meaningless; if t is greater than 30, the membership would be a negative number and again meaningless.

The range always describes the values of the dependent variable, in this case $M(t)$. Using the graph from Example 6, we find that the smallest value $M(t)$ takes on in quadrant I is 0; the largest value $M(t)$ takes on in quadrant I is 361.

Therefore the restricted range is $0 \le M(t) \le 361$.

The relevant portion of the graph is shown below.

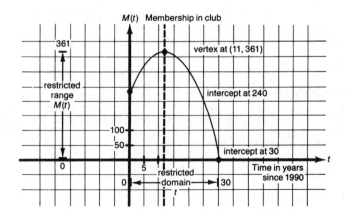

The graph of $M(t) = -t^2 + 22t + 240$ with $0 \le t \le 30$.

If you are able to answer the questions in Example 9 yourself, you have a good understanding of the material thus far.

EXAMPLE 9

Use the graph of Example 8 to answer the following:

a.   At what year between 1990 and the present was the membership of the country club the largest?

b.   What was the largest membership?

c.   How would you describe the future prospects for the country club?

Solution: a.   From the graph, the highest point on the parabola is the vertex which is located at $(11, 361)$. The interpretation of that point, as you probably recall, is that in the year 2001 the membership of the country club was 361.

Therefore the membership was the largest when $t = 11$, which corresponds to 2001.

b.   The largest membership was 361, which occurred in 2001.

c.   The club's prospects are poor; it is estimated that the country club will cease to have any members by the year 2020.

Are you starting to see the value of being able to graph quadratics?

You now should be able to interpret all points, especially the vertex and the intercepts, of a quadratic as they relate to various applications.

Here are some comprehensive exercises to try before continuing on to the next unit.

EXERCISES

1.   Let $y = f(x) = 0.01x^2 - 8x$

where $f(x) = $ the supply function per unit for the Albemarle Manufacturing Company's stress-reducing comfort recliner,

and $x = $ the price per unit, in dollars.

a.   Graph the supply function $f(x)$, including the x-intercepts.

b.   Interpret $f(800)$.

c.   Determine the restricted domain.

2.   The number of pounds of seed supplied by farmers at a price of p dollars per pound is given by

$$S(p) = p^2 - 100.$$

a.   Graph the supply function $S(p)$.

b.   Calculate and interpret $S(18.50)$.

c.   What would the price have to be in order for the farmers to be willing to supply exactly 800 pounds?

3. Assume you are the manager of a gourmet food shop. From past records you determined that the demand function for a rare blend of imported coffee is defined by

$$y = D(p) = -p^2 - 5p + 1,400$$

where p = the price, in dollars, for a pound of coffee

and y = the number of pounds sold (the demand) at price p.

a. Sketch the function $D(p)$, showing all the key points.

Hint: $p^2 + 5p - 1,400 = (p + 40)(p - 35)$.

b. Interpret the y-intercept.

c. Calculate and interpret $D(15.75)$.

d. Interpret the p-intercept located in the restricted domain.

e. State the restricted range.

f. Suppose you are to receive a shipment of 1,334 pounds of the imported coffee for the coming weekend. This particular brand is highly perishable, so you want to price the coffee so as to sell all 1,334 pounds within a reasonable time. What is the *highest* price you can charge per pound and be reasonably sure of selling all of it before it loses its flavor?

4. Suppose that $D(p) = 2p^2 - 297p + 10,960$

and $S(p) = 2p^2 + 5p - 250$

where $D(p)$ = the demand function per unit at price p,

$S(p)$ = the supply function per unit at price p,

and p = the price per unit, in dollars.

Determine the equilibrium price and corresponding quantity.

5. Cathy's Creatures are small driftwood sculptures handcrafted by a local artist, Cathy Morton. Cathy spends the winter months making hundreds of Creatures, which she then sells during the summer months at various arts and crafts fairs. She has found that the demand function for Cathy's Creatures is given by

$$q = D(p) = -2p^2 - 44p + 480$$

where p = the price, in dollars, for one of Cathy's Creatures

and q = the number of Creatures sold at price p.

a. Sketch the function, showing all the key points.

b. Calculate and interpret $D(5)$.

c. Determine the y-intercept (vertical intercept) and interpret its meaning.

d. State the restricted domain.

e. Interpret the x-intercept in the restricted domain.

C6.   The Dooley Company manufactures one product, a trivet. The company has determined that its total cost of producing x trivets is given by

$$C(x) = 8{,}135 + 4.75x + 0.02x^2$$

   a.   Calculate and interpret $C(312)$.

   b.   Interpret the y-intercept, if it exists.

   7.   A manufacturer of T-shirts has a fixed cost of $30,000 and a variable cost per shirt of $9. The selling price per shirt is $12.

   a.   How many shirts must be sold to break even?

   b.   How many T-shirts must be sold to make a profit of $6,000?

   8.   If a rubber ball is thrown vertically upward from the top of a building with an initial speed of 128 feet per second, its height, in feet, t seconds later is given by

$$H(t) = -16t^2 + 128t + 320$$

   a.   Graph the function $H(t)$.

   b.   Point $(2, 512)$ is on the graph of $H(t)$. Interpret its meaning with respect to the problem.

   c.   Compute and interpret $H(9)$.

   d.   Use the graph to determine when the ball will hit the ground.

   e.   Use the graph to determine how high the ball will rise.

   f.   Determine the height of the building.

C9.   The Smith Company's profits, P, in thousands of dollars, are related to the time, t, in years, since the company's incorporation by the function

$$P(t) = 0.4t^2 - 0.8t + 8.44 \quad t > 0$$

   a.   Graph the function.

   b.   Interpret the y-intercept (vertical intercept), if it exists.

   c.   Determine the coordinates of the vertex and interpret the meaning.

   d.   Find the company's annual profit (or loss) for the third year.

   e.   Based on the graph, how would you describe the future prospects for the company?

   10.   Jim recently opened a restaurant specializing in home-style cooking. His accountant has advised him that his profit is related to the number of customers as defined by

$$P(t) = -x^2 + 140x - 4{,}000$$

   where $P(x)$ = the profit (or loss), in dollars, per day

   and x = the number of customers per day.

   a.   Graph $P(x)$.

   b.   Interpret the graph and all key points.

   If you can do this one, you've got it made!

If additional practice is needed:

Barnett, Ziegler, and Byleen, pages 65–67, problems 45–53
Harshbarger and Reynolds, pages 165–167, problems 33–51; page 174, problems 21–27
Hoffmann and Bradley, pages 26–27, problems 25–29; pages 55–61, problems 1–47
Waner and Costenoble, pages 77–79, problems 15–28

UNIT 8

Polynomial Functions

In this unit you will learn to recognize polynomial functions of one variable and to determine their degrees. If the polynomial function is a cubic, you will be able to determine some, but not all, of the characteristics of its graph.

CUBIC FUNCTIONS

Definition: $y = f(x) = ax^3 + bx^2 + cx + d$, with a, b, c, and d being real numbers, $a \neq 0$, is called a **cubic** or **third-degree function of one variable.**

In other words, a third-degree function of one variable must contain a cubed term, x^3, and no higher powered term.

Definition: $f(x) = ax^3 + bx^2 + cx + d$ is called the **standard form** of a cubic function of one variable.

Note: *All* terms are on the right side of the equal sign with *only y* or $f(x)$ on the left.

Recall that the graph of a function was defined as the set of all points, (x, y) whose coordinates satisfy the function.

One way to graph cubic functions is to find and plot a large number of points. We will do one problem together to illustrate this approach.

Problem 1

Graph: $y = f(x) = 2x^3 - 7x^2 - 7x + 5$.

Solution: Find and plot a large number of points whose coordinates satisfy the function.

Let $x = -2$; then $y = 2(-2)^3 - 7(-2)^2 - 7(-2) + 5$

$$= 2(-8) - 7(4) + 14 + 5$$

$$= -16 - 28 + 14 + 5$$

$$= -25, \text{ which gives point } (-2, -25).$$

Let $x = -1$; then $y = 2(-1)^3 - 7(-1)^2 - 7(-1) + 5$

$$= 2(-1) - 7(1) + 7 + 5$$

$$= -2 - 7 + 5 + 5$$

$$= 3, \text{ which gives point } (-1, 3).$$

Find the remaining points yourself. Find all the points listed; otherwise you may miss the shape of the curve.

Let $x = 0$; then $y =$

Let $x = 1$; then $y =$

Let $x = 2$; then $y =$

Let $x = 3$; then $y =$

Let $x = 4$; then $y =$

Let $x = 5$; then $y =$

Plot the points on the grid provided and connect them with a smooth curve. Locate as many points as you need to be sure of the curve's basic shape.

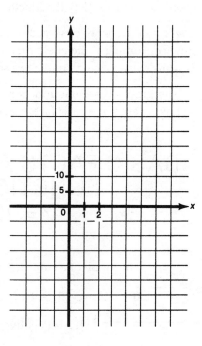

Answer:

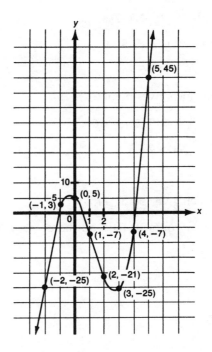

Again notice that every ordered pair whose coordinates satisfy the function, in this case $y = f(x) = 2x^3 - 7x^2 - 7x + 5$, must correspond to some point on the curve. And every point on the curve must have coordinates that satisfy the function.

As you probably observed, unless you use a graphing calculator plotting points could become quite time-consuming if the cubic function was more complicated, and the key points could be easily missed. For this reason we will move to a second method.

The graph of a cubic function, like that of a linear or a quadratic function, is very predictable. The following are some basic facts that can be used when graphing a cubic function of one variable.

BASIC FACTS ABOUT THE GRAPH OF A CUBIC FUNCTION IN STANDARD FORM: $y = f(x) = ax^3 + bx^2 + cx + d$

1. The graph of a cubic function is a smooth, ⌢⌣-shaped curve.

2. If a, the coefficient of the cubed term, is positive, the curve's right-hand tail is going up: ⌢⌣. If a is negative, the curve's right-hand tail is going down: ⌣⌢.

3. The y-intercept is d, the constant. Remember that the y-intercept is the value of y when $x = 0$; thus $y = a(0)^3 + b(0)^2 + c(0) + d = d$.

4. There are at most **three x-intercepts**, "at most" meaning that there can be three or fewer. The x-intercepts are the values of x when $y = 0$; thus the solution to $0 = ax^3 + bx^2 + cx + d$ yields the x-intercepts.

On the basis of the above information about cubic functions of one variable, it is possible to do nothing more than a rough sketch of the function at this time. In later units we will refine the approach so as to include more detail on how high the peak is and how low the dip is.

SUGGESTED APPROACH FOR SKETCHING CUBIC FUNCTIONS OF ONE VARIABLE

1.   Write the cubic function in standard form.

2.   Determine whether the curve's right-hand tail is going up or down.

3.   Find the y-intercept at d.

4.   Find the x-intercepts, if any, by solving $0 = ax^3 + bx^2 + cx + d$. Unfortunately most cubic equations are not easily factored and we will be unable to find the x-intercepts analytically.

5.   Plot the above points and connect them with a smooth, ⌢‿↑ -shaped curve.

Examples 1–4 will illustrate this approach.

EXAMPLE 1

Sketch: $y = f(x) = x^3 - x^2 - 2x$.

Solution: 1.   Write in standard form and compare:

$$y = f(x) = \boxed{a}\, x^3 \boxed{+b}\, x^2 \boxed{+c}\, x \boxed{+d}$$
$$y = f(x) = \boxed{}\, x^3 \boxed{-}\, x^2 \boxed{-2}\, x \boxed{}$$

Thus $a = 1, b = -1, c = -2, d = 0$.

Graph will be a cubic.

2.   Since a is positive, <u>curve's right-hand tail goes up</u>: ⌢‿↑.   ⌢‿↑

3.   The y-intercept is $d = 0$.

4.   Find the x-intercepts by solving:

$$0 = x^3 - x^2 - 2x$$
$$0 = x(x^2 - x - 2)$$
$$0 = x(x - 2)(x + 1)$$
$$x = 0, \quad x = 2, \quad x = -1$$

The x-intercepts are –1, 0, and 2.

5. Plot the above points and connect them with a smooth curve.

Answer:

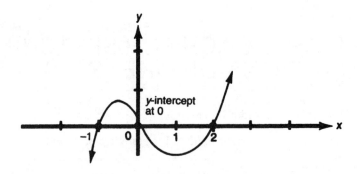

Comment: Remember that the intent is to do a rough sketch. We do not know how high the peak on the left goes up or how low the dip on the right goes down. That information will come later.

If greater detail was needed, we could plot more points.

EXAMPLE 2

Sketch: $y = f(x) = -x^3 - 5x^2 - 6x$.

Solution: 1.   Write in standard form and compare:

$$y = f(x) = \boxed{a}\,x^3 \boxed{+b}\,x^2 \boxed{+c}x\boxed{+d}$$
$$y = f(x) = \boxed{-}\,x^3 \boxed{-5}\,x^2 \boxed{-6}x\;\boxed{}$$

Thus $a = -1$, $b = -5$, $c = -6$, $d = 0$.

Graph will be a cubic.

2.   Since a is negative, curve's right-hand tail goes down: ∿.

3.   The y-intercept is $d = 0$.

4.   Find the x-intercepts by solving:

$$0 = -x^3 - 5x^2 - 6x$$
$$0 = -x(x^2 + 5x + 6)$$
$$0 = -x(x + 2)(x + 3)$$
$$x = 0, \quad x = -2, \quad x = -3$$

The x-intercepts are -3, -2, and 0.

5.   Plot the above points and connect them with a smooth curve. Let me repeat: this is merely a sketch of the general shape. All we know at this point is that the curve is ∿-shaped and goes through the three intercepts.

Answer:

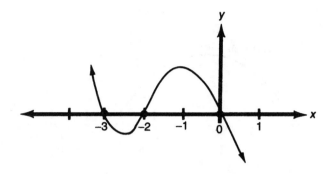

EXAMPLE 3

Sketch $y = f(x) = (x - 3)(x + 2)(x - 5)$.

Solution: Believe it or not, this one is easier because it is factored already.

1. Write in standard form and compare.

   As you probably have noticed, the only values that are actually used are a, the coefficient of the cubed term, and d, the constant. Therefore, rather than waste time multiplying everything out, determine only the cubed term and the constant.

$$y = f(x) = (x - 3)(x + 2)(x - 5)$$

$$= (x)(x)(x) + \cdots + (-3)(2)(-5)$$

$$= x^3 + \cdots + 30$$

$$y = f(x) = a\,x^3 + bx^2 + cx + d$$

   Thus $a = 1$ and $d = 30$.

   Graph will be a cubic.

2. Since a is positive, curve's right-hand tail goes up: .

3. The y-intercept is $d = 30$.

4. Find the x-intercepts by solving:

$$0 = (x - 3)(x + 2)(x - 5)$$

$$x = 3, \quad x = -2, \quad x = 5$$

   The x-intercepts are 3, –2, and 5.

   Wasn't that easy? But had the function been stated as $y = f(x) = x^3 - 6x^2 - x + 30$, it is unlikely that we would have been able to determine the x-intercepts unless you are far better than I at factoring cubics.

5. Plot the above points and connect them with a smooth curve.

Answer:

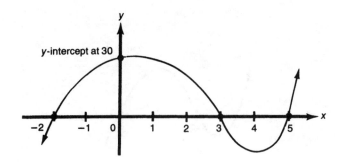

EXAMPLE 4

Sketch: $y = f(x) = -x^3 + 1$.

Solution: 1.   Write in standard form and compare:

$$y = f(x) = \boxed{a}\, x^3 + bx^2 + cx \boxed{+ d}$$
$$y = f(x) = \boxed{-}\, x^3 \qquad\qquad \boxed{+ 1}$$

Thus $a = -1, b = 0, c = 0, d = 1$.

Graph will be a cubic.

2.   Since a is negative, curve's right-hand tail goes down: ∿.

3.   The y-intercept is 1.

4.   Find the x-intercepts by solving:

$$0 = -x^3 + 1$$
$$x^3 = 1$$
$$x = 1$$

The x-intercept is 1.

5.   Several possibilities exist for the sketch, but they must all have three things in common: there is exactly one x-intercept and it is 1, the y-intercept is 1, and the curve is shaped like

this: ∿.

Answer:

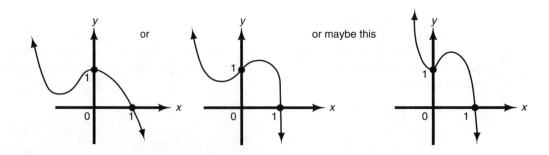

Ready to try one?

Problem 2

Sketch: $y = f(x) = x^3 - 6x^2 + 9x$.

Solution: 1.   Write in standard form and compare.

2.   Curve's right-hand tail will go _____.

3.   The y-intercept is _____.

4.   The x-intercepts are _____ because $\begin{aligned}0 &= x^3 - 6x^2 + 9x \\ 0 &= x(x-3)(x-3)\end{aligned}$

Notice there are three factors, but two of them are identical, $(x-3)$ and $(x-3)$. Thus, there are only two distinct x-intercepts, 0 and 3. The 3 is called a double root. At a double root, the graph will "bounce" rather than go through the intercept. I think the sketch below will illustrate why the word *bounce* is used to describe the graph at a double root.

5.   Plot the points on the grid provided and connect them with a smooth curve.

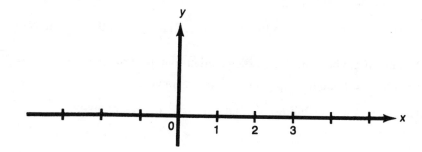

Answer:

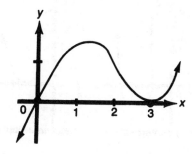

What happens if we are unable to factor the cubic equation to find the x-intercepts? In that case, no formula, such as the quadratic formula, exists to help us. Therefore, without plotting points, there is little that can be done. Example 5 will illustrate what I mean.

EXAMPLE 5

Sketch: $y = f(x) = \dfrac{1}{3}x^3 - 3x^2 + 5x + 2$.

Solution: 1.   Write in standard form and compare:

$$y = f(x) = \boxed{a}\, x^3 \boxed{+ b}\, x^2 \boxed{+ c}\, x \boxed{+ d}$$

$$y = f(x) = \boxed{\tfrac{1}{3}}\, x^3 \boxed{- 3}\, x^2 \boxed{+ 5}\, x \boxed{+ 2}$$

Thus $a = \dfrac{1}{3}$, $b = -3$, $c = 5$, $d = 2$.

Graph will be a cubic.

2.   Since a is positive, curve's tail goes up: .

3.   The y-intercept is 2.

4.   Find the x-intercepts by solving:

$$0 = \frac{1}{3} x^3 - 3x^2 + 5x + 2$$

I could spend days and still never be able to factor, thus solve, the equation. This inability doesn't mean there aren't any x-intercepts; it only means I am unable to factor the cubic and find them analytically.

The best that can be said is that the function has at most three x-intercepts.

5.   Not much can be done with a sketch other than to show that the curve is   -shaped, righ-hand tail opening up, and the y-intercept is 2.

Here are some possibilities of what the curve might look like.

Answer:

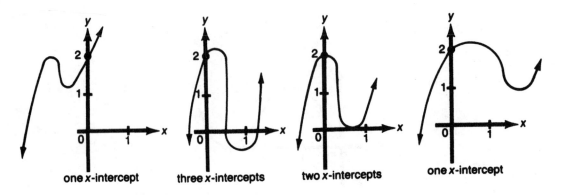

one x-intercept   three x-intercepts   two x-intercepts   one x-intercept

Of course, if we had nothing better to do, we could always plot points.

In Unit 15, we will come back and finish this example.

IDENTIFICATION OF POLYNOMIAL FUNCTIONS

Linear, quadratic, and cubic functions are all part of a larger group of functions called polynomial functions. Next we will consider what is meant by a polynomial function, how we can identify such functions, and what procedure we can use to graph them.

Definition: A function $y = f(x) = a_n x^n + a_{n-1} x^{n-1} + \cdots + a_2 x^2 + a_1 x + a_0$, where n is a counting number, $a_n \neq 0$, and $a_0, a_1, \ldots, a_n$ are real numbers, is called an ***n*th-degree polynomial function** involving one independent variable x and a dependent variable y. Sometimes this definition is shortened to read "a polynomial function of one variable," meaning there is only one independent variable.

In other words, in an nth-degree polynomial function:

1. There are one independent variable and one dependent variable.

2. The dependent variable is raised only to the first power.

3. One term contains the independent variable raised to the nth power.

4. The highest exponent is n.

5. All exponents are counting numbers.

6. Neither the independent nor the dependent variable appears in any denominator.

7. No term contains a product of the independent and dependent variables.

Here are some examples of polynomial functions of one variable and their degrees:

$$y = f(x) = x^4 + 3x - 1 \qquad \text{fourth degree}$$

$$y = g(x) = 1 - 5x^3 \qquad \text{third degree}$$

$$y = h(x) = 27x^{19} - 10x^7 + x \quad \text{nineteenth degree}$$

$$y = f(x) = x^2 - 3x + 4 \qquad \text{second degree}$$

$$y = F(x) = 2x - 7 \qquad \text{first degree}$$

Here are some examples that are not polynomial functions of one variable:

$$y = f(x) = \sqrt{x^2 + x - 1}$$

$$y = g(x) = \frac{3 - x}{x}$$

$$y^2 = x^3 + 3x^2 + 2x + 7$$

$$\frac{1}{y} = x^4 + 3x - 1$$

Before proceeding, determine why each of the above is not a polynomial function.

GRAPHING POLYNOMIAL FUNCTIONS

First-, second-, and third-degree polynomial functions are nothing more than linear, quadratic, and cubic functions with the coefficients of the x-terms changed to different letters of the alphabet. And we know a lot about graphing linear, quadratic, and cubic functions.

To summarize what we know so far:

First-degree polynomial functions $y = f(x) = mx + b$

They are also called linear functions.

The graph is a straight line.

The slope is m.

The y-intercept is b, the constant.

Second-degree polynomial functions: $y = f(x) = ax^2 + bx + c$

They also are called quadratic functions.

The graph is a $\smile$-shaped curve called a parabola.

If a is positive, the curve opens up: $\smile$.

If a is negative, the curve opens down: $\frown$.

The y-intercept is c, the constant.

There are at most two x-intercepts.

The solution to $0 = ax^2 + bx + c$ yields the x-intercepts.

The low point on the curve (or the high point if the curve opens down) is called the vertex.

The vertex is located at point

$$\left(\frac{-b}{2a}, \frac{4ac - b^2}{4a}\right) \quad \text{or} \quad \left(\frac{-b}{2a}, f\left(\frac{-(b)}{2a}\right)\right)$$

The curve is symmetric to a vertical line (called the axis of symmetry) through the vertex.

Third-degree polynomial functions: $y = f(x) = ax^3 + bx^2 + cx + d$

They also are called cubic functions.

The graph is a $\sim$-shaped curve.

If a is positive, the curve's right-hand tail goes up: $\sim$.

If a is negative, the curve's right-hand tail goes down: $\sim$.

The y-intercept is d, the constant.

There are at most three x-intercepts.

The solution to $0 = ax^3 + bx^2 + cx + d$ yields the x-intercepts.

The predictability ends with third-degree polynomial functions. The graphs of higher-order polynomial functions, starting with fourth-degree functions, take on a variety of shapes. The best we can do at this stage is to offer the following generalizations.

BASIC FACTS ABOUT THE GRAPHS OF POLYNOMIAL FUNCTIONS

1. The graph is always a smooth curve without any breaks or sharp corners.

2. The number of x-intercepts is at most equal to the degree of the polynomial function.

3. The domain is the set of real numbers.

In other words, if you were to make up a fifth-degree polynomial function, there is no way to predict what its graph would look like other than to say it would be a nice smooth curve without any breaks or sharp corners, and it would have at most five x-intercepts. Graphing would require locating an adequate number of points and connecting them with a smooth curve.

The graph of your fifth-degree polynomial function might look like one of the following:

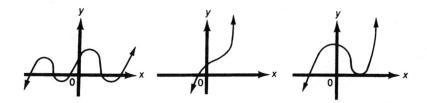

It could not look like any of these:

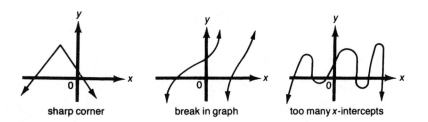

We will not attempt to graph any higher order polynomial functions in this unit.

You should now be able to identify a polynomial function and to distinguish it from functions that are not polynomial functions. Also you should be able to describe the basic shapes of the graphs for first-, second-, and third-degree polynomial functions. Remember that:

the graph of a first-degree polynomial function, also called a linear function, is a straight line;

the graph of a second-degree polynomial function, also called a quadratic function, is a ⌣-shaped curve called a parabola;

the graph of a third-degree polynomial function, also called a cubic function, is a ⌒⌣-shaped curve.

CALCULUS AND THE CALCULATOR (Optional)

Changing the Dimensions of the Viewing Window

Thus far, most of the examples were selected so that the graph could be seen using the standard viewing window (**ZOOM**) (**6**). Recall that its dimensions are [–10, 10] by [–10, 10] as shown here, which can be seen by pressing (**WINDOW**).

```
WINDOW
 Xmin=-10
 Xmax=10
 Xscl=1
 Ymin=-10
 Ymax=10
 Yscl=1
 Xres=1
```

What to do when the standard viewing window is not satisfactory is the subject of this section. In Unit 3 two approaches for enlarging the viewing window for a linear function were introduced. In this unit we will basically use the same two methods but apply them to functions in general and explain them in greater detail. The two approaches are presented by way of examples.

The first approach relies on your knowledge of the function to establish more appropriate dimensions. I assume by now that displaying all the keystrokes is no longer necessary, so they will be listed only when we are doing something new.

EXAMPLE 6

Graph: $y = f(x) = x^2 - 53x + 350$.

Solution: Create a table of values for the function and view the graph. In the standard viewing window only an "almost vertical" line off to the right is visible, which is not very useful.

Rather than using trial and error to change the viewing dimensions, we first will do some analytical work. What do we know about this function?

This is a quadratic function with $a = 1$, $b = -53$, and $c = 350$.

The y-intercept is 350. It is a parabola, which opens up. Return to the table of values, use the formula for the x-coordinate of the vertex, $\frac{-b}{2a}$, and key in (**5**) (**3**) (**÷**) (**(**) (**2**)

(**×**) (**1**) (**)**) to find that the vertex is located at (26.5, –352.3).

Now we have some useful information to establish more appropriate dimensions for our viewing window.

For the x-values, –50 to 50 would more than accommodate the vertex's x-coordinate of 26.5. For the y-values, –400 to 400 would accommodate the vertex's y-coordinate of –352.3 and the y-intercept of 350.

Press (**WINDOW**). Set Xmin, Xmax, Ymin, and Ymax accordingly.

Upon viewing the graph, you might decide that using a window with dimensions [–25, 75] by [–400, 500] would be even better. I did, but the final choice is yours.

```
WINDOW
 Xmin=-50
 Xmax=50
 Xscl=1
 Ymin=-400
 Ymax=400
 Yscl=1
 Xres=1
```

Answer:

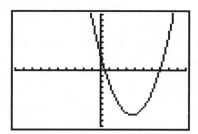

When analytical methods are unavailable or too time-consuming, a second approach for changing the viewing window is to use the (**TRACE**) and (**ZOOM**) keys.

EXAMPLE 7

Graph $y = f(x) = 2x^3 + 7x^2 - 15x + 20$.

Solution: Create a table of values for the function and view the graph. This time in the standard viewing window you see only an "almost vertical" line off to the left.

Press (**TRACE**), and the value of the y-intercept, which is 20, appears at the bottom of the screen. Press (**ZOOM**) (**3**) (**ENTER**). More of the graph is visible; but it is difficult to see much detail.

Press (**TRACE**) and move the cursor along while noting the various x- and y-values at the bottom of the screen. These values provide the clues as to what the new dimensions for the viewing window should be.

It appears that dimensions of [–10, 10] by [–100, 100] should be sufficient.

Press (**WINDOW**), change dimensions, and view the graph.

Answer:

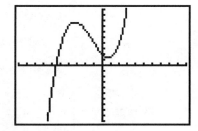

Before leaving this example completely, press (**TRACE**) and move the cursor to some point close to the x-intercept. Press (**ZOOM**) (**2**) (**ENTER**) and notice what happens. The viewing window changes to magnify that portion of the graph centered around the position of the cursor. You can get as close to a point as desired by repeating the process of alternately pressing (**TRACE**) with (▶) or (◀) to move the cursor followed by (**ZOOM**) (**2**) (**ENTER**).

I suggest you do some experimenting with all these features. Set various viewing window dimensions, practice using 2:ZOOM In, 3:ZOOM Out, and TRACE. And if you really want to be creative, try 1:Zbox and 0:ZOOMFit. Determine which options are best for the way you approach graphing functions. I have favorites, and I expect you will too.

Before beginning the next unit, you should work the following exercises.

EXERCISES

By means of sketches, show the pertinent characteristics of each of the following functions, including *all* the x-intercepts:

1. $f(x) = x^3 - 4x^2 - 5x$

2. $f(x) = x^3 - 9x^2 + 18x$

3. $f(x) = -x^2 + 3x + 4$

4. $f(x) = 2x^3 + 6x^2 - 20x$

5. $f(x) = -x + 3$

6. $f(x) = 12x^3 - 24x^2 + 12x$

7. $f(x) = (x - 2)(x - 3)$

8. $f(x) = (x - 2)(x + 3)(x - 1)$

9. $f(x) = x(x - 2.5)(x + 4)$

10. $f(x) = -x^3 + 3x^2$

11. $f(x) = (x - 1)(x^2 + 4x - 12)$

12. $f(x) = x^3 - 4x$

C13. Return to Example 6 and find the two x-intercepts.

C14. Return to Example 7 and find the x-intercept.

15. The Triangle Company manufactures expensive watches. The company estimates that, if the watches are priced at \$90, then 500 watches will be sold, whereas at a price of \$130 only 300 watches will be sold. Graph the demand function,

$$y = D(x),$$

where x = the price of a watch, in dollars,

and y = the estimated number of watches that will be sold.

16. Graph $f(x) = \begin{cases} 3 & \text{if } 5 < x \\ x^2 & \text{if } x \leq 2 \\ x+1 & \text{if } 3 \leq x \leq 5 \end{cases}$

If additional practice is needed:

Barnett, Ziegler, and Byleen, page 88, problems 1–14
Harshbarger and Reynolds, pages 185–188, problems 1–25, 30–34
Hoffmann and Bradley, page 26, problems 1–8

UNIT 9

The Derivative

This unit starts the beginning of what is traditionally called differential calculus. The purpose of this unit is to provide you with an understanding of a derivative. An intuitive approach will be used rather than formal, rigorous mathematical definitions. When you have finished the unit, you will be able to differentiate polynomial functions.

Recall the following definition from Unit 4:

$$\text{Definition:} \quad \textbf{Slope of a line} = m = \frac{\Delta y}{\Delta x} = \frac{y_2 - y_1}{x_2 - x_1}$$

Recall that the slope of a line is the rate at which y changes with respect to x. And a fundamental property of a straight line is that the slope is constant. As a review, consider the function $f(x) = \frac{2}{3}x + 5$, with slope $= m = \frac{\Delta y}{\Delta x} = \frac{2}{3}$. In other words, if x increases by 3, the y value will increase by 2 and this ratio will remain constant for the entire line.

Differential calculus deals with finding the slope of a curve.

When we speak of the slope of a curve, we really mean the slope of the tangent line to the curve at a point.

What do we mean by a tangent line? Think back to your days in geometry class. Remember a tangent to a circle? You learned that it is a line that touches the circle at only one point. That is close to what is meant here.

Consider $f(x)$ with point P as shown on the following page. The tangent line at P is the limiting position of line PQ as Q moves closer and closer to P. Notice how the position of the line is changing as Q is drawn closer to P. (PQ, PQ_1, PQ_2, PQ_3, and PQ_4). The limiting position of PQ is the line marked T. This same limiting position results whether P is approached with lines from the left or the right of P. The sequence of lines PS, PS_1, and PS_2 has the same limiting position as S is drawn closer to P. Since line T is the limiting position whether the approach is from the left or the right, line T is defined as the **tangent line at point P.**

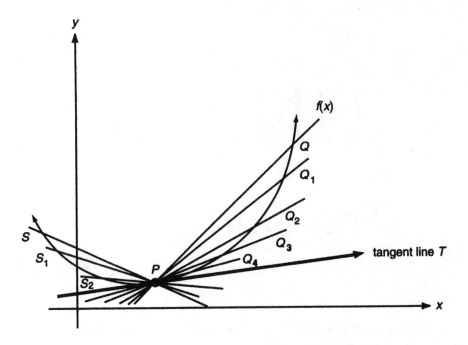

Not every function has a unique tangent line at each point on its graph. For example, the graph below does not have a tangent at point A. The sequence of lines AS, AS_1, and AS_2 on the left do not converge to the same limiting position as Q is drawn closer to A on the right.

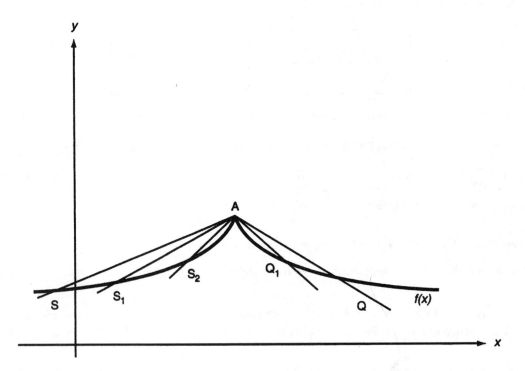

Here we go with another picture. In this case it will be helpful to think of the tangent line as being a very short line segment rather than a line.

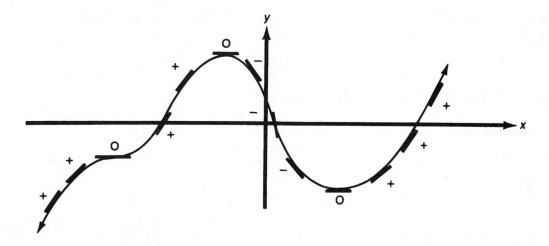

The short line segments represent tangent lines at various points along the curve.

Notice that the slopes of the tangent lines are changing. In some places the slopes are positive, in others negative; at still other points, the slope is zero. Unlike a straight line, a curve does *not* have a constant slope.

Definition: $f'(x)$ represents the slope of the tangent line to the graph of $y = f(x)$, provided that the tangent line exists.

Definition: $f'(x)$ is called the **derivative**. $f'(x)$ is read "f prime of x."

RULES FOR DETERMINING DERIVATIVES

Next we will learn four rules for determining derivatives. In this unit we will concentrate on the mechanics of finding the derivatives of polynomials, and we will wait until Unit 10 to illustrate the relationship of the derivative to the slope of the curve.

Let me explain the notation to be used: $f(x)$, $u(x)$, and $v(x)$ denote functions of x, such as various polynomials, and c denotes a constant, such as a number.

Here is the first rule.

Power Rule

If $f(x) = x^n$, then $f'(x) = nx^{n-1}$ for all n, n being a real number.

In words, if x is raised to some exponent, the derivative is equal to the exponent times x raised to the
exponent minus 1.

Of all the rules for determining derivatives, the Power Rule is the one used most frequently. Memorize it!

We will do several examples. For the moment, leave Example 1 blank; we will come back to it later.

Power Rule:	If $f(x) = x^n$, then $f'(x) = nx^{n-1}$

EXAMPLE 1

EXAMPLE 2 — If $f(x) = x^2$, then $f'(x) = 2x^{2-1} = 2x$

EXAMPLE 3 — If $f(x) = x^3$, then $f'(x) = 3x^{3-1} = 3x^2$

EXAMPLE 4 — If $f(x) = x^4$, then $f'(x) = 4x^{4-1} = 4x^3$

EXAMPLE 5 — If $f(x) = x^5$, then $f'(x) = 5x^{5-1} = 5x^4$

Do you see the pattern? To find the derivative, you take the exponent times x raised to the exponent minus 1. Before returning to Example 1, try a few problems yourself.

Problem 1 — If $f(x) = x^6$, then $f'(x) =$

Problem 2 — If $f(x) = x^{34}$, then $f'(x) =$

Problem 3 — If $f(x) = x^{88}$, then $f'(x) =$

Problem 4 — If $f(x) = x^{27}$, then $f'(x) =$

Answers: $6x^5$; $34x^{33}$; $88x^{87}$; $27x^{26}$

By now you should be comfortable enough with the Power Rule that we can return to Example 1 and finish it. It should be obvious what function I want in there.

EXAMPLE 1 (Continued)

$$\text{If } f(x) = x$$

$$= x^1 \qquad \text{because if no exponent is written, the exponent is understood to be 1.}$$

$$\text{Then } f'(x) = 1x^{1-1}$$

$$= 1x^0$$

$$= x^0$$

$$= 1 \qquad \text{because the definition of a zero exponent is } b^0 = 1 \text{ if } b \neq 0.$$

We will refer to Examples 1–5 for the rest of the unit, so go back and write the following information in the blank space provided in Example 1: If $f(x) = x$, then $f'(x) = 1$.

We're ready for the second rule.

Derivative of a Constant Function

If $f(x) = c$, then $f'(x) = 0$, where c is a constant.

The derivative of a constant term is 0.

If $f(x) = c$, then $f'(x) = 0$.

EXAMPLE 6
If $f(x) = 5$, then $f'(x) = 0$.

EXAMPLE 7
If $f(x) = -10$, then $f'(x) = 0$.

EXAMPLE 8
If $f(x) = 2$, then $f'(x) = 0$.

And the third.

Derivative of a Constant Times a Function

If $f(x) = c \cdot u(x)$, then $f'(x) = c \cdot u'(x)$.

The derivative of a constant times a function equals the constant times the derivative of the function.

EXAMPLE 9

If $f(x) = 5x^2$, find $f'(x)$.

Solution: If $f(x) = 5x^2$,   This is a constant times a function.

From the Power Rule,

then $f'(x) = 5(2x)$ the derivative of $x^2 = 2x$.

= $10x$

EXAMPLE 10

If $f(x) = -3x^4$, find $f'(x)$.

Solution: If $f(x) = -3x^4$,   This is a constant times a function.

From the Power Rule,

then $f'(x) = -3(4x^3)$ the derivative of $x^4 = 4x^3$.

= $-12x^3$

EXAMPLE 11

If $f(x) = 7x$, find $f'(x)$.

Solution: If $f(x) = 7x$,

From the Power Rule,

then $f'(x) = 7(1)$ the derivative of x is 1.

= 7

EXAMPLE 12

If $f(x) = cx^n$, find $f'(x)$.

Solution: If $f(x) = cx^n$,

From the Power Rule,

then $f'(x) = cnx^{n-1}$ the derivative of $x^n = nx^{n-1}$.

Example 12 is a special case of a constant times a function where the function is x raised to some power, and can be used as a rule by itself. In words, the derivative of a constant times x raised to some exponent equals the constant times the exponent times x raised to the exponent minus 1.

Special Case: If $f(x) = cx^n$, then $f'(x) = cnx^{n-1}$

Rule 4, the last rule of this unit, combines the three preceding rules.

Derivative of Sums and/or Differences

If $f(x) = u(x) + v(x)$, then $f'(x) = u'(x) + v'(x)$.

If $f(x) = u(x) - v(x)$, then $f'(x) = u'(x) - v'(x)$.

The derivative of the sum (or difference) of two or more functions equals the sum (or difference) of their respective derivatives.

The first three rules dealt with taking the derivative of a single term. Rule 4 defines how to take the derivative of several terms; the derivative is taken term by term.

EXAMPLE 13

Find the derivative of $f(x) = x^3 + 7x^2 - 3x + 10$.

Solution: If $f(x) = x^3 + 7x^2 - 3x + 10$,

then $f'(x) = 3x^2 + 7(2x) - 3(1) + 0$ from the Power Rule.

$= 3x^2 + 14x - 3$

EXAMPLE 14

Find the derivative of $f(x) = 15x^4 - 2x^3 + 11x^2 + 3x - 1$.

Solution: If $f(x) = 15x^4 - 2x^3 + 11x^2 + 3x - 1$,

then $f'(x) = 15(4x^3) - 2(3x^2) + 11(2x) + 3(1) - 0$

$= 60x^3 - 6x^2 + 22x + 3$

Not too bad is it?

Keep in mind what we are finding. The geometric interpretation is that the derivative is a formula, expressed in terms of x, for finding the slope of the tangent line at any point on the graph of a function, $f(x)$, provided that the tangent line exists. We will explore what all of that means in Unit 10 after you have had an opportunity to practice taking derivatives.

Through the combined use of the following rules, you should now be able to take the derivative of any polynomial function.

If $f(x) = x^n$, then $f'(x) = nx^{n-1}$ for all n, n being a real number.

If $f(x) = c$, then $f'(x) = 0$, where c is a constant.

If $f(x) = c \cdot u(x)$, then $f'(x) = c \cdot u'(x)$.

If $f(x) = cx^n$, then $f'(x) = cnx^{n-1}$.

If $f(x) = u(x) + v(x)$, then $f'(x) = u'(x) + v'(x)$.

If $f(x) = u(x) - v(x)$, then $f'(x) = u'(x) - v'(x)$.

Before going on to the next unit, try the following exercises.

EXERCISES

Find $f'(x)$ for each of the following:

1. $f(x) = 3x^{15}$

2. $f(x) = -7x^2$

3. $f(x) = 21x^7$

4. $f(x) = x$

5. $f(x) = -11x^9$

6. $f(x) = 10$

7. $f(x) = -8$

8. $f(x) = 4x + 3$

9. $f(x) = -7x$

10. $f(x) = -3x + 17$

11. $f(x) = \dfrac{2}{5}x$

12. $f(x) = 3x^2 + 2x - 6$

13. $f(x) = -8x^3 - 7x^2 + 11x - 5$

14. $f(x) = 5x^3 - 4x^2 + 3x + 20$

15. $f(x) = -21x^2 - 7x^3 + 10x^4$

16. $f(x) = 34x^2 + x^5$

17. $f(x) = \dfrac{1}{4}x^4 - \dfrac{1}{2}x^2 - x$

18. $f(x) = \frac{1}{3}x^3 + \frac{1}{2}x^2 - 2x + 13$

19. $f(x) = \frac{1}{3}x^{12} - 3x^2 + \sqrt{5}\,x - \pi$

20. $f(x) = ax^2 + bx + c$

21. $f(x) = ax^5 + \frac{1}{3}x^3 - 2x + 3$

22. $f(x) = mx + b$

23. $f(x) = ax^3 + bx^2 + cx + d$

For each of the following functions, find $f'(x)$ and $f'(2)$:

24. $f(x) = x^2 - 3x + 1$

25. $f(x) = -x^3 - x$

26. $f(x) = ax^3 + x$

27. $f(x) = 5x$

For each of the following functions, find $f'(x)$, $f(3)$, and $f'(3)$:

28. $f(x) = x^3 - 2x + 1$

29. $f(x) = 2 - 5x^2$

30. $f(x) = x^2 - 6x + 12$

If additional practice is needed:

Hoffmann and Bradley, page 106, problems 1–4
Waner and Costenoble, pages 149–150, problems 1–14

UNIT 10

More About the Derivative

In this unit we will continue our introduction to a derivative by illustrating the relationship between the slope of the curve and the derivative. At the end of the unit, alternative notation for denoting a derivative will be described. In this unit and the next few, we will limit our discussion to polynomial functions.

To illustrate the relationship between the derivative and the slope of the tangent line to a curve, let's return to a familiar function, the second-degree polynomial.

EXAMPLE 1

Graph: $y = f(x) = x^2 - 6x + 5$.

Solution: 1. This is a quadratic with $a = 1$, $b = -6$, $c = 5$.

2. The parabolic curve opens up.

3. The y-intercept is 5.

4. The x-intercepts are 1 and 5 because

$$0 = x^2 - 6x + 5$$
$$0 = (x - 1)(x - 5)$$
$$x - 1 = 0 \quad x - 5 = 0$$
$$x = 1 \qquad x = 5$$

what numbers when added result in 6 but when multiplied yield 5 (1 & 5)

$(x-1)(x-5)$

5. The vertex is located at $(3, -4)$ because
the axis of symmetry is $x = 3$
and $y = f(3) = (3)^2 - 6(3) + 5 = 9 - 18 + 5 = -4$.

Answer:

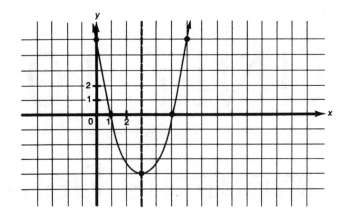

In Unit 9, we stated that from a geometric interpretation the derivative is a formula, expressed in terms of x, for finding the slope of the tangent line at any point on the graph of a function $f(x)$.

EXAMPLE 2

Find the derivative for $f(x) = x^2 - 6x + 5$.

Solution: $f(x) = x^2 - 6x + 5$

$f'(x) = 2x - 6(1) + 0$

$= 2x - 6$

The sooner you are able to skip that first step, the better.

Definition: $f'(a)$ is the slope of the tangent line of the curve at the specific point where $x = a$.

EXAMPLE 3

Compute and interpret: $f'(5), f'(3),$ and $f'(2)$ from Examples 1 and 2.

Solution: From Example 1: $f(x) = x^2 - 6x + 5$

From Example 2: $f'(x) = 2x - 6$

$f'(5) = 2(5) - 6 = 4$

From the graph, when $x = 5, y = 0$, which is point $(5, 0)$.

Answer: The slope of the tangent line to the curve at point $(5, 0)$ is 4.

Remember that tangent lines are represented by *very short* line segments. On the grid on the next page, I have drawn just such a line segment at point $(5,0)$ with a slope of 4. I have tried to aid your understanding by showing, off to the side, what a line looks like with a slope of 4. The lines are meant to be parallel.

$$f'(3) = 2(3) - 6 = 0$$

From the graph, when $x = 3$, $y = -4$, which is point $(3, -4)$.

Answer: The slope of the tangent line to the curve at point $(3, -4)$ is 0.

A line with 0 slope is horizontal and the tangent is shown as such.

$$f'(2) = 2(2) - 6 = 4 - 6 = -2$$

In this instance, the y value for the point needs to be determined.

$$\text{If } x = 2, \text{ what is } y? \quad f(2) = (2)^2 - 6(2) + 5$$

$$= 4 - 12 + 5$$

$$= -3, \text{ which is point } (2, -3).$$

Answer: The slope of the tangent line to the curve at point $(2, -3)$ is -2.

The tangent line is shown as a very short line segment with slope of -2. Again I have tried to aid you by showing, off to the side, what a line looks like with a slope of -2.

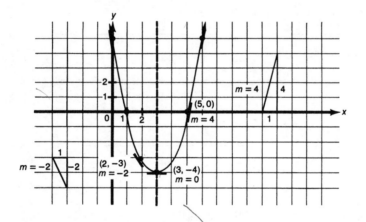

There was nothing special about the points selected for Example 3. Indeed, any value can be used for x, as Problem 1 is intended to show.

Problem 1

Using the information provided in Examples 1 and 2, find and interpret $f'(10)$.

Solution:

Answer: $f'(10) = 14$. The slope of the tangent line to the curve at $x = 10$ is 14. Better still, the slope of the tangent line to the curve at point $(10, 45)$ is 14.

Sometimes the derivative is referred to as the **instantaneous rate of change** of y with respect to x. It is an estimate or approximation of the change in y for a unit change in x. Notice the similarity of the wording to that for the slope of a line.

In Problem 1 you found the slope of the tangent to be 14. Another interpretation of the number 14 would be that, as x increases by 1 from point $(10, 45)$, the y value will increase by approximately 14. This is an approximation because 14 represents the slope of the tangent line, whereas the function itself is a curve.

EXAMPLE 4

Find and interpret $f'(x)$ for $f(x) = 5x - 2$.

Solution: $f(x) = 5x - 2$

$f'(x) = 5$

The derivative is a formula for finding the slope of the tangent line to the curve at any point. In this example, the formula equals the constant 5, regardless of the values of x. But that is consistent with what we already know. $f(x) = 5x - 2$ is a linear function with slope of 5 and the slope is constant everywhere on the line. To say that $f'(x) = 5$ is the same thing as $m = 5$, but in different symbols.

Answer: $f'(x) = 5$ means that the slope of the line is 5.

Consider another, similar situation.

EXAMPLE 5

Find and interpret $f'(x)$ for $f(x) = 3$.

Solution: $f(x) = 3$

$f'(x) = 0$

Answer: $f'(x) = 0$ means that the slope of the line is 0.

Comments: Again this is consistent with what we know. The graph of $f(x) = 3$ is a horizontal line with y-intercept of 3 and slope of 0. The graph of $f(x) = c$ always represents a horizontal line, and the slope of a horizontal line is 0. That is the reasoning behind the rule that the derivative of a constant function is zero.

And now for some unfamiliar examples.

EXAMPLE 6

Find and interpret $f'(0)$ for $f(x) = 7x^5 + 3x^4 - 11x^3 + x^2 + 25x + 10$.

Solution: $y = f(x) = 7x^5 + 3x^4 - 11x^3 + x^2 + 25x + 10$

$y = f(0) = 0 + 0 - 0 + 0 + 0 + 10$

$= 10$, which is point $(0, 10)$ on the curve.

$$f'(x) = 35x^4 + 12x^3 - 33x^2 + 2x + 25$$

$$f'(0) = 0 + 0 - 0 + 0 + 25$$

$$= 25$$

Answer: The slope of the tangent line to the curve at point (0, 10) is 25.

Comment: This is a fifth-degree polynomial function. I have no more idea of what the graph of this curve looks like than you do. What we have determined is that at point (0, 10) the curve is increasing at the approximate rate of 25; this means that if x increases by 1 (*from* 0), the y-value will increase (*from* 10) by approximately 25. In other words, the next point will be at about (0 + 1, 10 + 25) = (1, 35). If need be, we could calculate the next point by substituting 1 into the function to determine y:

$$y = f(1) = 7(1)^5 + 3(1)^4 - 11(1)^3 + (1)^2 + 25(1) + 10$$

$$= 7 + 3 - 11 + 1 + 25 + 10$$

$$= 35$$

In this example, our approximation turned out to be the exact answer.

The next example will use the derivative as an aid in curve sketching.

EXAMPLE 7

Find $f(0)$, $f'(0)$, $f(1)$, $f'(1)$, $f(2)$, $f'(2)$, $f(3)$, $f'(3)$, $f(4)$, and $f'(4)$, where $y = f(x) = x^3 - 6x^2 + 9x$, and use the results to sketch the graph of $f(x)$.

Solution: $y = f(x) = x^3 - 6x^2 + 9x$

$$y = f(0) = 0 - 0 + 0$$

$\qquad = 0$ which means (0, 0) is a point on the curve.

$$f'(x) = 3x^2 - 12x + 9$$

$$f'(0) = 0 - 0 + 9$$

$$= 9$$

The slope of the tangent line to the curve at point (0, 0) is 9.

$$y = f(x) = x^3 - 6x^2 + 9x$$

$$y = f(1) = 1 - 6 + 9$$

$\qquad = 4$ which means (1, 4) is a point on the curve.

$$f'(x) = 3x^2 - 12x + 9$$

$$f'(1) = 3 - 12 + 9$$

$$= 0$$

The slope of the tangent line to the curve at point (1, 4) is 0.

$y = f(x) = x^3 - 6x^2 + 9x$

$y = f(2) = (2)^3 - 6(2)^2 + 9(2)$

$\quad = 8 - 24 + 18$

$\quad = 2$ which means (2, 2) is a point on the curve.

$f'(x) = 3x^2 - 12x + 9$

$f'(2) = 3(2)^2 - 12(2) + 9$

$\quad = 12 - 24 + 9$

$\quad = -3$

The slope of the tangent line to the curve at point (2, 3) is –3.

$y = f(x) = x^3 - 6x^2 + 9x$

$y = f(3) = (3)^3 - 6(3)^2 + 9(3)$

$\quad = 27 - 54 + 27$

$\quad = 0$ which means (3, 0) is a point on the curve.

$f'(x) = 3x^2 - 12x + 9$

$f'(3) = 3(3)^2 - 12(3) + 9$

$\quad = 27 - 36 + 9$

$\quad = 0$

The slope of the tangent line to the curve at point (3, 0) is 0.

$y = f(x) = x^3 - 6x^2 + 9x$

$y = f(4) = (4)^3 - 6(4)^2 + 9(4)$

$\quad = 64 - 96 + 36$

$\quad = 4$ which means (4, 4) is a point on the curve.

$f'(x) = 3x^2 - 12x + 9$

$f'(4) = 3(4)^2 - 12(4) + 9$

$\quad = 48 - 48 + 9$

$\quad = 9$

The slope of the tangent line to the curve at point (4, 4) is 9.

The preceding information is plotted on the grid below. Can you see in your own mind the shape of the curve?

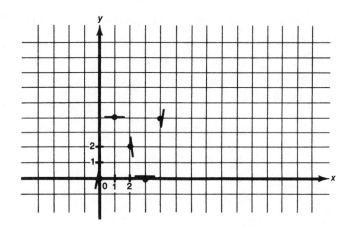

Now combine the above information with our previous knowledge. The function $y = f(x) = x^3 - 6x^2 + 9x$ is a cubic with $a = 1$. The curve is opening like this: .

Answer:

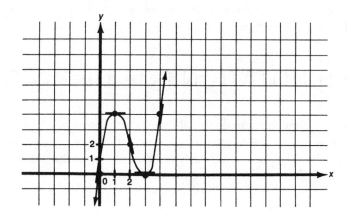

Return to Problem 2 of Unit 8, and compare the results. What important information has been added?

Two important items have been determined. One, the basic location of the graph now has been established. Two, the high point of the left peak is at $(1, 4)$. Do you see why?

It is time to reverse the procedure slightly.

EXAMPLE 8

If $f(x) = x^2 - 3x - 4$, find the value of x such that the slope is 11.

Solution: $f(x) = x^2 - 3x - 4$

$f'(x) = 2x - 3$

The derivative is a formula for finding the slope of the tangent line to the curve at a point. This problem asks for the value of x when the slope is 11. In other words, when $f'(x) = 11$, what is x? The answer is the solution to the following equation:

$$11 = 2x - 3$$

$$14 = 2x$$

$$x = 7$$

Answer: When $x = 7$, the slope of the tangent line to the curve is 11. Or, shortened, when $x = 7$, the slope of the curve is 11. You may wish to verify the answer.

Problem 2

If $f(x) = x^2 - 3x - 4$, find the value of x such that the slope is 0.

Solution:

Answer: $x = \dfrac{3}{2}$

Do you recognize what you have found? Can you relate the value of x to anything that is already known about this parabola? If not, I suggest you graph the function.

OTHER NOTATION FOR DERIVATIVES

Thus far, $f'(x)$ has been used exclusively to denote the derivative. For a variety of reasons, which we will not discuss here, numerous other notations are used as well. Some have more advantages than others. The important thing is that you be able to recognize each and to understand the notation.

Consider the function: $y = f(x) = 3x^2 - 7x + 2$

We know that:

1. $f'(x)$ denotes the derivative and $f'(x) = 6x - 7$.

Other notations that also denote the derivative are:

2. y' read "y prime" and $y' = 6x - 7$.

3. f' read "f prime" and $f' = 6x - 7$.

In addition there are some new symbols that denote a derivative:

4. $\dfrac{dy}{dx}$ read "dy dx" and $\dfrac{dy}{dx} = 6x - 7$.

$\dfrac{dy}{dx}$ is a recognized symbol in itself. In other words, it is *not* a fraction whereby the d's may be cancelled.

$\dfrac{dy}{dx}$ stresses the rate of change concept, much as the notation $\dfrac{\Delta y}{\Delta x}$ does for the slope of a line.

One of the major advantages of this notation is that by definition the dependent variable appears in the numerator, the independent variable appears in the denominator, and any other variables appearing in the function are to be considered constants.

5. $\dfrac{df}{dx}$ read "*df dx*" uses the name of the function rather than the dependent variable.

And then there are a few variations, the necessity for which you will not see until Unit 11.

6. $\dfrac{d(3x^2 - 7x + 2)}{dx}$ read "the derivative of $3x^2 - 7x + 2$ with respect to x" and again

   $\dfrac{d(3x^2 - 7x + 2)}{dx} = 6x - 7.$

7. $D_x(3x^2 - 7x + 2)$ read the same as number 6 and $D_x(3x^2 - 7x + 2) = 6x - 7.$

All total, then, there are seven common notations that are used to represent the derivative. Although I have favorites, I will try to use all seven so that you can get used to them.

EXAMPLE 9

If $y = f(x) = 5x^3$, find each of the following:

a. $f'(x)$

b. y'

c. f'

d. $\dfrac{dy}{dx}$

e. $\dfrac{d(5x^3)}{dx}$

f. $D_x(5x^3)$

Solution: All of the above notations denote the derivative.

Answer: The derivative of $5x^3$ is $15x^2$.

VOCABULARY LESSON

Derivative is a noun. *Differentiate* is a verb. And *differentiable* is an adjective. Therefore, the following two sentences mean exactly the same thing:

a.   Find the derivative of $y = f(x) = 3x^2$.

b.   Differentiate the function $y = f(x) = 3x^2$.

And to say that the function $y = f(x) = 3x^2$ is differentiable means that the derivative exists. Examples 10 and 11 pull together the concepts of this unit with the writing of linear functions.

EXAMPLE 10

Write an equation of the tangent line to the curve, $y = f(x) = x^2$, at point $(-3, 9)$.

Solution: To write an equation of a line, the point-slope formula is used.

In order to use the formula, we need two items, the slope and a given point on the line. The point was given as $(-3, 9)$. What is the slope?

$$y = f(x) = x^2 \text{ is the function.}$$

$$f'(x) = 2x \text{ is the formula for the slope of the tangent line.}$$

$$f'(-3) = 2(-3) = -6 \text{ is the slope of the tangent line at the point.}$$

Use point-slope formula. $y - y_1 = m(x - x_1)$

Substitute. $y - 9 = -6(x - (-3))$

Remove parentheses. $y - 9 = -6(x + 3)$

$y - 9 = -6x - 18$

Solve for y. $y = -6x - 9$

Answer: An equation of the tangent line to the curve at point $(-3, 9)$ is $y = -6x - 9$.

EXAMPLE 11

Write an equation of the tangent line to the curve, $y = f(x) = x^2$, where $x = 5$.

Solution: To write an equation of a line, the point-slope formula is used.

In order to use the formula, we need two items, the slope and a given point on the line. All that is given is that $x = 5$. What is the slope? What is the y-coordinate?

$$y = f(x) = x^2 \text{ is the function.}$$

$$y = f(5) = (5)^2 = 25 \text{ means that } (5, 25) \text{ is the point on the curve.}$$

$$f'(x) = 2x \text{ is the formula for the slope of the tangent line.}$$

$$f'(5) = 2(5) = 10 \text{ is the slope of the tangent line at the point.}$$

Use point-slope formula. $y - y_1 = m(x - x_1)$

Substitute: $y - 25 = 10(x - 5)$

Remove parentheses. $y - 25 = 10x - 50$

Solve for y. $y = 10x - 25$

Answer: An equation of the tangent line to the curve at point $(5, 25)$ is $y = 10x - 25$.

CALCULUS AND THE CALCULATOR (Optional)

Evaluating the Derivative at a Point

I somewhat hesitate to introduce this feature of the graphing calculator; but it does provide you with an easy way to check your analytical work. Notice that I used the word *check*, which implies you first have done the problem with paper and pencil. Though this time we will do it on the calculator first.

EXAMPLE 12

If $y = f(x) = x^4 + 2x^3 - x + 1$, find $f'(-2.5)$.

Solution: Notice the problem does not ask for the derivative but only for the value of the derivative when $x = -2.5$.

Key in the function, $f(x)$, and view the graph.

To evaluate the derivative, press (**2nd**) (**CALC**) (**6**) followed by (**−**) (**2**) (**·**) (**5**)

(**ENTER**) for the specified x-value. $\frac{dy}{dx} = -26.00001$ appears at the bottom of the screen.

Answer: $f'(-2.5) = -26$

For verification of our answer, in the next problem you will be asked to rework Example 12 analytically.

I expect you understand why I hesitated to introduce this feature. It certainly saves considerable time and effort if the actual derivative is not needed. But, and I cannot stress this enough, you need to be able to differentiate functions manually. For the time being, unless directed otherwise, use this particular feature for **checking** answers.

Problem 3

Given: $y = f(x) = x^4 + 2x^3 - x + 1$

Find $\frac{dy}{dx} = f'(x)$ and verify that $f'(-2.5) = 26$.

Solution:

Answer: $\frac{dy}{dx} = f'(x) = 4x^3 + 6x^2 - 1$

You should now be able to evaluate and explain expressions such as $f'(a)$, where $x = a$. In addition you should be able to write an equation of the tangent line to the curve at the point where $x = a$.

Further, you should be able to recognize the seven common notations used to denote a derivative: $f'(x)$, f', y', $\frac{dy}{dx}$, $\frac{df}{dx}$, $\frac{df(x)}{dx}$, and D_x.

Before going on to the next unit, do the following exercises on finding derivatives.

EXERCISES

1. If $f(x) = -5x^2 + 2x - 1$, find $\dfrac{df}{dx}$.

2. Find $\dfrac{dy}{dx}$ if $y = 11 + 2x - x^3$.

3. Find $D_x(7x + 2 - 3x^5)$.

4. Find $D_x(x^7 - 7x)$.

5. If $y = f(x) = x^4 - 8x^3 - 3x + 15$, find $\dfrac{dy}{dx}$.

6. If $f(x) = 3 - 7x^2 + 21x + 2x^5$, find $f'(1)$.

7. Find and interpret f' for $f(x) = 11x - 2$.

8. Find $\dfrac{d(1 - 3x^2)}{dx}$.

9. Find $\dfrac{d(x^2 - 5)}{dx}$.

10. If $y = \dfrac{x^6}{3} - 2x$, find $\dfrac{dy}{dx}$. Hint: $\dfrac{x^6}{3} = \dfrac{1}{3}x^6$.

11. If $y = f(x) = 3x - \dfrac{x^2}{2}$, find $f'(x)$.

12. Differentiate: $f(x) = \dfrac{x^4}{4} - \dfrac{x^3}{3} + 5x^2$.

13. Differentiate: $f(x) = \dfrac{3 - x^2}{2}$.

14. Let $f(x) = 4x^2 - 2x + 9$.

    a. Find $f'(x)$.

    b. Determine the slope at $x = 0$.

    c. Find $f'(2)$.

    d. Determine the value of x such that the slope is 0.

15. Let $f(x) = x^2 + 48x + 2{,}000$.

    a. Find $f'(x)$.

    b. Determine the slope of the tangent line at $x = 0$.

    c. Find $f'(1)$.

    d. Determine the value(s) of x such that the slope of the tangent line to the curve is 0.

16.　　Let $y = f(x) = -x^2 + 140x - 10$.

　　　a.　Find $f'(x)$.

　　　b.　Find $f(0)$.

　　　c.　Find $f'(0)$.

　　　d.　Find x such that $f'(x) = 0$.

17.　　Given the function $f(x) = 4x^2 - 2x + 9$, find:

　　　a.　$f'(x)$

　　　b.　$f(1)$

　　　c.　An equation of the line tangent to the graph at $x = 1$

18.　　Find an equation of the line tangent to the graph of $y = x^9 - 5x^8 + x + 12$ at point $(-1, 5)$.

C19.　Find and interpret $f(-1.25)$ and $f'(-1.25)$ for the function given in Exercise 18.

20.　　Given the function $y = f(x) = x^4 - 3x^3 + 2x^2 - 6$, find:

　　　a.　$f'(x)$

　　　b.　An equation of the line tangent to the graph at $x = 2$

21.　　Let $y = f(x) = x^3 - 6x^2$ and find $f'(-1), f'(0), f'(2), f'(4)$, and $f'(6)$, Use the results to sketch the graph.

C22.　Given the function $y = f(x) = 2x^4 - 5x^3 + x^2 - 4x + 1$, find $f'(x)$ and $f(0), f'(0), f(2), f'(2), f(4), f'(4), f(-3)$, and $f'(-3)$.

　　　Hint:　First enter $f(x)$ for $\backslash Y_1$ and $f'(x)$ for $\backslash Y_2$; then create a table of x-values of 0, 2, 4, and –3.

C23.　Let $y = f(x) = 10x^9 - 5x^5 + 7x^3 + 5x - 9$ and find $\dfrac{dy}{dx}$ when $x = 0.25$.

C24.　Given $y = 12x^{20} + 8x^{10} - 2x^7 + 17x - 6$, find $f'(0)$ and $f'(-2)$.

If additional practice is needed:

　Barnett, Ziegler, and Byleen, pages 139–142, problems 1–32; pages 170–171, problems 15–42
　Harshbarger and Reynolds, page 680, problems 1–10

UNIT 11

Instantaneous Rates of Change

In preceding units the geometric interpretation of a derivative as a formula that expresses the slope of the tangent to the curve as a function of x was used. This unit will show how the derivative can be interpreted as a rate of change.

As stated in Unit 10, the derivative also is referred to as the instantaneous rate of change of y with respect to x. The derivative is used to estimate the change in y for a unit change in x. This is an approximation because the slope of the tangent line is used to provide the estimate of the change, rather than the function itself.

A sampling of various application problems illustrating the derivative as a rate of change will be introduced throughout the unit. Specific examples include the concepts of marginal cost, marginal revenue, marginal profit, and velocity.

EXAMPLE 1

Mr. Smith is president of the board of trustees for the city's Repertory Theatre. The board's current effort is a major fund-raising project being conducted by direct-mail solicitations. Next week, a letter requesting donations to support the Repertory Theatre will be sent to a group of people whose names appear on a mailing list compiled by the board. Mr. Smith has formulated that, if x letters, in thousands, are sent, the amount of money donated will be

$$f(x) = 0.5x^2 + 14x + 90$$

where x = the number, in thousands, of letters sent

and $f(x)$ = the money donated, in thousands of dollars.

a. Find and interpret $f(10)$.

b. Express the rate of change in money donated with respect to the number of letters sent.

c. Find the rate of change of money donated at $x = 10$.

d.   Interpret the answer to part c.

e.   Compute the actual change in the amount of money donated that will result if 1,000 additional letters
     are mailed.

Solution: a. $f(x) = 0.5x^2 + 14x + 90$

$f(10) = 0.5(10)^2 + 14(10) + 90$

$= 0.5(100) + 140 + 90$

$= 50 + 140 + 90$

$= 280$

Remember that both variables are measured in thousands.

If 10,000 letters of solicitation are sent, it is estimated that \$280,000 will be donated.

b.   The rate of change in money donated, $f(x)$, with respect to the number, x, of letters sent is
     the derivative.

$$\text{Rate of change} = f'(x) = 0.5(2x) + 14$$

$$= x + 14$$

For any value of x, this derivative is an approximation of the *additional* amount of money,
in thousands of dollars, that will be donated if an *additional* 1,000 letters are mailed out.

Note: Recall that the derivative is stated as the change in y for a unit change in x. In this
         example a unit change in x represents a change of 1,000 letters.

c.   Rate of change $= f'(x) = x + 14$

$f'(10) = 10 + 14$

$= 24$

d.   Currently 10,000 letters are being sent. If an additional 1,000 letters are sent, it is estimated
     that an additional \$24,000 will be donated.

e.   The actual change is the difference between the amount donated if 11,000 letters are sent
     and the amount donated if 10,000 letters are sent.

$$f(x) = 0.5x^2 + 14x + 90$$

$$f(11) = 0.5(11)^2 + 14(11) + 90$$

$$= 0.5(121) + 154 + 90$$

$$= 304.5$$

$$f(10) = 280$$

$$\text{Actual change} = f(11) - f(10)$$

$$= 304.5 - 280$$

$$= 24.5$$

Currently 10,000 letters are being sent. If an additional 1,000 letters are sent, the actual
change will result in an additional \$24,500 being donated.

The reason for the difference between the actual change and the estimated change is that the rate of change of the money donated varied with the number of letters sent. The instantaneous rate of change in part c can be thought of as the change in money donated that would occur if the rate of change remained constant.

In geometric terms, the difference between these two quantities is the difference between using the tangent line to estimate the next point one unit away and using the actual curve to determine the point.

A sketch of the graph for $f(x)$ and the tangent line at (10, 280) will help to clarify why there is a difference.

The graph of $f(x)$ is a parabola opening up with its vertex located to the left of the y-axis. The portion in quadrant I is shown below.

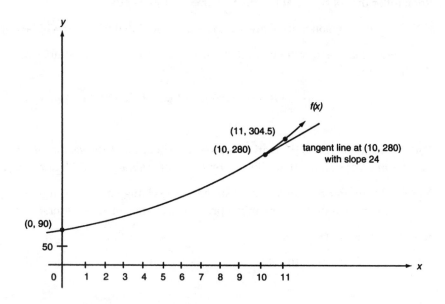

Problem 1

Continue with Mr. Smith's fund-raising project.

a.   Find $f(16)$.

b.   Find the rate of change of money donated at $x = 16$. Are the donations increasing or decreasing at this point?

c.   Interpret $f'(16)$.

d.   Compute the actual change in the amount of money donated if 1,000 additional letters are sent.

Solution:

Answers: a. 402.   b. 30 or $30,000, increasing.   c. If the number of letters sent is increased from 16,000 to 17,000, it is estimated that an additional $30,000 will be collected.   d. $30,500.

Here is another example.

EXAMPLE 2

It is estimated that t months from now the number, in thousands, of out-of-state vehicles on North Carolina ferries will be $N(t) = -t^2 + 13t$.

a. Derive an expression for the rate at which the number of out-of-state vehicles will be changing with respect to time t months from now.

b. At what rate will the number of vehicles be changing 4 months from now?

c. Will the number of vehicles be increasing or decreasing?

d. By how much will the number of vehicles actually change during the fifth month?

Solution: a. The rate of change in the number of out-of-state vehicles with respect to time is the derivative of the function $N(t)$:

$$\text{Rate of change} = N'(t) = -2t + 13.$$

b. The rate of change in the number of out-of-state vehicles 4 months from now will be

$$N'(4) = -2(4) + 13 = -8 + 13 = 5 \text{ or } 5,000 \text{ cars per month.}$$

It is an estimate of the change in $N(t)$ for a unit change in t; that is, as t increases by 1 (from 4 to 5), the value of $N(t)$ will increase by approximately 5.

c. The number of vehicles will be increasing.

d. The actual change in the number of vehicles during the fifth month is the difference between the number at the end of 5 months and the number at the end of 4 months. Or

$$\text{Change in number of vehicles} = N(5) - N(4)$$

$$= [-(5)^2 + 13(5)] - [-(4)^2 + 13(4)]$$

$$= [-25 + 65] - [-16 + 52]$$

$$= -25 + 65 + 16 - 52$$

$$= 4 \text{ or } 4,000 \text{ cars}$$

Why the difference between the actual change and the monthly rate of change? The situation is illustrated on the following page. In part b the tangent line is being used to approximate a unit change of 1 on the curve. The instantaneous rate of change can be thought of as the change in the number of vehicles that would occur during the fifth month if the rate of change remained constant.

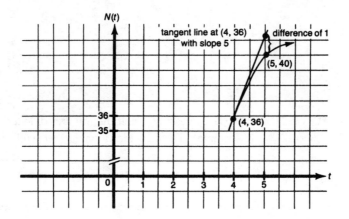

INSTANTANEOUS VELOCITY

A formula for the **instantaneous velocity** of an object at any point in time is given by the instantaneous rate of change of distance with respect to time. Or, simply, velocity equals the derivative of the distance function expressed in terms of time.

For our example we will return to the rubber ball being thrown in Exercise 8 of Unit 7. It will be helpful if you review your answers to questions a–f before continuing.

EXAMPLE 3

Unit 7, Exercise 8—Revisited

If a rubber ball is thrown vertically upward from the top of a building, its height, in feet, t seconds later is given by

$$H(t) = -16t^2 + 128t + 320.$$

a.   Find the formula giving the ball's instantaneous velocity.

b.   Find the ball's instantaneous velocity at $t = 5$.

c.   What was the initial velocity of the ball when it was thrown upward?

d.   What is the velocity of the rubber ball at the instant it hits the ground?

Solution: a.   Height $= H(t) = -16t^2 + 128t + 320$

Velocity $= H'(t) = -32t + 128$

b.   To determine the instantaneous velocity at $t = 5$, the derivative, $H'(t)$, must be evaluated at $t = 5$.

$$\text{Velocity} = H'(t) = -32t + 128$$

$$H'(5) = -32(5) + 128$$

$$= -160 + 128$$

$$= -32 \text{ feet per second}$$

The minus sign indicates the direction of the velocity (down).

An instantaneous rate of 32 feet per second at the end of 5 seconds means that, if the rate were to remain constant for the next second, the ball would **fall** an additional 32 feet. If the ball is accelerating or decelerating (that is, if the rate does not remain constant), then the instantaneous rate is an approximation of what actually happens during the next second.

c. To determine the initial velocity, which would be the velocity at $t = 0$, the derivative is evaluated at time $t = 0$.

$$\text{Velocity} = H'(t) = -32t + 128$$

$$H'(0) = 0 + 128$$

$$= 128 \text{ feet per second}$$

The rubber ball was thrown vertically upward with an initial velocity of 128 feet per second.

d. In order to determine the velocity of the ball when it hits the ground, we must know when it will hit the ground. The ball will hit the ground when $H(t) = 0$, or

$$0 = -16t^2 + 128t + 320.$$

From Unit 7, Exercise 8, the solution was found to be 10.

$$\text{Velocity} = H'(t) = -32t + 128$$

$$H'(10) = -32(10) + 128$$

$$= -320 + 128$$

$$= -192 \text{ feet per second}$$

A negative sign implies a downward direction.

The rubber ball will have a velocity of −192 feet per second when it hits the ground.

Problem 2

Continue with Example 3.

a. Sketch the graph of $H(t)$.

b. Determine the ball's instantaneous velocity at $t = 2$.

c. Is the ball's height increasing or decreasing at this point in time?

d. How many seconds will it take for the ball to reach its maximum height?

e. What is its maximum height?

f. Determine the ball's instantaneous velocity at $t = 4$.

g. How do you explain your answer to part f?

Solution:

Answers:

a.

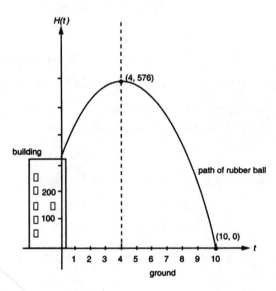

b.   64 feet per second.   c. increasing.   d. 4 seconds.   e. 576 feet.   f. 0 foot per second.   g. The ball has reached its maximum height and is about to start falling back to the ground.

MARGINAL ANALYSIS

In economics, the term **marginal analysis** is used to denote the approximation procedure of using the derivative of a function to estimate the change in the function produced by a one unit increase in the size of the variable.

Examples 4–7 will introduce the concepts of marginal cost, marginal revenue, and marginal profit.

EXAMPLE 4

Suppose the total cost of producing x funky digital watches is given by the cost function

$$C(x) = 20 + 2x + 0.01x^2 \quad x \geq 0$$

where x = the number of watches,

and $C(x)$ = the total cost, in dollars, to produce x watches.

a.   Determine the cost of producing 100 watches.

b.   Determine the cost of producing 101 watches.

c.   Determine the cost of producing the 101st watch.

Solution: a. $C(100) = 20 + 2(100) + 0.01(100)^2$

$= 20 + 200 + 0.01(10,000)$

$= 20 + 200 + 100$

$= 320$

The total cost of producing 100 watches is $320.

b. $C(101) = 20 + 2(101) + 0.01(101)^2$

$= 20 + 202 + 0.01(10,201)$

$= 20 + 202 + 102.01$

$= 324.01$

The cost of producing 101 watches is $324.01.

c.   The cost of producing the 101st watch is found by subtracting:

$$C(101) - C(100) = 324.01 - 320$$

$$= 4.01$$

The cost of producing one additional watch after 100, or the cost of producing the 101st watch, is $4.01.

In Example 4 we computed the additional cost of producing one more unit, here when $x = 100$. In practice, this quantity is usually approximated by the derivative $C'(x)$, which is called the **marginal cost**. Marginal cost is the rate of change in the total production cost with respect to the number of units produced. It is a good approximation of the cost of producing one additional unit.

> Definition: **Marginal cost** $= MC = C'(x)$,
>
> where $C(x) =$ the total production cost of producing x units.

EXAMPLE 5

Continue with the total cost function for producing funky digital watches given in Example 4 of $C(x) = 20 + 2x + 0.01x^2$ $x \geq 0$.

a.   Find the equation for marginal cost.

b.   Find the marginal cost at $x = 100$.

c.   Interpret the answer to part b.

Solution: a.   By definition, marginal cost equals the derivative of cost; therefore $MC = C'(x) = 2 + 0.02x$

b. $MC = C'(x) = 2 + 0.02x$

$MC = C'(100) = 2 + 0.02(100)$

$= 2 + 2$

$= 4$

c. At a production level of $x = 100$, one more watch costs approximately $4 to produce. In other words, the marginal cost of the 101st watch is $4.

Did you notice how very close the approximation is to the actual cost of producing the 101st unit? The answers vary by only $0.01.

Problem 3

Use the information provided in Example 5 to answer the following:

a. Find the marginal cost at $x = 200$.

b. Interpret your answer to part a.

Solution:

Answers: a. 6.   b. At production level $x = 200$, one more watch costs approximately $6 to produce. Or, the marginal cost of the 201st watch is $6.

By a similar line of reasoning, given a revenue function $R(x)$, its derivative, $R'(x)$, is called the **marginal revenue**. Marginal revenue is the rate of change of total revenue with respect to the number of units sold. It is a good approximation of the revenue derived from selling one additional unit.

Definition: **Marginal revenue** $= MR = R'(x)$,

where $R(x) =$ the total revenue derived from selling x units.

EXAMPLE 6

The total sales revenue function, $R(x)$, for selling x funky digital watches is given by

$$R(x) = 8x \quad x \ge 0$$

where $x =$ the number of units

and $R(x) =$ the total revenue, in dollars, for selling x units.

a. Find the equation for marginal revenue.

b. Find the marginal revenue at $x = 100$.

c. Interpret the answer to part b.

Solution: a.   By definition, marginal revenue equals the derivative of the revenue function,

$$MR = R'(x).$$

Therefore $MR = R'(x) = 8.$

In this example, the marginal revenue formula is a constant, meaning that revenue increases by $8 for each additional watch sold.

b. $MR = R'(100) = 8$

c.   At production level $x = 100$, approximately $8 of additional sales revenue is derived from selling one more watch. Or, in other words, the marginal revenue derived from selling the 101st watch would be $8.

Again, by a similar line of reasoning, given a profit function $P(x)$, its derivative, $P'(x)$, is called the **marginal profit**. Marginal profit is the rate of change of total profit with respect to the number of units produced and sold. It is an approximation of the profit derived from producing and selling one additional unit.

Definition: **Marginal profit** $= MP = P'(x)$,

where $P(x) =$ the total profit derived from producing and selling x units.

Note: Many economics texts use the symbol π (pi) for profit to avoid confusion with P, which often stands for price. To help you become familiar with the notation, the symbol will be used with some of the problems in this book as well.

EXAMPLE 7

Use the information provided in Examples 5 and 6 to answer the following:

a.   Write the total profit function, $\pi(x)$, for the funky digital watches.

b.   Find the marginal profit function $\pi'(x)$.

c.   Compute $\pi'(200)$ and interpret your result.

Solution: a.   Profit for an organization is the difference between total revenue and total cost. If $\pi(x)$ represents the total profit from the production and sale of funky digital watches, then

$$\pi(x) = R(x) - C(x)$$

$$= 8x - (20 + 2x + 0.01x^2)$$

$$= 6x - 20 - 0.01x^2$$

b.   By definition, marginal profit equals the derivative of profit; therefore

$$MP = \pi'(x) = 6 - 0.02x$$

c. $MP = \pi'(x) = 6 - 0.02x$

$\pi'(200) = 6 - 0.02(200)$

$= 6 - 4$

$= 2$

At a production level $x = 200$, approximately $2 of additional profit is derived from producing and selling one more funky digital watch. In other words, the marginal profit derived from producing and selling the 201st watch would be $2.

You should now have a workable understanding of the derivative as a rate of change and be able to use it to estimate the change in y for a unit change in x.

Furthermore, given an expression for distance as a function of time, you should be able to determine the instantaneous velocity of an object at any given point in time. And you should be able to determine marginal cost, marginal revenue, and marginal profit equations, as well as to interpret their meanings.

Before beginning the next unit, complete the following exercises and pay particular attention to the interpretations. When you have completed the exercises, check your answers against those at the back of the book.

EXERCISES

1.   A city's population x years after 2002 is predicted to be

$$P(x) = 50{,}000 + 200x^2.$$

   a.   What was the population in 2002?

   b.   Find the rate, in terms of people per year, at which the city is growing.

   c.   Find and interpret $P'(10)$.

2.   An ant colony was treated with an insecticide, and the number of survivors, $N(t)$, in hundreds, after t hours was found to be given approximately by

$$N(t) = -t^2 - 5t + 750.$$

   a.   Find $N'(t)$.

   b.   Find the rate of change in the colony at $t = 10$.

   c.   How many ants were in the colony at the start of the experiment?

C3.   The weekly total cost function of the ABC Company is given by

$$C(x) = 0.258x^2 + 1{,}500$$

      where cost is measured in the number of units produced.

   a.   Calculate and interpret $C(0)$.

   b.   Find the marginal cost equation.

   c.   Use the marginal cost equation to estimate the cost of producing the 25th unit.

   d.   Compute the actual cost of producing the 25th unit.

4.   The total cost, $C(x)$, in dollars, of manufacturing x units is given by

$$C(x) = 10 - 2.5x^2 + x^3 \quad x \geq 2.$$

   a.   Find the marginal cost equation.

   b.   Find the marginal cost at $x = 10$ level of production.

   c.   Interpret the answer to part b.

5.   The Triangle Company manufactures only one item, large clocks. The Accounting Department has determined that the total cost, $C(x)$, in dollars, and total revenue, $R(x)$, in dollars, for x clocks manufactured and sold are given by:

$$C(x) = 50x + 5{,}000$$

$$R(x) = 260x - 2x^2.$$

    a.   Write the equation that defines the total profit function, $P(x)$.

    b.   Find the marginal profit equation.

    c.   Use the marginal profit equation to estimate the profit of producing and selling the 45th clock.

    d.   Compute the actual cost of producing and selling the 45th unit.

6.   A company's total cost, in dollars, of manufacturing q units is given by

$$C(q) = 25q^2 + q + 100 \quad q > 0.$$

    a.   Use the marginal cost equation to estimate the cost of manufacturing the 101st unit.

    b.   Compute the actual cost of manufacturing the 101st unit.

C7.   The T. C. Mellott Company finds that its profits on day x of an advertising campaign is given by

$$y = P(x) = -2.5x^2 + 50x + 30{,}000$$

    where $P(x)$ = profit, in dollars, for day x of the campaign.

    a.   Determine the formula for the instantaneous rate of change for profit with respect to the days of the campaign.

    b.   Calculate and interpret $P(0)$.

    c.   Calculate and interpret $P(5)$.

    d.   Calculate and interpret $P'(5)$.

8.   In April 1999 the number of manufactured housing sales for one state was found to be 1,757. At that time it was projected that t months from then ($t = 0$ for April 1999), the number of manufactured housing sales for the state would be

$$N = \frac{1}{3}t^3 - 6t^2 + t + 1{,}757 \,.$$

    a.   Determine the formula for the instantaneous rate of change for the number of manufactured housing sales with respect to time.

    b.   At what rate was the number of housing sales changing with respect to time 6 months from then? Were sales increasing or decreasing at that point in time?

    c.   At what rate was the number of housing sales changing with respect to time 12 months from then? Were sales increasing or decreasing at that point in time?

9.   The revenue derived from selling x units of some item is given by the revenue function, defined by

$$R(x) = -2x^2 + 80x \quad 0 \le x \le 40 \,.$$

    a.   Write the equation for marginal revenue.

    b.   Find the marginal revenue at $x = 5$.

    c.   Interpret the answer to part b.

10.   It is estimated that the daily output at a certain plant is

$$Q(x) = 0.1x^3 + 3x^2 + 1$$

where x is the number of workers employed at the plant. Currently 10 workers are employed.

    a.   Estimate the effect that one additional worker will have on the daily output.

    b.   Compute the actual change in the daily output that will result if one additional worker is hired.

11.   The revenue derived from selling x units of some item is given by the revenue function, defined by

$$R(x) = -x^2 + 30x.$$

    a.   Calculate and interpret $R(20)$.

    b.   Calculate and interpret $R'(20)$.

    c.   Find $R(21)$.

12.   Unit 7, Exercise 10—Revisited

      Jim's profit at his restaurant is related to the number of customers as defined by

$$P(x) = -x^2 + 140x - 4{,}000$$

      where $P(x) =$ the profit (or loss), in dollars per day

      and $x =$ the number of customers served per day.

    a.   Find the equation for marginal profit.

    b.   Find the marginal profit at $x = 50$.

    c.   Is profit increasing or decreasing at this point?

13.   At a medium-sized college in the South, the number of students enrolled is given by the function

$$N(t) = 5{,}000 - 25t - 9t^2$$

      where $t =$ time, in years, and $t = 0$ corresponds to the year 2001.

    a.   Find and interpret $N(2)$.

    b.   Find and interpret $N'(2)$.

    c.   What are the future prospects for the college?

14.   A ball is tossed upward from the top of a tower. The function that describes the height above the ground of the ball is

$$y = H(t) = -16t^2 + 64t + 80$$

      where $H(t) =$ the height, in feet, above the ground

      and $t =$ the time, in seconds, from when the ball was tossed.

    a.   What is the instantaneous velocity at $t = 1$?

    b.   What is the height of the ball after 4 seconds?

    c.   What is the velocity of the ball after 4 seconds?

       A negative sign implies a downward direction.

    d.   What was the initial velocity with which the ball was tossed?

    e.   When will the ball reach its maximum height?

         Hint: Graph the function.

    f.   How high will the ball rise?

    g.   When will the ball hit the ground?

    h.   What is the velocity of the ball at the instant it hits the ground?

If additional practice is needed:

Barnett, Ziegler, and Byleen, pages 172–173, problems 61–72; pages 210–212, problems 1–18
Harshbarger and Reynolds, pages 721–723, problems 1–34
Hoffmann and Bradley, pages 118–121, problems 35–57; page 143, problems 1–6; page 168, problems 24–25
Waner and Costenoble, pages 169–172, problems 1–30

UNIT 12

More Differentiation Rules

This unit will continue our discussion of finding derivatives. Specifically, you will learn three more rules that will enable you to take the derivative of additional groups of functions besides polynomial functions.

Recall from Unit 9 the following rules:

Power: If $f(x) = x^n$, then $f'(x) = nx^{n-1}$

Constant: If $f(x) = c$, then $f'(x) = 0$

Constant $\cdot$ Function: If $f(x) = c \cdot u(x)$, then $f'(x) = c \cdot u'(x)$

Special Case: If $f(x) = c \cdot x^n$, then $f'(x) = cnx^{n-1}$

Sum: If $f(x) = u(x) + v(x)$, then $f'(x) = u'(x) + v'(x)$

Difference: If $f(x) = u(x) - v(x)$, then $f'(x) = u'(x) - v'(x)$

PRODUCT RULE

The Product Rule allows us to take the derivative of a product of two functions without necessarily multiplying them out first.

Product Rule

If $f(x) = F(x) \cdot S(x)$, then $f'(x) = F(x) \cdot S'(x) + S(x) \cdot F'(x)$

In words, the derivative of a product of two functions is the first function times the derivative of the second function plus the second function times the derivative of the first function.

The above statement can be shortened somewhat to the following: the derivative of a product of two functions is the first times the derivative of the second plus the second times the derivative of the first. I used $F(x)$ and $S(x)$ to denote the first and second functions so as to make the rule easier to follow.

In all of the examples for this unit, an extra step will be added using the $\frac{df(x)}{dx}$ notation to indicate how the rule is applied before actually taking the derivative. You may wish to omit the step after you become more familiar with the rules, but at the beginning you should leave it in.

EXAMPLE 1

If $f(x) = x^3(x - 5)$, find the derivative.

Solution: One approach would be to remove the parentheses and then take the derivative of the polynomial function term by term. But I want to illustrate the use of the Product Rule.

The original function: $f(x) = x^3(x - 5)$

Product Rule: $f'(x) = F(x) \cdot S'(x) + S(x) \cdot F'(x)$

Extra step to show
what I intend to do: $f'(x) = x^3 \dfrac{d(x - 5)}{dx} + (x - 5)\dfrac{d(x^3)}{dx}$

Take the derivatives. $= x^3(1) \qquad + (x - 5)(3x^2)$

Remove the parentheses. $= x^3 + 3x^3 - 15x^2$

Combine like terms. $= 4x^3 - 15x^2$

Problem 1

If $f(x) = x^3(x - 5)$, remove the parentheses and take the derivative term by term to verify that the answer is the same.

Solution:

Which method was easier? I prefer proceeding term by term, but in this unit we will practice using the Product Rule. After that, the choice is yours, that is, until we reach the units on exponential and logarithmic functions. To differentiate products involving either exponential or logarithmic functions, the use of the Product Rule will be required.

EXAMPLE 2

Use the Product Rule to find the derivative of $f(x) = (x^4 + 3)(3x^3 + 1)$.

Solution: Start with the function. $f(x) = (x^4 + 3)(3x^3 + 1)$

               Product Rule: $f'(x) = F(x) \cdot S'(x) + S(x) \cdot F'(x)$

               Extra step to show
                  what is intended: $f'(x) = (x^4 + 3)\dfrac{d(3x^3 + 1)}{dx} + (3x^3 + 1)\dfrac{d(x^4 + 3)}{dx}$

               Take the derivatives. $= (x^4 + 3)(9x^2) + (3x^3 + 1)(4x^3)$

               Remove the parentheses. $= 9x^6 + 27x^2 + 12x^6 + 4x^3$

               Combine like terms. $= 21x^6 + 4x^3 + 27x^2$

Here's a word of warning before you try Problem 2. Be sure to keep each factor in parentheses until the very end; otherwise factors easily become mistaken for separate terms.

Problem 2

Use the Product Rule to find y' if $y = (x^2 + 1)(3x - 2)$.

Solution: Start with the function.

               Product Rule: $f'(x) = F(x) \cdot S'(x) + S(x) \cdot F'(x)$

               Extra step to show
                  what is intended:

               Take the derivatives.

               Remove the parentheses.

               Combine like terms.

                                                    Answer: $y' = 9x^2 - 4x + 3$

In case you did not get the correct answer for that problem, we will do another example using the Product Rule before continuing.

EXAMPLE 3

Use the Product Rule to find $\dfrac{dy}{dx}$ if $y = (x - 3)(x^2 + 7x + 10)$.

Solution: $y = (x - 3)(x^2 + 7x + 10)$

           $\dfrac{dy}{dx} = F(x) \cdot S'(x) + S(x) \cdot F'(x)$

               $= (x - 3)\dfrac{d(x^2 + 7x + 10)}{dx} + (x^2 + 7x + 10)\dfrac{d(x - 3)}{dx}$

               $= (x - 3)(2x + 7) + (x^2 + 7x + 10)(1)$

$$= 2x^2 + 7x - 6x - 21 + x^2 + 7x + 10$$

$$= 3x^2 + 8x - 11$$

QUOTIENT RULE

Thus far all that has happened since the initial introduction of the Power Rule is that the functions have become progressively more complicated. All of the functions, however, have been polynomial functions. And the derivative remained a formula for finding the slope of the tangent line to the curve at any point.

The only part that is about to change with the next rule is that we are no longer necessarily dealing with polynomial functions. The function may be, but probably is not, a polynomial function. The next rule is used to find the derivative of a quotient of two functions.

Quotient Rule

If $f(x) = \dfrac{N(x)}{D(x)}$, then $f'(x) = \dfrac{D(x) \cdot N'(x) - N(x) \cdot D'(x)}{[D(x)]^2}$

In words, the derivative of a quotient of two functions is the denominator times the derivative of the numerator minus the numerator times the derivative of the denominator, all over the denominator squared.

The Quotient Rule is not really as bad as it looks. In fact, in my opinion it is easier to use than the Product Rule once you become familiar with it. Again, I have tried to make the rule easier to follow by letting $N(x)$ denote the function in the numerator and $D(x)$ denote the function in the denominator.

We will do a variety of examples.

EXAMPLE 4

If $f(x) = \dfrac{x}{x^2 + 5}$, find $f'(x)$.

Solution: Start with the function. $f(x) = \dfrac{x}{x^2 + 5}$

Quotient Rule: $f'(x) = \dfrac{D(x) \cdot N'(x) - N(x) \cdot D'(x)}{[D(x)]^2}$

Extra step to show
what is intended: $f'(x) = \dfrac{(x^2 + 5)\dfrac{d(x)}{dx} - x\dfrac{d(x^2 + 5)}{dx}}{(x^2 + 5)^2}$

Take the derivatives.
$$= \frac{(x^2 + 5)(1) - x(2x)}{(x^2 + 5)^2}$$

Remove the parentheses but leave the denominator as is.
$$= \frac{x^2 + 5 - 2x^2}{(x^2 + 5)^2}$$

Combine like terms.
$$= \frac{-x^2 + 5}{(x^2 + 5)^2}$$

Comment: The preferred form for the answer is to leave the denominator in factored form; that is, not multiplied.

EXAMPLE 5

If $y = \dfrac{3}{x}$, find $\dfrac{dy}{dx}$.

Solution: Start with the function.
$$y = \frac{3}{x}$$

Quotient Rule:
$$f'(x) = \frac{D(x) \cdot N'(x) - N(x) \cdot D'(x)}{[D(x)]^2}$$

Extra step:
$$\frac{dy}{dx} = \frac{(x)\dfrac{d(3)}{dx} - 3\dfrac{d(x)}{dx}}{x^2}$$

Take the derivatives.
$$= \frac{(x)0 - 3(1)}{x^2}$$

Remove the parentheses.
$$= \frac{0 - 3}{x^2}$$

Combine like terms.
$$= \frac{-3}{x^2}$$

Here is a rather short problem for you to try, and then I'll do two more examples.

Problem 3

If $f(x) = \dfrac{5}{x^2}$, find $f'(x)$. Be sure to simplify the final answer.

Solution: Start with the function.

Quotient Rule: $$f'(x) = \frac{D(x) \cdot N'(x) - N(x) \cdot D'(x)}{[D(x)]^2}$$

Extra step to show
 what is to be done:

$f(x) = \frac{5}{x^2} \; ; f'(x) = \frac{x^2 \cdot 0 - 5(2x)}{(x^2)^2} = \frac{0 - 10x}{x^4} = -\frac{10}{x^3}$

Take the derivatives.

Remove the parentheses.

Combine like terms.

Reduce to lowest terms.

Answer: $f'(x) = \dfrac{-10}{x^3}$

Recall that fractions should always be simplified, that is, reduced to lowest terms. To simplify an algebraic fraction:

 a.   factor the numerator and denominator,

 b.   cancel all common factors.

EXAMPLE 6

$f'(x) = \frac{(x+2) \cdot 3x^2 - x^3(1)}{(x+2)^2} = \frac{3x^3 + 6x^2 - x^3}{(x+2)^2} = \frac{2x^3 + 6x^2}{(x+2)^2}$

If $f(x) = \dfrac{x^3}{x+2}$, find $f'(x)$.

$\frac{2x^2(x+3)}{(x+2)}$

Solution: Start with the function. $f(x) = \dfrac{x^3}{x+2}$

Quotient Rule: $$f'(x) = \frac{D(x) \cdot N'(x) - N(x) \cdot D'(x)}{[D(x)]^2}$$

Extra step: $$f'(x) = \frac{(x+2)\dfrac{d(x^3)}{dx} - x^3\dfrac{d(x+2)}{dx}}{(x+2)^2}$$

Take the derivatives. $$= \frac{(x+2)(3x^2) - x^3(1)}{(x+2)^2}$$

Remove the parentheses. $$= \frac{3x^3 + 6x^2 - x^3}{(x+2)^2}$$

Combine like terms. $$= \frac{2x^3 + 6x^2}{(x+2)^2}$$

Simplify. $$= \frac{2x^2(x+3)}{(x+2)^2}$$

Here are two warnings concerning the Quotient Rule: (1), at the beginning of the problem don't fail to put both the numerator and the denominator in parentheses; (2), don't attempt to simplify by canceling common factors until the very **last** step. A common error is to cancel a factor in the denominator with a term, rather than a factor, in the numerator.

EXAMPLE 7

$$y' = \frac{(x-5)(1) - (x+2)(1)}{(x-5)^2} \quad \frac{(x-5) - x - 2)}{(x-5)2} = \frac{-7}{(x-5)^2}$$

If $y = \dfrac{x+2}{x-5}$, find $\dfrac{dy}{dx}$.

Solution: Start with the function.

$$y = \frac{x+2}{x-5}$$

Quotient Rule:

$$f'(x) = \frac{D(x) \cdot N'(x) - N(x) \cdot D'(x)}{[D(x)]^2}$$

Extra step:

$$\frac{dy}{dx} = \frac{(x-5)\dfrac{d(x+2)}{dx} - (x+2)\dfrac{d(x-5)}{dx}}{(x-5)^2}$$

Take the derivatives.

$$= \frac{(x-5)(1) - (x+2)(1)}{(x-5)^2}$$

Remove the parentheses,
   except in the denominator.

$$= \frac{x-5-x-2}{(x-5)^2}$$

Combine like terms.

$$= \frac{-7}{(x-5)^2}$$

Now it's your turn to try a more complicated problem.

Problem 4

$$\frac{(1+x^2)4 - (2x)(4x)}{(1+x^2)^2} = \frac{4+4x^2-8x^2}{(1+x^2)^2} = \frac{4-4x^2}{(1+x^2)^2}$$

If $y = \dfrac{4x}{1+x^2}$, find $\dfrac{dy}{dx}$.

$$\frac{4(1-x^2)}{(1+x^2)^2}$$

Solution: Start with the function.

Quotient Rule:

$$f'(x) = \frac{D(x) \cdot N'(x) - N(x) \cdot D'(x)}{[D(x)]^2}$$

Extra step:

Take the derivatives.

Remove the parentheses.

Combine like terms.

Simplify:

Answer: $\dfrac{dy}{dx} = \dfrac{4 - 4x^2}{(1 + x^2)^2}$

CHAIN RULE/POWER OF A FUNCTION

The last rule of the unit is for taking the derivative of a function raised to some power, n. This rule has two different names, the Chain Rule and the Power of a Function; the latter is more descriptive.

> **Chain Rule/Power of a Function**
>
> If $f(x) = [u(x)]^n$, then $f'(x) = n[u(x)]^{n-1} u'(x)$, where n is a real number.
>
> In words, the derivative of a function raised to a power is equal to the exponent times the function raised to the exponent minus 1, and all of that is multiplied times the derivative of the function.

An easy way to think of this rule is that it usually takes the form of having a function in parentheses raised to some power.

In symbols, the Chain Rule looks like this:

$$\text{If } f(x) = (\quad)^n, \text{ then } f'(x) = n(\quad)^{n-1}\frac{d(\quad)}{dx}$$

In words, the derivative of a quantity raised to a power is equal to the exponent times the quantity raised to the exponent minus 1 times the derivative of what is in the parentheses.

After the last two rules, this one is really short and fast.

EXAMPLE 8

$f'(x) = 5(7x^2 - 2x + 13)^4 \cdot (14x - 2)$

Find the derivative of $f(x) = (7x^2 - 2x + 13)^5$.

Solution: If we had nothing better to do with our time, we could multiply the expression out and take the derivative term by term. Obviously the Chain Rule is meant to be the better way.

Start with the function. $f(x) = (7x^2 - 2x + 13)^5$

Use the Chain Rule. $f'(x) = n[u(x)]^{n-1} u'(x)$

Extra step to show
what is intended: $f'(x) = 5(7x^2 - 2x + 13)^4 \dfrac{d(7x^2 - 2x + 13)}{dx}$

Take the derivative. $= 5(7x^2 - 2x + 13)^4(14x - 2)$

Nothing more need be done unless you prefer to rewrite the answer as

$$f'(x) = 5(14x - 2)(7x^2 - 2x + 13)^4$$

or

$$f'(x) = 10(7x^2 - 1)(7x^2 - 2x + 13)^4$$

EXAMPLE 9

$f'(x) = 10(3x+1)^9(3)$
$30(3x+1)^9$

Find the derivative of $f(x) = (3x + 1)^{10}$.

Solution: Start with the function. $f(x) = (3x + 1)^{10}$

   Chain Rule: $f'(x) = n[u(x)]^{n-1}u'(x)$

   Extra step: $f'(x) = 10(3x + 1)^9 \dfrac{d(3x + 1)}{dx}$

   Take the derivative. $= 10(3x + 1)^9(3)$

                    $= 30(3x + 1)^9$ would be the preferred way to write it.

Since these problems are so short, try two.

Problem 5

$f'(x) = 12(x^2-3x+5)^{11}\cdot(2x-3)$

If $f(x) = (x^2 - 3x + 5)^{12}$, find $f'(x)$.

$24x - 36(x^2-3x+5)$

Solution: Start with the function.

   Chain Rule: $f'(x) = n[u(x)]^{n-1}u'(x)$

   Take the derivative.

                                     Answer: $f'(x) = 12(x^2 - 3x + 5)^{11}(2x - 3)$

Problem 6

$D_x = 14(3x-5)^{13}\cdot 3$
$42(3x-5)^{13}$

Find $D_x(3x - 5)^{14}$.

Solution: Start with the function.

   Chain Rule. $f'(x) = n[u(x)]^{n-1}u'(x)$

                                     Answer: $D_x(3x - 5)^{14} = 42(3x - 5)^{13}$

CALCULUS AND THE CALCULATOR (Optional)

Writing Equations of Tangent Lines

In previous units you were often asked to find an equation of the line tangent to the graph of some function at a specified x-value. To do so meant finding the corresponding y-coordinate, finding the derivative, evaluating it at the given point, and then using the point-slope formula to write an equation of the line.

Using still another feature of the graphing calculator, we are able to draw a tangent line to the graph of a function at a given point and write its equation, without our having to do much of anything. One example should be sufficient, as the keystroke sequence is quite short.

EXAMPLE 10

Given the function $f(x) = x^2 - 2x - 6$, find an equation of the line tangent to the graph at $x = 2$.

Solution: Key in the function and view the graph.

To write an equation of the line tangent to the graph at $x = 2$,

Press (**2nd**) (**DRAW**) (**5**) followed by (**2**) (**ENTER**) for the specified x-value.

On the screen, the tangent line is drawn to the graph, and at the bottom appears $x = 2$ and $y = 2x - 10$.

Answer: An equation of the line tangent to the graph at $x = 2$ is given by $y = 2x - 10$.

You should now be able to take the derivative of the product of two functions, of the quotient of two functions, and of a function raised to a power.

Here is a summary of the rules thus far:

Power:	If $f(x) = x^n$, then $f'(x) = nx^{n-1}$
Constant:	If $f(x) = c$, then $f'(x) = 0$
Constant · Function:	If $f(x) = c \cdot u(x)$, then $f(x) = c \cdot u'(x)$
Special Case:	If $f(x) = c \cdot x^n$, then $f'(x) = cnx^{n-1}$
Sum:	If $f(x) = u(x) + v(x)$, then $f'(x) = u'(x) + v'(x)$
Difference:	If $f(x) = u(x) - v(x)$, then $f'(x) = u'(x) - v'(x)$
Product Rule:	If $f(x) = F(x) \cdot S(x)$, then $f'(x) = F(x)S'(x) + S(x)F'(x)$
Quotient Rule:	If $f(x) = \dfrac{N(x)}{D(x)}$, then $f'(x) = \dfrac{D(x)N'(x) - N(x)D'(x)}{[D(x)]^2}$
Chain Rule:	If $f(x) = [u(x)]^n$, then $f'(x) = n[u(x)]^{n-1}u'(x)$

Before beginning the next unit, use the appropriate rules to differentiate the various functions listed in the exercises below. Final answers should be simplified.

EXERCISES

1.   Find $f'(x)$ for $f(x) = 3x(x^2 + x - 5)$.

2.   Use the Product Rule to find $f'(x)$ for $f(x) = (2x + 1)(x^2 - 3)$.

3.   Find the derivative of $f(x) = (x - 1)(2x^3 - x^2 + 7x + 2)$.

4.   Use the Product Rule to find $\dfrac{dy}{dx}$ for $y = (x^2 + 1)(2 - x^3)$.

5.   Differentiate $y = f(x) = 3(x^7 + 2x^5 - x^3 + 6x - 1)$.

6.   Find the derivative of $f(u) = (u^3 + 5)^7$.

7.   Given $y = (3x^7 - x + 13)^5$, find y'.

8.   If $f(x) = \dfrac{-1}{x^2}$, find $f'(x)$ and simplify.

9.   Find f' when $f(x) = \dfrac{4x + 100}{x}$.

10.   Find $D_x\!\left(\dfrac{3}{x + 4}\right)$.

11.   Differentiate $y = \dfrac{x - 2}{3x^2 + 1}$.

12.   Find y' for $y = \dfrac{x^2 + 2}{x - 3}$.

13.   Find $\dfrac{dy}{dx}$ for $y = (11x^3 - 7x + \pi)^2$.

14.   Given $N = \dfrac{100t}{t - 9}$, find the rate of change of N with respect to t.

15.   If $f(x) = \dfrac{x + 2}{x^3}$, find $f'(x)$.

16.   Given $f(x) = (6x^2 + 12x + 1)^5$:

   a.   Find $f(-2)$ and interpret.

   b.   Find $f'(-2)$ and interpret.

17.   Given $f(x) = \dfrac{2x + 4}{x + 6}$:

   a.   Find f(–2) and interpret.

   b.   Find $f'(-2)$ and interpret.

18. Consider the revenue function defined by

$$R(x) = (x + 1)(2x^2 + x + 3) \quad x \geq 0$$

where x = the number of units sold

and $R(x)$ = the total revenue, in dollars, obtained by selling x units.

a. Find the equation for marginal revenue.

b. Find the marginal revenue at $x = 10$. Is revenue increasing or decreasing at this point?

19. If $C(x)$ = the total cost of producing x units of some product, then the average cost per unit is given by

$$A(x) = \frac{C(x)}{x} \quad x > 0$$

Given the cost function, in dollars, defined by

$$C(x) = 10x + 5,000:$$

a. Determine the equation for average cost, $A(x)$.

b. Determine the formula for $A'(x)$.

c. Compute $C(100)$ and interpret.

d. Compute $A(100)$ and interpret.

e. Compute $C'(100)$ and interpret.

f. Compute $A'(100)$ and interpret.

20. Suppose that the number of pounds of jelly beans people are willing to buy per day in a small candy store at a price of p is given by

$$D(p) = 6 + p - p^2 \quad 0 \leq p \leq 3$$

where p = the price, in dollars, per pound of jelly beans

and $D(p)$ = the number of pounds sold per day at price p.

a. Find $D'(p)$, the rate of change of demand with respect to price.

b. Find $D(1.5)$ and $D'(1.5)$ and interpret.

21. The concentration of alcohol in the bloodstream t hours after drinking 1 ounce of alcohol is about

$$A(t) = \frac{3t}{100 + t^2}.$$

a. Find $A'(t)$.

b. Find the approximate change in concentration as t changes from 2 to 3.

22. It is estimated that t years from now the population of people aged 65 and over in Japan will be

$$P(t) = 15 - \frac{6}{t + 1} \text{ million.}$$

a. Derive a formula for the rate at which the population will be changing with respect to time t years from now.

    b.   At what rate will the population be growing 1 year from now?

    c.   By how much will the population actually increase during the second year?

    d.   At what rate will the population be growing 9 years from now?

    e.   What will happen to the rate of population growth in the long run?

C23. Given $f(x) = (1.3x^4 - 0.02x^3 + 0.25x^2 - x + 5)^3$, find $f'(-1)$, $f'(0)$, $f'(1)$, $f'(2)$, and $f'(3)$.

C24. Suppose that the temperature T of food placed in a freezer is given by the function

$$T = \frac{600}{x^2 + 4x + 10}$$

       where x = the time in hours after it is placed in the freezer

       and T = the temperature reading in degrees on the Fahrenheit scale.

    a.   Find T'.

    b.   Find $T(0)$, $T'(0)$, $T(1)$, $T'(1)$, $T(10)$, and $T'(10)$ and interpret.

    c.   Find $T(24)$, $T'(24)$, $T(48)$, $T'(48)$, $T(72)$, and $T'(72)$.

    d.   What will happen to the temperature of the food in the long run?

C25. Given the function $f(x) = 4x^2 - 2x + 9$, find an equation of the line tangent to the graph at $x = 1$. Verify that your answer agrees with Unit 10, Exercise 17.

C26. Given the function $y = f(x) = x^4 - 3x^3 + 2x^2 - 6$, find an equation of the line tangent to the graph at $x = 2$. Verify that your answer agrees with Unit 10, Exercise 20.

If additional practice is needed:

   Barnett, Ziegler, and Byleen, pages 192–194, problems 1–48 and 65–71; pages 199–200, problems 1–16
   Harshbarger and Reynolds, pages 690–693, problems 1–59; pages 698–701, problems 1–53
   Hoffmann and Bradley, pages 128–131, problems 1–39; pages 156–160, problems 1–63
   Waner and Costenoble, pages 207–208, problems 1–66

UNIT 13

The Power Rule Revisited

In this unit you will learn how to take the derivative of functions with negative or fractional exponents and of functions involving radicals.

NEGATIVE EXPONENTS

Thus far all of our experience in taking derivatives has involved exponents that were whole numbers. That is about to change. First we will consider negative exponents. As a review, the definition of a negative exponent is repeated below, along with some sample problems.

$$b^{-n} = \frac{1}{b^n}$$

$$\frac{1}{b^n} = b^{-n}$$

Definition: **Negative Exponent**

$$b^{-n} = \frac{1}{b^n} \text{ and } \frac{1}{b^{-n}} = b^n$$

Therefore a negative exponent has the effect of moving a factor from top to bottom (or vice versa). Consider Examples 1–8, and be certain you understand the simplification.

EXAMPLE 1 $2^{-1} = \dfrac{1}{2}$

EXAMPLE 2 $x^{-1} = \dfrac{1}{x}$

EXAMPLE 3 $q^{-3} = \dfrac{1}{q^3}$

EXAMPLE 4 $2x^{-1} = \dfrac{2}{x}$

EXAMPLE 5 $3t^{-2} = \dfrac{3}{t^2}$

163

EXAMPLE 6 $\dfrac{1}{x^{-2}} = x^2$

EXAMPLE 7 $\dfrac{5}{x^{-3}} = 5x^3$

EXAMPLE 8 $\dfrac{2}{7p^{-1}} = \dfrac{2p}{7}$ $\dfrac{2}{7p^{-1}}$ $\dfrac{2}{7}p$

TAKING DERIVATIVES OF FUNCTIONS WITH NEGATIVE EXPONENTS

We are now ready to apply the Power Rule to take derivatives of functions containing x raised to a negative power. Keep in mind that derivatives always should be simplified, meaning that final answers should be written using positive integral exponents only—no fractional, negative, or zero exponents.

The basic procedure is as follows:

1. Rewrite the function as a power of x if it is not already written that way.

2. Use the Power Rule to take the derivative.

3. Simplify by removing the parentheses and doing whatever else will help.

4. If any negative exponents exist, use the definition of a negative exponent to rewrite the expression with positive exponents.

The Power Rule is repeated here for reference.

Power Rule

If $f(x) = x^n$, then $f'(x) = nx^{n-1}$ for all n, n being a real number.

EXAMPLE 9

If $f(x) = x^{-3}$, find $f'(x)$.

Solution: Given: $f(x) = x^{-3}$

Power Rule: If $f(x) = x^n$, then $f'(x) = nx^{n-1}$

Use the Power Rule. $f'(x) = -3x^{-3-1}$

Simplify. $= -3x^{-4}$

Rewrite with a positive exponent. $= \dfrac{-3}{x^4}$

EXAMPLE 10

If $f(x) = 2x^{-5}$, find $f'(x)$.

Solution: Given:

$$f(x) = 2x^{-5}$$

Use the Power Rule.

$$f'(x) = 2(-5x^{-5-1})$$

Simplify.

$$= -10x^{-6}$$

Rewrite with a positive exponent.

$$= \frac{-10}{x^6}$$

(handwritten: $f(x) = -5(2x^{-6})$
$-10 \frac{1}{x^6}$)

EXAMPLE 11

Find $f'(x)$ if $f(x) = \dfrac{1}{x^2}$.

Solution: We could use the Quotient Rule, but there is a faster method.

Given:

$$f(x) = \frac{1}{x^2}$$

(handwritten: $f(x) = x^{-2}$, $f'(x) = -2x^{-3} = \dfrac{2}{x^3}$)

Rewrite as a power of x.

$$= x^{-2}$$

Use the Power Rule.

$$f'(x) = -2x^{-2-1}$$

Simplify.

$$= -2x^{-3}$$

Rewrite with a positive exponent.

$$= \frac{-2}{x^3}$$

For comparison purposes, rework Example 11.

Problem 1

Use the Quotient Rule to find the derivative of $f(x) = \dfrac{1}{x^2}$.

Solution:

(handwritten:)
$$\frac{x^2 \cdot 0 - 2x \cdot 1}{(x^2)^2} = \frac{-2x}{x^4} = \frac{-2}{x^3}$$

Answer: $f'(x) = \dfrac{-2}{x^3}$

Which method was shorter? Which method did you like better? I prefer rewriting and doing the problem with negative exponents, but the choice is yours.

I'll do one more example before having you try a problem.

EXAMPLE 12

(handwritten: $2x^{-1} - 7x^{-4}$)

Find $f'(x)$ if $f(x) = \dfrac{2}{x} - \dfrac{7}{x^4}$.

(handwritten: $f'(x) = -2x^{-2} + 28(x)^{-5}$)

(handwritten: $\dfrac{-2}{x^2} + \dfrac{28}{x^5}$)

Solution: Given:

$$f(x) = \frac{2}{x} - \frac{7}{x^4}.$$

Rewrite as powers of x.

$$= 2x^{-1} - 7x^{-4}$$

Use the Power Rule.

$$f'(x) = 2(-1x^{-1-1}) - 7(-4x^{-4-1})$$

Simplify.

$$= -2x^{-2} + 28x^{-5}$$

Rewrite with positive exponents.

$$= \frac{-2}{x^2} + \frac{28}{x^5}$$

Problem 2

(handwritten: $f(x) = x^{-1}$ $f'(x) = -x^{-2}$)

(handwritten: $= \dfrac{-1}{x^2}$)

Use the Power Rule to find the derivative of $f(x) = \dfrac{1}{x}$.   *(handwritten: F(}*

Solution: Given:

Rewrite as a power of x.

Use the Power Rule.

Simplify.

Rewrite with a positive exponent.

Answer: $f'(x) = \dfrac{-1}{x^2}$

Try another problem, with no hints this time.

Problem 3

(handwritten: $f(x) = 3x^{-2}$)

(handwritten: $f' = -6x^{-3} = \dfrac{-6}{x^3}$)

Use the Power Rule to find the derivative of $f(x) = \dfrac{3}{x^2}$.

Solution:

Answer: $f'(x) = \dfrac{-6}{x^3}$

FRACTIONAL EXPONENTS

Now it is time to turn our attention to fractional exponents. For several reasons, fractional exponents are more troublesome to work with, but I'll try to make the procedure as straightforward as possible. Don't get discouraged.

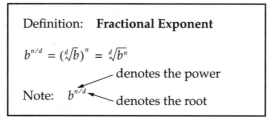

Note that two forms of the definition are given.

$$b^{n/d} = \left(\sqrt[d]{b}\right)^n = \sqrt[d]{b^n}$$

numerical   algebraic

The first form is useful in numerical calculations, provided that the dth root of b is a known integer. It is then convenient to take the root before raising to the nth power in order to work with smaller numbers.

The second form is the more common way of rewriting algebraic expressions with fractional exponents.

As the two forms suggest, we can do either the root or the power first.

We will do four numerical problems first and use the numerical form of the definition. For a more detailed explanation see the unit entitled "Fractional Exponents" in *Forgotten Algebra*.

$$\text{Numerical form:} \quad b^{n/d} = \left(\sqrt[d]{b}\right)^n$$

(with **power** indicated by the exponent n and **root** indicated by the index d)

EXAMPLE 13 $8^{1/3} = \left(\sqrt[3]{8}\right)^1 = (2)^1 = 2$

EXAMPLE 14 $27^{2/3} = \left(\sqrt[3]{27}\right)^2 = (3)^2 = 9$

EXAMPLE 15 $-8^{4/3} = \left(\sqrt[3]{-8}\right)^4 = (-2)^4 = 16$

EXAMPLE 16 $9^{-1/2} = \left(\sqrt{9}\right)^{-1} = (3)^{-1} = \dfrac{1}{3}$

or alternatively,

$$= \frac{1}{9^{1/2}} = \frac{1}{\sqrt{9}} = \frac{1}{3}$$

Now we will consider two algebraic problems and use the algebraic form of the definition to simplify the expressions.

$$\text{power}$$

Algebraic form: $b^{n/d} = \sqrt[d]{b^n}$

$$\text{root}$$

EXAMPLE 17 $x^{1/2} = \sqrt{x}$

EXAMPLE 18 $p^{2/3} = \sqrt[3]{p^2}$

Three problems should be enough for you to try because we want to get on to the real task of taking derivatives with fractional exponents. Simplify each of the following.

Problem 4

Rewrite the following expressions without using fractional exponents.

a. $w^{1/3}$

b. $z^{2/5}$

c. $x^{-1/2}$

Answers: a. $\sqrt[3]{w}$ b. $\sqrt[5]{z^2}$ c. $\dfrac{1}{\sqrt{x}}$

TAKING DERIVATIVES OF FUNCTIONS WITH FRACTIONAL EXPONENTS

We are now ready to use the Power Rule to take derivatives of functions with fractional exponents.

The basic procedure remains as stated earlier with a fifth step added:

1.   Rewrite the function as a power of x if it is not already written that way.

2.   Use the Power Rule to take the derivative.

3.   Simplify by removing parentheses and doing whatever else will help.

4.   If any negative exponents exist, use the definition of a negative exponent to rewrite the expression with positive exponents.

5.   If any fractional exponents exist, use the definition of a fractional exponent to rewrite the expression with radical signs.

The Power Rule is restated here one more time.

Power Rule

If $f(x) = x^n$, then $f'(x) = nx^{n-1}$ for all n, n being a real number.

To repeat in words, if x is raised to some exponent, the derivative is equal to the exponent times x raised to the exponent minus 1. In Examples 19 and 20, the exponents are fractions, but the rule remains unchanged. The only difference is that the calculations become more troublesome.

EXAMPLE 19

$f'(x) = \frac{1}{2} \cdot x^{-1/2} \qquad \frac{1}{2} \cdot \frac{1}{x^{1/2}} = \frac{1}{2\sqrt{x}}$

Find $f'(x)$ if $f(x) = x^{1/2}$.

Solution: Given: $f(x) = x^{1/2}$

Power Rule: If $f(x) = x^n$, then $f'(x) = nx^{n-1}$

Use the Power Rule. $f'(x) = \frac{1}{2}x^{(1/2)-1}$

Simplify. $= \frac{1}{2}x^{-1/2}$

Rewrite with a positive exponent. $= \frac{1}{2} \cdot \frac{1}{x^{1/2}}$

Rewrite without a fractional exponent. $= \frac{1}{2\sqrt{x}}$

EXAMPLE 20

$f'(x) = \frac{5}{3}x^{-\frac{2}{3}} = \frac{5}{3\sqrt[3]{x^2}}$

If $f(x) = 5x^{1/3}$, find $f'(x)$.

Solution: Given: $f(x) = 5x^{1/3}$

Use the Power Rule. $f'(x) = 5\left(\frac{1}{3}x^{(1/3)-1}\right)$

Simplify. $= 5 \cdot \frac{1}{3}x^{-2/3}$

Rewrite with a positive exponent. $= \frac{5}{3} \cdot \frac{1}{x^{2/3}}$

Rewrite without a fractional exponent. $= \frac{5}{3\sqrt[3]{x^2}}$

TAKING DERIVATIVES OF FUNCTIONS WITH RADICALS

The basic procedure remains as stated; however, step 1 may be reworked as follows:

1.   If the function contains a radical sign, rewrite the function as a power of x.

The remaining steps are the same, as Examples 21–24 will illustrate.

EXAMPLE 21

handwritten: $f(x) = 7x^{1/2}, f'(x) = \frac{7}{2}x^{-1/2} \quad \frac{7}{2} \cdot \frac{1}{\sqrt{2}}$

If $f(x) = 7\sqrt{x}$, find $f'(x)$.

handwritten: $\frac{7}{2\sqrt{2}}$

Solution: Given:

$$f(x) = 7\sqrt{x}$$

Rewrite as a power of x.

$$= 7x^{1/2}$$

Use the Power Rule.

$$f'(x) = 7\left(\frac{1}{2}x^{(1/2)-1}\right)$$

Simplify.

$$= 7 \cdot \frac{1}{2} \cdot x^{-1/2}$$

Rewrite with a positive exponents.

$$= \frac{7}{2} \cdot \frac{1}{x^{1/2}}$$

Rewrite without a fractional exponent.

$$= \frac{7}{2\sqrt{x}}$$

EXAMPLE 22

handwritten: $f(x) = 3x^{\frac{2}{3}}$
$f'(x) = 2x^{-1/3}$

If $f(x) = 3\sqrt[3]{x^2}$, find $f'(x)$.

handwritten: $= \frac{2}{\sqrt[3]{x}}$

Solution: Given:

$$f(x) = 3\sqrt[3]{x^2}$$

Rewrite as a power of x.

$$= 3x^{2/3}$$

Use the Power Rule.

$$f'(x) = 3\left(\frac{2}{3}x^{(2/3)-1}\right)$$

Simplify.

$$= \cancel{3} \cdot \frac{2}{\cancel{3}}x^{-1/3}$$

Rewrite with a positive exponent.

$$= 2 \cdot \frac{1}{x^{1/3}}$$

Rewrite without a fractional exponent.

$$= \frac{2}{\sqrt[3]{x}}$$

Problems 4 and 5 are for you to do.

Problem 5

handwritten: $f'(x) = \frac{1}{4} x^{-\frac{3}{4}}, \quad \frac{1}{4} \cdot \frac{1}{\sqrt[4]{x^3}}$

If $f(x) = x^{1/4}$, find $f'(x)$

Solution: Given: $f(x) =$

Use the Power Rule. $f'(x) =$

Simplify. $=$

Rewrite with a positive exponent. $=$

Rewrite without a fractional exponent. $=$

Answer: $\dfrac{1}{4\sqrt[4]{x^3}}$

Problem 6

handwritten: $f(x)\ 6x^{\frac{1}{2}}, \quad f'(x) = 3x^{-1/2} \quad \frac{3}{\sqrt{x}}$

If $f(x) = 6\sqrt{x}$, find $f'(x)$.

Solution: Given: $f(x) =$ *(handwritten:* $6\sqrt{x} = 6x^{1/2}$*)*

Rewrite as a power of x. $=$ *(handwritten:* $6x^{1/2}$*)*

Use the Power Rule. $f'(x) =$ *(handwritten:* $3x^{-1/2}$*)*

Simplify. $=$ *(handwritten:* $3\frac{1}{\sqrt{x}}$*)*

Rewrite with a positive exponent. $=$ *(handwritten:* $\frac{3}{\sqrt{x}}$*)*

Rewrite without a fractional exponent. $=$

Answer: $\dfrac{3}{\sqrt{x}}$

Examples 23 and 24 illustrate how to handle negative fractional exponents.

EXAMPLE 23

handwritten: $f(x) = x^{-1/2} \quad f'(x) = -\frac{1}{2} x^{-\frac{3}{2}} = -\frac{1}{2} \cdot \frac{1}{\sqrt{x^3}} x^{?} = -\frac{1}{2} \frac{1}{\sqrt{x^2 \cdot x}}$
$= -\frac{1}{2} \frac{1}{x\sqrt{x}}$

If $f(x) = \dfrac{1}{\sqrt{x}}$, find $f'(x)$.

Solution: $$f(x) = \frac{1}{\sqrt{x}}$$

    Rewrite as a power of x. $$= \frac{1}{x^{1/2}}$$

                                                        $$= x^{-1/2}$$

    Use the Power Rule. $$= -\frac{1}{2}x^{(-1/2)-1}$$

    Simplify. $$= -\frac{1}{2}x^{-3/2}$$

    Rewrite with a positive exponent. $$= \frac{-1}{2} \cdot \frac{1}{x^{3/2}}$$

    Rewrite without a fractional exponent. $$= \frac{-1}{2\sqrt{x^3}} \text{ or } \frac{-1}{2x\sqrt{x}}$$

EXAMPLE 24

$f(x)\ 3 \cdot x^{-\frac{2}{5}},\ F'(x) = -\frac{2}{5} \cdot 3\, x^{-\frac{7}{5}}\ )\ -\frac{6}{5} \cdot \frac{1}{5\sqrt{x^7}}\ )\ \frac{-6}{5\,5\sqrt{x^5\,x^2}}$

If $f(x) = \dfrac{3}{\sqrt[5]{x^2}}$, find $f'(x)$. $\dfrac{-6}{5\,x\sqrt[5]{x^2}}$

Solution: Given: $$f(x) = \frac{3}{\sqrt[5]{x^2}}$$

    Rewrite as a power of x. $$= 3x^{-2/5}$$

    Use the Power Rule. $$f'(x) = 3\left(-\frac{2}{5}x^{(-2/5)-1}\right)$$

    Simplify. $$= 3 \cdot \frac{-2}{5}x^{-7/5}$$

    Rewrite with a positive exponent. $$= \frac{-6}{5} \cdot \frac{1}{x^{7/5}}$$

    Rewrite without a fractional exponent. $$= \frac{-6}{5\sqrt[5]{x^7}} \text{ or } \frac{-6}{5x\sqrt[5]{x^2}}$$

Before concluding this unit, there is one additional type of problem to be considered—cases in which functions, not just x, are under radical signs. Now that you are able to use the Power Rule with fractional exponents, the same procedure carries over to functions requiring the use of the Chain Rule.

Recall the Chain Rule.

If $f(x) = [u(x)]^n$, then $f'(x) = n[u(x)]^{n-1}u'(x)$ for all n.

EXAMPLE 25

If $f(x) = \sqrt{x^2 + 5x - 3}$, find $f'(x)$.

[handwritten: $f(x) = (x^2 + 5x - 3)^{1/2}, f'(x) = \frac{1}{2}(x^2+5x-3)^{-1/2} \cdot (2x+5)$
$= \frac{1}{2} \cdot \frac{2x+5}{(x^2+5x-3)^{1/2}} = \frac{2x+5}{2\sqrt{x^2+5x-3}}$]

Solution:	Given:	$f(x) = \sqrt{x^2 + 5x - 3}$
	Rewrite as a power of x.	$= (x^2 + 5x - 3)^{1/2}$
	Use the Chain Rule.	$f'(x) = \frac{1}{2}(x^2 + 5x - 3)^{(1/2)-1}\frac{d}{dx}(x^2 + 5x - 3)$
	Simplify.	$= \frac{1}{2}(x^2 + 5x - 3)^{-1/2}(2x + 5)$
	Rewrite with a positive exponent.	$= \frac{1}{2} \cdot \frac{1}{(x^2 + 5x - 3)^{1/2}} \cdot (2x + 5)$
	Rewrite without a fractional exponent.	$= \frac{2x + 5}{2\sqrt{x^2 + 5x - 3}}$

Let's make the final example easier.

EXAMPLE 26

If $f(x) = \sqrt{3x}$, find $f'(x)$.

[handwritten: $f(x) = (3x)^{1/2}, f' = \frac{1}{2}(3x)^{-1/2} \cdot 3 = \frac{3}{2\sqrt{3x}}$]

Solution:	Given:	$f(x) = \sqrt{3x}$
	Rewrite as a power of x.	$= (3x)^{1/2}$
	Use the Chain Rule.	$f'(x) = \frac{1}{2}(3x)^{(1/2)-1}\frac{d(3x)}{dx}$
	Simplify.	$= \frac{1}{2}(3x)^{-1/2}3$
	Rewrite with a positive exponent.	$= \frac{1}{2} \cdot \frac{1}{(3x)^{1/2}} \cdot 3$
	Rewrite without a fractional exponent.	$= \frac{3}{2\sqrt{3x}}$

You should now be able to take the derivatives of functions with negative and/or fractional exponents. Implied within that statement is that you should also be able to take the derivatives of radical functions. In all cases, the final answer should be simplified, that is, written without the use of negative, fractional, or zero exponents.

Briefly, the basic procedure for taking derivatives of such functions is:

1. Rewrite the function as a power of x.

2. Use either the Power Rule or the Chain Rule to take the derivative.

3. Simplify by removing parentheses and doing whatever else is needed.

4. Use the definition of a negative exponent to replace any negative exponents with positive exponents.

5. Use the definition of a fractional exponent to rewrite any fractional exponents with radical signs.

The following exercises should be worked before starting the next unit.

EXERCISES

Differentiate each of the following and simplify:

1. $f(x) = 7x^{-1}$

2. $f(x) = 3x^{-4}$

3. $f(x) = -5x^{-2}$

4. $f(x) = \dfrac{11}{x^3}$

5. $f(x) = \dfrac{3}{x^{-2}}$

6. $y = 12x^{-3} + 3x$

7. $y = x^{1/2}$

8. $f(x) = 2x^{3/2}$

9. $y = x^{2/5}$

10. $f(x) = x^{-1/3}$

11. $f(x) = -2x^{2/3}$

12. $f(x) = \dfrac{5}{x^2} - \dfrac{1}{x}$

13. $y = 2\sqrt{x}$

14. $y = \sqrt{2x}$

15. $f(x) = \sqrt[3]{8x}$

16. $y = \dfrac{2}{\sqrt{x}}$

17. $f(x) = \dfrac{-2}{\sqrt[3]{x^2}}$

18. $g(x) = \dfrac{1}{3x^2}$

19. $f(x) = (2x)^{-1}$ *Chain rule*

20. $y = \sqrt{1 - 3x^2}$ — *chain rule*

21. $D(x) = \sqrt{4x - 3}$

22. $D(p) = \sqrt[3]{p^2 + 1}$

23. $f(x) = \sqrt{x^5 - 8}$

24. If $g(u) = 2u - 4u^{-1}$, find $g'(2)$.

25. If $f(x) = \sqrt{2x - 1}$, find $f'(13)$.

The last two exercises are for practice keying functions containing radicals. The keystrokes are provided along with the answers in the back of the book.

C26. If $f(x) = \sqrt[3]{3x^2 - 1}$, find $f'(x), f'(0), f'(1), f'(2)$, and $f'(3)$.

C27. If $f(x) = \sqrt{x^2 + 5x - 2}$, find $f'(x), f(-10), f'(-10), f(-5), f'(-5), f(5), f'(5), f(10)$, and $f'(10)$.

If additional practice is needed:

   Barnett, Ziegler, and Byleen, pages 183–186, problems 1–88; pages 200–202, problems 17–83

   Harshbarger and Reynolds, pages 681–682, problems 11–58

   Hoffmann and Bradley, page 117, problems 1–24; pages 143–146, problems 7–37; pages 166–167, problems 1–19

   Waner and Costenoble, pages 160–161, problems 1–92; pages 217–219, problems 1–66

UNIT 14

Maximum and Minimum Points

The major objectives of Units 14–18 are to develop methods for determining where a function achieves maximum or minimum values and what they are. In preparation for learning those methods, this unit consists solely of terminology and definitions. It is brief, but it will provide the foundation for the techniques that follow in Units 15–18.

HIGHER-ORDER DERIVATIVES

Thus far $f'(x)$ has been discussed as the derivative of the function $f(x)$. Actually it is the first derivative of the function. There are second derivatives, third derivatives, and so on.

> Definition: The **second derivative** of a function $f(x)$ is the derivative of $f'(x)$ and is denoted by $f''(x)$.

Finding a second derivative is a straightforward matter. You differentiate the first derivative.

EXAMPLE 1

If $f(x) = 5x^4 - 7x^3 + x^2 + 34x - 1$, find $f'(x)$ and $f''(x)$.

Solution: $f(x) = 5x^4 - 7x^3 + x^2 + 34x - 1$

$$f'(x) = 5(4x^3) - 7(3x^2) + (2x) + 34$$

$$= 20x^3 - 21x^2 + 2x + 34$$

$$f''(x) = 20(3x^2) - 21(2x) + 2$$

$$= 60x^2 - 42x + 2$$

I am confident that, if I were to ask, you could correctly find the third derivative of the above function without even seeing an example.

Only first and second derivatives will be of interest to us.

OTHER NOTATION FOR SECOND DERIVATIVES

There are just as many different notations for second derivatives as for first derivatives. Again, the important thing is to be able to recognize and to understand each notation.

Consider this function: $y = f(x) = x^3 - 7x^2 + x - 6$.

We know the following:

- $f'(x)$ denotes the first derivative and $f'(x) = 3x^2 - 14x + 1$.

- $f''(x)$ denotes the second derivative and $f''(x) = 6x - 14$.

Other notations that denote second derivatives are:

- y'' read "y double prime" and $y'' = 6x - 14$.

- f'' read "f double prime" and $f'' = 6x - 14$.

- $\dfrac{d^2y}{dx^2}$ read simply as "the second derivative of y with respect to x."

Do notice the location of the 2's. In the numerator, the 2 is between the d and y. In the denominator, the 2 is after the d and x.

Problem 1

If $f(x) = x^5 - 3x^2 + 27$, find $f'(x)$ and $f''(x)$.

Solution: $f(x) =$

$f'(x) =$

$f''(x) =$

Answers: $f'(x) = 5x^4 - 6x; f''(x) = 20x^3 - 6$

We will not use second derivatives until Unit 17, but the concept of a first derivative was needed for a definition in this unit.

CRITICAL POINTS

As before, this will be an intuitive approach. We will use the accompanying picture to explain the words. In this picture everything possible is happening. Some of it is beyond the scope of this book, but I want the definitions to be complete in the event that you continue on with additional studies in calculus.

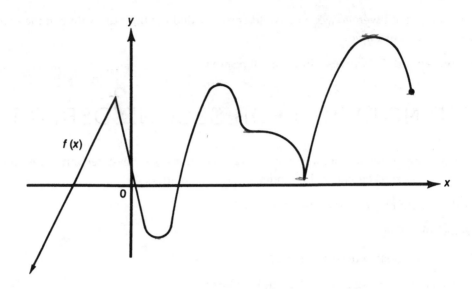

Definition: A **critical point** of a function is a point in its domain where the first derivative is zero or undefined.

What does that mean in terms of our picture?

To say that the first derivative is zero really means that the slope of the tangent line is zero, and that in turn means that the tangent line is horizontal and the curve "levels off."

To say that the first derivative is undefined has two possible meanings: (1), the slope of the tangent line is undefined, and therefore the tangent line is vertical; or (2), the tangent line fails to exist and the curve is "pointed."

Before reading the answer below, can you say how many critical points there are in the drawing?

CLASSIFICATION OF CRITICAL POINTS

All critical points can be classified into one of three categories.

Relative Maximum: The words *relative maximum* are used to describe a critical point that is a peak, a point that is higher than any other, nearby point on the graph. *Local maximum* is used frequently in place of *relative maximum*.

Relative Minimum: The words *relative minimum* are used to describe a critical point that is a low point, a point that is lower than any other, nearby point on the graph. Again *local minimum* and *relative minimum* are used interchangeably.

Stationary Inflection Point: The words *stationary inflection point* are used to describe critical points that are neither high points nor low points.

A word of caution: **first, the point must be a critical point.** Only a critical point is classified further as to whether it is a relative maximum, a relative minimum, or a stationary inflection point.

To return to the question posed earlier, there are *six* critical points in the above drawing. On the next page, tangent lines have been added to the drawing to help you "see" the critical points. Also, coordinates have been added so that we can distinguish among the various points.

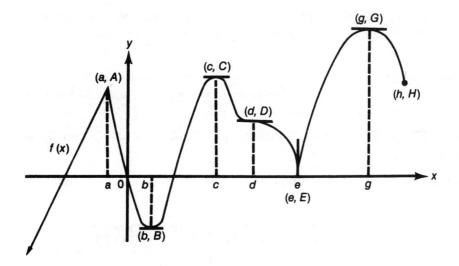

Recall that a critical point exists where the tangent line is either horizontal, vertical, or nonexistent. Be sure you understand why each of the points indicated is a critical point.

Critical points: (a, A); (b, B); (c, C); (d, D); (e, E); and (g, G).

Using the drawing as our guide, we can classify each of the critical points further as to whether it is a relative maximum, a relative minimum, or a stationary inflection point. Relative maxima (*maxima* is the plural of maximum) are peaks or high points. Relative minima (*minima* is the plural of minimum) are low points. Stationary inflection points are what is left over, neither high points nor low points. Try classifying the critical points yourself, before reading the answers.

1. (a, A) relative maximum

2. (b, B) relative minimum

3. (c, C) relative maximum

4. (d, D) stationary inflection point

5. (e, E) relative minimum

6. (g, G) relative maximum

In words:

1. A relative maximum value for y is A and it occurs when $x = a$.

2. A relative minimum value for y is B and it occurs when $x = b$.

3. A relative maximum value for y is C and it occurs when $x = c$.

4. Although (d, D) is a critical point, it is neither a relative maximum nor a relative minimum. It is a stationary inflection point.

5. A relative minimum value for y is E and it occurs when $x = e$.

6. A relative maximum value for y is G and it occurs when $x = g$.

Although (h, H) is a low point it is *not* a relative minimum. Why? Because it is not a critical point, and therefore, by definition, cannot be a relative minimum.

In practical applications, you will sometimes see functions that fail to be differentiable, that is, do not have a derivative at certain values in the domain. It can be shown that a continuous function f fails to be differentiable at a point when the graph of *f* makes an abrupt change of direction at that point. A function also fails to be differentiable at a point where the tangent line is vertical, since the slope of a vertical line is undefined.

Both of these cases are illustrated in our drawing. The derivative of the function does not exist at $x = a$ because the graph makes an abrupt change of direction at that point. Finally, the function is not differentiable at $x = e$ because the tangent line is vertical at that point.

Future discussion will be restricted to only differentiable functions, so sharp points and vertical tangent lines will be disallowed.

Maximum and minimum values are sometimes called extreme values, or extrema.

ABSOLUTE EXTREMA

Absolute Maximum: The term *absolute maximum* is used to denote the largest value of the function, if it exists. The absolute maximum is located at the highest point on the graph. It need not be a critical point.

Absolute Minimum: The term *absolute minimum* is used to denote the smallest value of the function, if its exists. The absolute minimum is located at the lowest point on the graph. It need not be a critical point.

In the drawing on the preceding page, (*g*, *G*) is the highest point on the graph. Thus *G* is the absolute maximum value and it occurs when $x = g$. There is no absolute minimum. Even though *B* is the smallest (lowest) critical point, the arrow on the left side of the graph indicates that the curve continues downward and thus has values of *y* that would be smaller (lower) than *B*. Had the arrow not been there, with an endpoint instead, we would have had an absolute minimum.

INCREASING AND DECREASING FUNCTIONS

There is one last concept to be mentioned before leaving our picture. Remember that we always read graphs from left to right, exactly the same as we read a sentence.

> Definition: If the curve is rising as you move from left to right, the function is said to be **increasing.**

> Definition: If the curve is falling as you move from left to right the function is said to be **decreasing.**

Starting at the arrow on the left side of the curve and moving to the right, I will verbally describe the graph, using the words increasing and decreasing.

The curve is increasing until you get to $x = a$,

at $x = a$, there is a sharp corner;

then it is decreasing until you get to $x = b$,

at $x = b$, the curve levels off;

then it increases until $x = c$,

at $x = c$, the curve levels off again;

then it decreases until $x = d$,

at $x = d$, the curve levels off again;

then it decreases until $x = e$,

at $x = e$, there is a sharp point;

then it increases until $x = g$,

at $x = g$, the curve levels off again;

then it decreases until $x = h$.

and at $x = h$, the curve ends.

In symbols, the above information on increasing and decreasing would look like this:

If $x < a$, $f(x)$ is increasing.

If $a < x < b$, $f(x)$ is decreasing.

If $b < x < c$, $f(x)$ is increasing.

If $c < x < d$, $f(x)$ is decreasing.

If $d < x < e$, $f(x)$ is decreasing.

If $e < x < g$, $f(x)$ is increasing.

If $g < x \leq h$, $f(x)$ is decreasing.

Notice that intervals of x are used to describe where the function is increasing and decreasing. Only the x-values, not the y-values, are used. Also notice that the values where the critical points are located are not included as part of the interval.

Critical points were defined in terms of derivatives. Can we do the same thing for increasing and decreasing? Yes. Consider the section of the graph from $b < x < c$. Sketch in three or four tangent lines along that section. What do they all have in common? Their slopes are all positive. Do the same thing for the section from $c < x < d$. This time all the slopes are negative. That leads us to the following theorems.

THEOREM: If $f'(x)$ is positive for each value of x in some open interval, then $f(x)$ is increasing on that interval.

THEOREM: If $f'(x)$ is negative for each value of x in some open interval, then $f(x)$ is decreasing on that interval.

To summarize what we have said so far:

If the derivative at some point is positive, the curve is increasing.

If the derivative is negative, the curve is decreasing.

If the derivative is zero, the curve levels off, and a critical point exists.

Critical points are points where the first derivative is zero or undefined. Graphically that definition implies that critical points occur where the tangent line is horizontal, vertical, or nonexistent.

A critical point can be classified as either a

1. relative (or local) maximum—a high point,

2. relative (or local) minimum—a low point, or

3. stationary inflection point—neither a high nor a low point.

An absolute maximum is the largest value (highest point on the curve), if it exists. Similarly, an absolute minimum is the smallest value (lowest point on the curve), if it exists. These maxima or minima need not be critical points.

Maximum and minimum values are sometimes called extrema. The term *relative extrema* refers to the relative maximum and relative minimum values of a function, and the term *absolute extrema* refers to the absolute maximum and absolute minimum values.

For the rest of this book, the only critical points we will consider are those where the derivative is zero; that is, the tangent line is horizontal. Keep in mind that others exist, but they are beyond the scope of this presentation.

Do the following exercises before continuing on to the next unit.

EXERCISES

Use the drawing provided to answer the following questions:

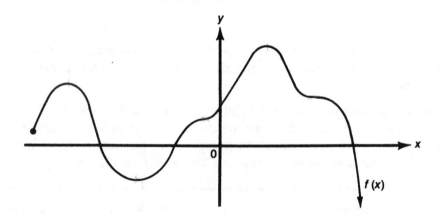

1. How many critical points are there?

2. Indicate where they are located.

3. Sketch the tangent line at each critical point.

4. Classify each critical point.

5. Does the function have an absolute maximum?

6. Does the function have an absolute minimum?

7. Over what intervals is the function decreasing?

8. Over what intervals is the function increasing?

C9. Given $f(x) = x^4 - 12x^3 + 48x^2 - 64x$.

    a. Graph using a viewing window [–5, 10] by [–30, 10].

    b. How many critical points are there?

    c. Using TRACE, determine approximately where they are located.

    d. Classify each critical point.

    e. Does the function have an absolute maximum?

    f. Does the function have an absolute minimum?

UNIT 15

Classifying Critical Points— One Procedure

In this unit you will learn one method for determining where a function achieves maximum or minimum values.

In Unit 14, you learned to classify critical points as either relative maxima, relative minima, or stationary inflection points by determining, on the basis of a drawing, whether they were high points, low points, or neither. But what happens when a drawing is not available? How can we determine where the relative maximum and relative minimum occur, or if they occur at all? There is a theorem to help us.

THEOREM: If a function has a relative maximum or minimum, it will occur at a critical point.

In an effort to locate the relative maximum or minimum points, the first step, then, is to locate all the critical points. Recall that a critical point is located where the first derivative is zero or undefined. From this unit on, we will consider only critical points in which the first derivative is zero.

LOCATING CRITICAL POINTS

Critical points occur where the first derivative equals zero, $f'(x) = 0$.

To locate critical points:

1. Take the first derivative.

2. Set it equal to zero.

3. Solve for x. (x^* is used frequently to denote a critical point.)

4. Find the y-coordinate, $f(x^*)$.

   Reminder: To find the y-coordinate, you must substitute x^* into the original function, $y = f(x)$.

EXAMPLE 1

Find all critical points for $y = f(x) = x^2 - 4x + 5$.

Solution:	Given:	$y = f(x) = x^2 - 4x + 5$

Critical points occur where $f'(x) = 0$.

Take the first derivative.	$f'(x) = 2x - 4$
Set it equal to zero.	$0 = 2x - 4$
Solve for x.	$4 = 2x$
	$x^* = 2$
Find the y-coordinate.	$y = f(2) = (2)^2 - 4(2) + 5$
Reminder: Substitute into the original function, $y = f(x)$.	$= 4 - 8 + 5$
	$= 1$

Answer: $f(x)$ has a critical point at $(2, 1)$.

EXAMPLE 2

Find all critical points for $y = f(x) = x^3 + 3x^2 + 1$.

Solution:	Given:	$y = f(x) = x^3 + 3x^2 + 1$

Critical points occur where $f'(x) = 0$.

Take the first derivative.	$f'(x) = 3x^2 + 6x$
Set it equal to zero.	$0 = 3x^2 + 6x$
	$0 = 3x(x + 2)$
Solve for x.	$3x = 0 \qquad x + 2 = 0$
	$x^* = 0 \qquad\quad x^* = -2$

Find the y-coordinates.

If $x^* = 0$,

$$f(0) = (0)^3 + 3(0)^2 + 1 = 1$$

and $(0, 1)$ is a critical point.

If $x^* = -2$,

$$f(-2) = (-2)^3 + 3(-2)^2 + 1$$

$$= -8 + 12 + 1$$

$$= 5$$

and $(-2, 5)$ is a critical point.

Answer: $f(x)$ has two critical points, one at $(0, 1)$ and another at $(-2, 5)$.

Here's an easy problem for you.

Problem 1

Find all critical points for the function $y = f(x) = x^{16}$.

Solution: Given:

Critical points occur where $f'(x) = 0$.

Take the first derivative.

Set it equal to zero.

Solve for x.

Find the y-coordinate.

Answer: $f(x)$ has a critical point at $(0, 0)$.

We are halfway to our objective. You now should be able to locate a function's critical points; but how do we classify them without the aid of a drawing? At the start, I stated you would learn one method in this unit. That is not really true. You will learn two, but they are related.

CLASSIFYING CRITICAL POINTS BY USING PRIOR KNOWLEDGE ABOUT GRAPHS

One way to classify critical points is to use your knowledge of the graph of the function. In other words, sketch or visualize your own drawing. Considering all the time we spent on first-, second-, and third-degree polynomial functions, this part should be easy.

We will return to our first example.

EXAMPLE 3

Find and classify all critical points for $y = f(x) = x^2 - 4x + 5$.

Solution: From Example 1, we found that $f(x)$ has a critical point at $(2, 1)$.

To classify it, we ask ourselves what we know about its graph.

We don't need everything, just enough for a very quick sketch.

Point $(2, 1)$ is plotted on the grid below, along with its horizontal tangent. The horizontal tangent is most important; that shows it is a critical point and reminds us that the curve levels off at that point.

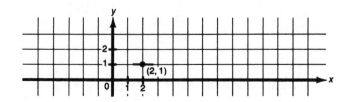

This function is a quadratic with $a = 1$.

The parabolic curve opens up: and we have enough information to complete our quick sketch.

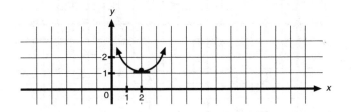

Answer: The critical point (2, 1) is a relative minimum. In fact, it is an absolute minimum because we know the vertex is the lowest point (or smallest value) on the curve.

In other words, the absolute minimum value for y is 1, and it occurs when $x = 2$.

EXAMPLE 4

Find and classify all critical points for $y = f(x) = x^2 - 12x + 1$.

Solution: Given: $y = f(x) = x^2 - 12x + 1$

Critical points occur where $f'(x) = 0$.

Take the first derivative. $f'(x) = 2x - 12$

Set it equal to zero. $0 = 2x - 12$

Solve for x. $12 = 2x$

$x^* = 6$

Find the y-coordinate. $y = f(6) = 36 - 72 + 1 = -35$

$f(x)$ has a critical point at (6, –35).

To classify the critical point, use your knowledge of the graph.

Plot the critical point (6, –35) along with its horizontal tangent to show that the curve levels off there.

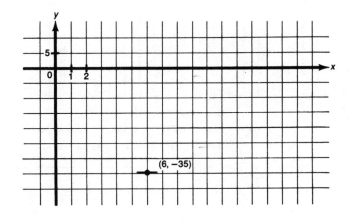

This is a quadratic function with $a = 1$, and the parabola opens up: 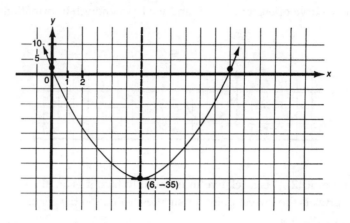.

Answer: The critical point at $(6, -35)$ is a relative minimum. Again, you could be more precise in that it is an absolute.

In other words, the absolute minimum value for y is -35, and it occurs when $x = 6$.

Ready to try a problem?

==========

Problem 2

Find and classify all critical points for the function $y = f(x) = 100 - 10x - x^2$.

Solution: Given:

Critical points occur where $f'(x) = 0$.

Take the first derivative.

Set it equal to zero.

Solve for x.

Find the y-coordinate.

Use your knowledge of the graph to classify the critical point.

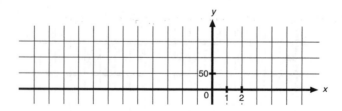

Answer: The critical point at $(-5, 125)$ is an absolute maximum. Or the maximum value for y is 125, and it occurs when $x = 5$.

EXAMPLE 5

Find all critical points for $y = f(x) = 5x - 1$.

Solution: Given: $y = f(x) = 5x - 1$

Critical points occur where $f'(x) = 0$.

Take the first derivative. $f'(x) = 5$

Set it equal to zero. $0 = 5$

Solve for x.   This equation has no solution.

Answer: $f(x)$ has no critical points.

Can you explain why there are no critical points? What does the graph of $f(x)$ look like? Does a straight line have any critical points? That is like asking if a straight line has any high or low points. The answer is no.

We have done examples with first- and second-degree functions; now it is time to consider third-degree. To save time, we will return to Example 2.

EXAMPLE 6

Find and classify all critical points for $y = f(x) = x^3 + 3x^2 + 1$.

Solution: From Example 2, we found that $f(x)$ had critical points at $(-2, 5)$ and $(0, 1)$.

To classify the critical points, use your knowledge of what the graph of $f(x)$ looks like.

First, plot both critical points. Be sure to include the horizontal tangents to remind yourself that the curve levels off at each of the points.

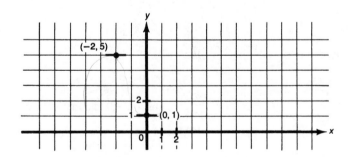

The function is a cubic with $a = 1$; therefore it opens like this: . Since we have already plotted our two critical points, we can sketch in the curve without much effort.

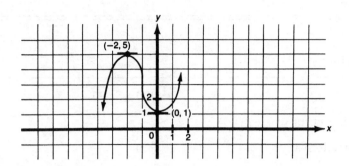

Notice from our sketch that the cubic has two critical points. The one on the left is a relative maximum, and the one on the right is a relative minimum. There are no absolutes because the curve continues downward on the left and upward on the right.

Answer: $(-2, 5)$ is a relative maximum and $(0, 1)$ is a relative minimum.

EXAMPLE 7

Find and classify all critical points for $y = f(x) = \dfrac{1}{3} x^3 - x^2 - 3x + 1$.

Solution: Given:

$$y = f(x) = \frac{1}{3} x^3 - x^2 - 3x + 1$$

Take the first derivative.

$$f'(x) = \frac{1}{3} (3x^2) - 2x - 3$$

$$= x^2 - 2x - 3$$

Set it equal to zero.

$$0 = x^2 - 2x - 3$$

Solve for x.

$$0 = (x - 3)(x + 1)$$

$$x^* = 3 \quad x^* = -1$$

Find the y-coordinates.

If $x^* = 3$,

$$f(3) = 9 - 9 - 9 + 1 = -8$$

and $(3, -8)$ is a critical point.

If $x^* = -1$,

$$f(-1) = \frac{1}{3} (-1)^3 - (-1)^2 - 3(-1) + 1$$

$$= \frac{1}{3} (-1) - (1) + 3 + 1$$

$$= -\frac{1}{3} - 1 + 3 + 1$$

and $\left(-1, 2\dfrac{2}{3}\right)$ is a critical point.

Plot both critical points along with their horizontal tangents. Using your knowledge of third-degree polynomial functions, you realize that the function is a cubic with $a = 1$, therefore it is shaped as ⟋⌣⟍. Finish the sketch.

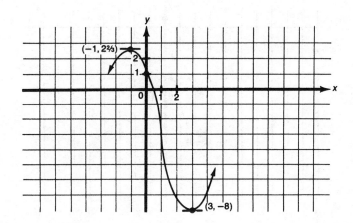

Answer: The critical point $\left(-1, 2\frac{2}{3}\right)$ is a relative maximum.

The critical point $(3, -8)$ is a relative minimum.

I will do one more example before having you try a problem.

EXAMPLE 8

Find and classify all critical points for $y = f(x) = -x^3 - 3x^2 + 1$.

Solution: Given: $y = f(x) = -x^3 - 3x^2 + 1$

Take the first derivative. $f'(x) = -3x^2 - 6x$

Set it equal to zero. $0 = -3x^2 - 6x$

Solve for x. $0 = -3x(x + 2)$

$x^* = 0 \quad x^* = -2$

Find the y-coordinates.   If $x^* = 0$,

$y = f(0) = 0 - 0 + 1 = 1$

and $(0, 1)$ is a critical point.

If $x^* = -2$,

$y = f(-2) = 8 - 12 + 1 = -3$

and $(-2, -3)$ is a critical point.

The critical points are plotted along with their horizontal tangents. The function is a cubic with $a = -1$; therefore it is shaped as 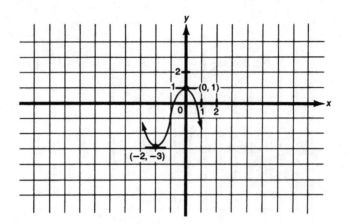.

Answer: The critical point $(-2, -3)$ is a relative minimum.

The critical point $(0, 1)$ is a relative maximum.

Here is your problem.

Problem 3

Find and classify all critical points for $y = f(x) = 7x^3 - 21x + 2$.

Solution: Given:

Take the first derivative.

Set it equal to zero.

Solve for x.

Find the y-coordinates.

To classify the critical points, use your knowledge of the graph. Plot both critical points along with their horizontal tangents. Finish the sketch.

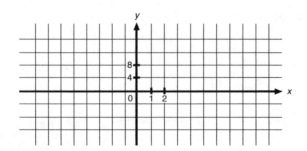

Answer: The critical point $(-1, 16)$ is a relative maximum.

The critical point $(1, -12)$ is a relative minimum.

The next example is a bit unusual, and I had best explain it.

EXAMPLE 9

Find and classify all critical points for $y = f(x) = 4x^3 + 1$.

Solution: Given: $y = f(x) = 4x^3 + 1$

   Take the first derivative. $f'(x) = 12x^2$

   Set it equal to zero. $0 = 12x^2$

   Solve for x. $0 = x^2$

                                               $x^* = 0$

   Find the y-coordinate.   If $x^* = 0$,

                                  $y = f(0) = 0 + 1 = 1$

                        and $(0, 1)$ is a critical point.

Plot the critical point along with its horizontal tangent. The function is a cubic with $a = 4$; therefore it is shaped as ∫∖◡◜. But this cubic function has only one critical point! How can it be ∫∖◡◜-shaped?

The best way I can explain it is to ask you to imagine that you are holding onto both ends of the curve and that you gently pull until both the peak and the valley disappear, but leave one leveling-off point in the middle, like the sketch below.

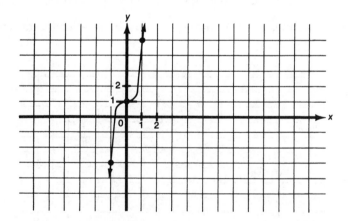

The above situation can be generalized to conclude that, if a cubic has only one critical point, that point will be a stationary inflection point.

Answer: The critical point $(0, 1)$ is a stationary inflection point.

The vast majority of functions that you will encounter in calculus as well as in various economics courses tend to be first-, second-, or third-degree polynomial functions. As you have seen, these functions are predictable and their maximum and minimum values rather easily determined. But, of course, there are other functions, and we need a method for classifying all critical points.

One method for classifying critical points is the **original-function test.**

THE ORIGINAL-FUNCTION TEST FOR CLASSIFYING CRITICAL POINTS

Let $f(x)$ be a function with a critical point at $x^* = a$. Determine the values of $f(x)$ to the left and right of the critical point.

1. If $f(x) > f(x^*)$ to the left of $x^* = a$ and if $f(x) > f(x^*)$ to the right of $x^* = a$, then the critical point at $x^* = a$ is a relative minimum.

2. If $f(x) < f(x^*)$ to the left of $x^* = a$ and if $f(x) < f(x^*)$ to the right of $x^* = a$, then the critical point at $x^* = a$ is a relative maximum.

3. If neither statement 1 nor statement 2 is satisfied, then the critical point at $x^* = a$ is a stationary inflection point.

Although the above test is a bit wordy, it is rather straightforward to use. Let me first explain the reasoning before doing any examples.

A relative minimum by definition is a low point. The original-function test examines what happens with the graph on either side of the critical point. If, as in statement 1 above, the points on the curve on either side of the critical point are both higher, then the critical point must be a relative minimum.

A relative maximum is defined as a high point. Therefore, if, as in statement 2 above, the points on the curve on either side of the critical point are both lower, then the critical point must be a relative maximum.

If neither of the two situations exist, the critical point must be a stationary inflection point.

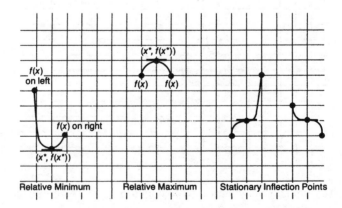

To use the original-function test, select an x value on either side of the critical point $x^* = a$ and quite close to it. Evaluate $f(x)$, the original function, at both values and compare the y-coordinates according to the test.

The examples that follow will involve functions whose graphs are unfamiliar to us. We will start by returning to Problem 1.

EXAMPLE 10

Find and classify all critical points for $y = f(x) = x^{16}$.

Solution: From Problem 1, $y = f(x) = x^{16}$ was found to have a critical point at $(0, 0)$.

To classify the critical point, use the original-function test. Select a value for x on either side of the critical point, substitute the values into the original function, and then compare.

On the left: Let $x = -1$; then $y = f(-1) = (-1)^{16} = 1$

At the critical point: $x^* = 0$, and $y = f(0)$ $= 0$

On the right: Let $x = 1$; then $y = f(1)$ $= (1)^{16}$ $= 1$

Do a quick sketch. The point on the left and the point on the right are both higher than the critical point.

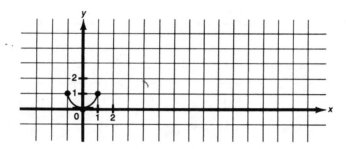

Answer: The critical point $(0, 0)$ is a relative minimum.

EXAMPLE 11

Find and classify all critical points for $y = f(x) = x^5$.

Solution: Given: $y = f(x) = x^5$

Take the first derivative. $f'(x) = 5x^4$

Set it equal to zero. $0 = 5x^4$

Solve for x. $x^* = 0$

Find the y-coordinate.   If $x^* = 0$,

$y = f(0) = 0$

and $(0, 0)$ is a critical point.

To classify the critical point, use the original-function test.

For point $(0, 0)$:

On the left: Let $x = -1$, then $y = f(-1) = (-1)^5 = -1$

At the point: $x^* = 0$,   and $y = f(0)$ =   0

On the right: Let $x = 1$,   then $y = f(1)$ $= (1)^5$ $= 1$

Do a quick sketch. Notice that the point on the left is lower than the critical point and the point on the right is higher than the critical point.

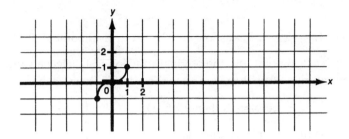

Answer: The critical point $(0, 0)$ is a stationary inflection point.

Examples 10 and 11 were selected with the critical point occurring at the origin, $(0, 0)$, so as to make our calculations easier. Obviously that will not always be the case, as Example 12 illustrates.

EXAMPLE 12

Find and classify all critical points for $y = f(x) = x^5 - 5x^4 + 1$.

Solution: Given: $y = f(x) = x^5 - 5x^4 + 1.$

Take the first derivative. $f'(x) = 5x^4 - 20x^3$

Set it equal to zero. $0 = 5x^4 - 20x^3$

Solve for x. $0 = 5x^3(x - 4)$

$$x^* = 0 \quad x^* = 4$$

Find the y-coordinates.   If $x^* = 0$,

$$y = f(0) = 0 - 0 + 1 = 1$$

and $(0, 1)$ is a critical point.

If $x^* = 4$,

$$y = f(4) = 1{,}024 - 1{,}280 + 1 = -255$$

and $(4, -255)$ is a critical point.

To classify the critical point, use the original-function test.

For point $(0, 1)$:

On the left: Let $x = -1$, then $y = f(-1) = -1 - 5 + 1 = -5$

At the point: $x^* = 0$, and $y = f(0) = $ 1

On the right: Let $x = 1$, then $y = f(1) = 1 - 5 + 1 = -3$

This is the second situation described in the test. Since the points on both sides of the critical point are lower than the critical point, it is a relative maximum. A sketch is provided on the following page.

For point (4, –255):

On the left: Let $x = 3$, then $y = f(3) = 243 - 405 + 1 = -161$

At the point: $x^* = 4$, and $y = f(4)$ $= -255$

On the right: Let $x = 5$, then $y = f(5) = 3{,}125 - 3{,}125 + 1 = 1$

This is the first situation described in the test. The point on the left of the critical point is higher and the point on the right is also higher; the result is that the critical point is a relative minimum. The points are shown on the sketch.

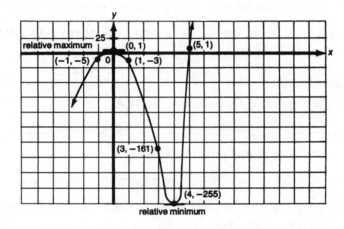

Answer: The critical point (0, 1) is a relative maximum.

The critical point (4, –255) is a relative minimum.

CALCULUS AND THE CALCULATOR (Optional)

Finding Maximum and Minimum Points

Critical points are located by taking the first derivative, setting it equal to zero, and solving for x. But what happens if you are unable to solve the resulting equation? That is exactly the situation in the next example, and we will examine how to locate a critical point using the minimum (or maximum) feature of the calculator.

EXAMPLE 13

Find and classify all critical points for $y = f(x) = x^4 + x^2 + x + 1$.

Solution: Use a viewing window [–2, 2] by [–5, 5] and graph the function.

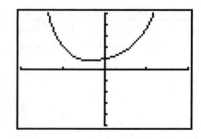

From the graph we can see there is one critical point, which is a relative minimum. In fact, it is an absolute minimum because it is the lowest point (or smallest value) on the curve. At this point in time, however, classifying it as a relative minimum is sufficient for our purposes. But where is it? Before trying to locate it, let's get a better picture. Press (TRACE), move the cursor to somewhere near the low point of the graph, and press (ZOOM) (2) (ENTER), which enlarges that section of the graph. To locate the critical point, which is a minimum, press (2nd) (CALC) (3).

Note: If the critical point was a maximum, the keystroke would be 4 rather than 3.

At the <u>Left Bound?</u> prompt, move the cursor to a point just to the left of where the critical point appears to be and press (ENTER).

At the <u>Right Bound?</u> prompt, move the cursor to a point just to the right of where the critical point appears to be and press (ENTER).

At the <u>Guess?</u> prompt, move the cursor between the boundaries shown on the screen by the two arrows, ▶ ◀, and press (ENTER).

Minimum $X = -0.38$ $Y = 0.79$ appears at the bottom of the screen.

Answer: The critical point $(-0.38, 0.79)$ is a relative minimum. In other words, the minimum value for y is 0.79, and it occurs when $x = -0.38$.

Comments: My answer was rounded to two decimal places. Answers may vary slightly depending on where your boundaries were set.

Could we have solved this problem algebraically without the use of a graphing calculator? Try it yourself and see what happens.

You should be able now to locate critical points and classify them.

Critical points are located by taking the first derivative, setting it equal to zero, solving for x, and then determining the corresponding y-coordinate.

To classify critical points, if the function is a first-, second-, or third-degree polynomial function, you should use your knowledge of the graph of the function.

1.   First-degree polynomial functions have no critical points.

2.   Second-degree polynomial functions have one critical point, either an absolute maximum or an absolute minimum, depending on which way the parabola is opening.

3.   Three possibilities exist with third-degree polynomial functions. If there are two critical points, one is a relative maximum and one is a relative minimum, again depending on which way the curve is opening. If there is only one critical point, it is a stationary inflection point. It is also possible that there are no critical points, in which case there is nothing to classify.

To classify points for functions other than first-, second-, or third-degree polynomial functions, one method is the original-function test. Points are examined on either side of the critical point for the original function; then, briefly:

1. If both points are higher than the critical point, the critical point is a relative minimum.

2. If both points are lower than the critical point, the critical point is a relative maximum.

3. If neither situation exists, the critical point is a stationary inflection point.

EXERCISES

Find and classify all critical points for each of the following: Optional—Confirm your answers using a graphing calculator before checking the answers in the back of the book.

1. $f(x) = 2x^2 - 4x$

2. $f(x) = -x^2 - 1$

3. $f(x) = 3x - 2$

4. $f(x) = 6x^2 + x - 1$

5. $f(x) = 5x^2 - 40x + 10$

6. $f(x) = x^3 + 3x^2$

7. $f(x) = x^3 + 15x^2 + 1,000$

8. $f(x) = -x^3 + 7$

9. $f(x) = 12x^{-1}$

10. $f(x) = x^3 - 3x^2 - 9x + 5$

11. $f(x) = 10x^3 - 25$

12. $f(x) = x^6 + 2$

13. $f(x) = 2x^3 - 3x^2 - 36x + 7$

14. $f(x) = 5x^7 - 35x^6 + 63x^5$ This one is a bit more difficult than the others, but also more interesting.

15. $f(x) = 4x^5 - 5x^4 + 5$

16. $f(x) = \frac{1}{3}x^3 - 3x^2 + 5x + 2$ When you have finished this problem, return to Example 5 of Unit 8 and see what you have found.

17. $f(x) = 2x^3 + 10.5x^2 - 12x$

18. $f(x) = \frac{-5}{x - 4}$

19. $P(x) = -3x^2 + 54x$

20. Graph: $f(x) = \begin{cases} 3 + x^2 & \text{if } x < 0 \\ 3 & \text{if } x = 0. \\ 3 + x & \text{if } x > 0 \end{cases}$

C21. Find and classify all critical points for $y = 2x^3 + 3x^2 - 12x + 10$. Use a viewing window $[-5, 5]$ by $[-30, 40]$. Verify your answers using analytical methods.

C22. This last problem is meant to reinforce the importance of analytical methods along with the graphing feature on a calculator.

Given $y = x^4 - 12x^3 + x^2 - 2$.

a.   Use a viewing window $[-5, 15]$ by $[-2,500, 1,000]$ with Xscl = 1 and Yscl = 500 and find and classify all critical points.

b.   Rework the problem using analytical methods. In other words, take the first derivative, set it equal to zero, solve for x, and then determine the corresponding y-coordinate.

If additional practice is needed:

Barnett, Ziegler, and Byleen, pages 251–253, problems 1–8 and 25–54
Waner and Costenoble, page 268, problems 1–20

UNIT 16

Classifying Critical Points—
A Second Procedure

In this unit you will learn a second method for determining where a function achieves maximum or minimum values.

In Unit 15, first you learned how to locate critical points, and then how to classify them based on your knowledge of what the graph looked like. If you were unfamiliar with the particular function, the original-function test was to be used.

The technique for locating critical points remains the same, but we will introduce a second procedure for classifying the points. It is called the **first-derivative test.**

The basis for the first-derivative test is a theorem from Unit 14, which is repeated here in a shortened version.

> THEOREM: If $f'(x) > 0$ on an open interval, then $f(x)$ is increasing.
>
> If $f'(x) < 0$ on an open interval, then $f(x)$ is decreasing.

Recall we said that, if the derivative is positive on an interval, the curve is increasing. If the derivative is negative, the curve is decreasing. And if the derivative is zero, the curve is leveling off, and you have a critical point.

The first-derivative test examines what is happening on either side of the critical point, much like the original-function test. With this second method, however, we use the first derivative to determine whether the curve is increasing or decreasing.

THE FIRST-DERIVATIVE TEST FOR CLASSIFYING CRITICAL POINTS

Let $y = f(x)$ be a function with a critical point at $x^* = a$; determine the values of $f'(x)$ to the left and right of the critical point.

1. If $f'(x) < 0$ to the left of $x^* = a$ and

    if $f'(x) > 0$ to the right of $x^* = a$, then

    the critical point at $x^* =$ is a relative minimum.

2. If $f'(x) > 0$ to the left of $x^* = a$ and

    if $f'(x) < 0$ to the right of $x^* = a$, then

    the critical point at $x^* = a$ is a relative maximum.

3. If neither statements 1 nor statement 2 is satisfied, the critical point at $x^* = a$ is a stationary inflection point.

In words:

1. If the derivative is negative on the left and positive on the right, the critical point is a relative minimum.

2. If the derivative is positive on the left and negative on the right, the critical point is a relative maximum.

3. If the derivative has the same sign on both sides, the critical point is a stationary inflection point.

Before doing any examples, let me explain and draw a picture of each situation.

1. $f'(x) < 0$ to the left of $x^* = a$ means the curve is decreasing to the left of the critical point. At the critical point the curve levels off. $f'(x) > 0$ to the right of $x^* = a$ means the curve is increasing to the right of the critical point. The sketch below reflects this situation and should illustrate why the critical point is a relative minimum.

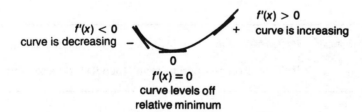

2. By similar reasoning, the second situation has the curve increasing to the left, leveling off at the critical point, and decreasing to the right. A sketch should help to convince you that the critical point is a relative maximum.

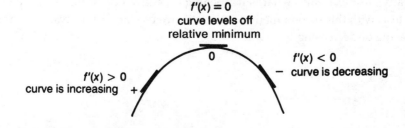

3. Two possibilities remain for the third situation.

    a. If $f'(x) > 0$ both to the left and the right, then the curve is increasing, leveling off, and increasing again.

    b. If $f'(x) < 0$ both to the left and the right, then the curve is decreasing, leveling off, and decreasing again.

    In either case, the critical point is a stationary inflection point.

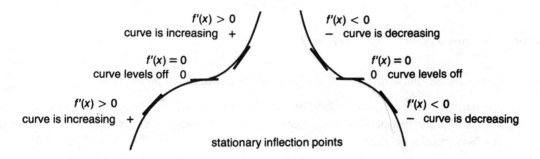

To use the first-derivative test, select a value for x on either side of the critical point $x^* = a$. As with the original-function test, the values should be quite close to a. This time, however, we use $f'(x)$ to calculate the two values and compare their signs. The actual numerical answer is unimportant. Only the sign, positive or negative, is of concern in the test.

Enough words; we will work through some examples.

EXAMPLE 1

Find all critical points for $y = f(x) = x^4 - 4x$.

Use the first-derivative test to classify the critical points.

Solution: $y = f(x) = x^4 - 4x$

$f'(x) = 4x^3 - 4$

Find all critical points. Critical points occur where $f'(x) = 0$.

$0 = 4x^3 - 4$

$4x^3 = 4$

$x^3 = 1$

$x^* = 1$

If $x^* = 1$,

$y = f(1) = 1 - 4 = -3$ and $(1, -3)$ is the only critical point.

Use the first-derivative test with $f'(x) = 4x^3 - 4$.

For point $(1, -3)$ the curve:

On the left: Let $x = 0$, then $f'(0) = 0 - 4 = -$ is decreasing

At point $(1, -3)$:   levels off

On the right: Let $x = 2$, the $f'(2) = 32 - 4 = +$ is increasing

This is the first situation described in the test. The first derivative is negative to the left of the critical point and positive to the right. Can you visualize the graph? It is decreasing, leveling off, then increasing. What must the critical point be?

decreasing   **increasing**

levels off

Answer: The critical point $(1, -3)$ is a relative minimum.

Be sure to notice the following two items regarding the use of the first-derivative test:

1. The values selected on either side of the critical point are substituted into the first derivative. That is how the test gets its name. Similarly, with the original-function test we substituted the numbers into the original function.

2. The actual numerical value of the derivative is unimportant; only the sign, positive or negative, need be determined.

EXAMPLE 2

Sketch the graph of $y = f(x) = x^4 - 4x$.

Solution: $f(x)$ is a fourth-degree polynomial function.

The y-intercept is 0.

The x-intercepts are 0 and $\sqrt[3]{4}$ because

$$0 = x^4 - 4x$$

$$0 = x(x^3 - 4)$$

$$x = 0 \qquad x^3 - 4 = 0$$

$$x^3 = 4$$

$$x = \sqrt[3]{4}$$

From Example 1, the function has a relative minimum at $(1, -3)$, which implies the curve is decreasing on the left and increasing on the right.

From the fact that the graph of a polynomial function is a smooth, continuous curve without any breaks or sharp corners, a sketch of the curve would appear as shown here.

Answer:

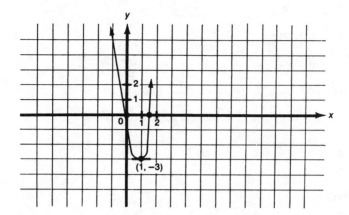

How do I know the curve is continuing upward on the right where I have the arrow? In order for the curve to turn downward again, there would need to be another critical point. But we found there was only one and it is shown. The same line of reasoning applies to the left-hand arrow.

EXAMPLE 3

Use the graph from Example 2 to do the following:

a.   Locate the absolute extrema of $f(x)$.

b.   State the interval(s) over which $f(x)$ is increasing.

c.   State the interval(s) over which $f(x)$ is decreasing.

Solution: a.   From the drawing, the absolute minimum value of $f(x)$ is –3 and occurs when $x = 1$. The function $f(x)$ does not have an absolute maximum.

b.   The function $f(x)$ is increasing over the interval $1 < x$.

c.   The function $f(x)$ is decreasing over the interval $x < 1$.

Remember that we read graphs from left to right. For this reason, although the arrow at the left is pointing upward, think of starting at the arrow and moving to the right. The curve is falling; hence the function is said to be decreasing.

Consider another fourth-degree polynomial function.

EXAMPLE 4

Find all critical points for $y = f(x) = x^4 - 8x^3 + 18x^2 - 27$.

Use the first derivative test to classify the critical points.

Solution: $y = f(x) = x^4 - 8x^3 + 18x^2 - 27$

$$f'(x) = 4x^3 - 24x^2 + 36x$$

$$= 4x(x^2 - 6x + 9)$$

$$= 4x(x - 3)(x - 3)$$

$$= 4x(x - 3)^2$$

Find all critical points. Critical points occur where $f'(x) = 0$.

$$0 = 4x^3 - 24x^2 + 36x$$

$$0 = 4x(x - 3)^2$$

$$4x = 0 \qquad x - 3 = 0$$

$$x^* = 0 \qquad x^* = 3$$

If $x^* = 0$,

$$y = f(0) = 0 - 0 + 0 - 27 = -27$$

and $(0, -27)$ is a critical point.

If $x^* = 3$,

$$y = f(3) = 81 - 8(27) + 18(9) - 27 = 81 - 216 + 162 - 27 = 0$$

and $(3, 0)$ is another critical point.

Use the first derivative test with $f'(x) = 4x(x - 3)^2$.

Since we need determine only the sign, not the actual value, I will shortcut some of the work.

For point $(0, -27)$ with $f'(x) = 4x^3 - 24x + 36x$ the curve:

On the left: Let $x = -1$, then $f'(-1) = -4 - 24 - 36 = -$ is decreasing

At point $(0, -27)$:   levels off

On the right: Let $x = 1$,   then $f'(1)$ $= 4 - 24 + 36 = +$ is increasing

Try to visualize the curve. It is decreasing, leveling off, then increasing. The critical point $(0, -27)$ is a relative minimum.

Since we need determine only the sign, not the actual value, of the derivative, I will show you how to shortcut some of the calculations. In order to do so, **you must use the first derivative in factored form rather than multiplied out.** Then only the sign of the factor is considered. For example, if $x = 2$, then $(x - 3)$ would be indicated as a minus factor, $(-)$, because $2 - 3$ is a negative number.

For point (3, 0) with $\qquad$ $f'(x) = 4x(x-3)^2$ $\qquad$ the curve:

On the left: Let $x = 2$, then $f'(2) = (+)(+)(-)^2 = +$ is increasing

At point (3, 0):   levels off

On the right: Let $x = 4$, then $f'(4) = (+)(+)(+)^2 = +$ is increasing

At this critical point, the curve is increasing, leveling off, then increasing again. Therefore, the critical point (3, 0) is a stationary inflection point.

I should add that this is my least favorite of the three procedures described in Units 15–17 because it is rather long. Nevertheless, it is important enough that we consider it and you become proficient with its use. You may even like it. After the next unit, the choice will be yours as to which of the three procedures you use to classify critical points.

EXAMPLE 5

Sketch the graph of $y = f(x) = x^4 - 8x^3 + 18x^2 - 27$.

a. Locate the absolute extrema of $f(x)$.

b. State the interval(s) over which $f(x)$ is increasing.

c. State the interval(s) over which $f(x)$ is decreasing.

Solution: 1.   The function is a fourth-degree polynomial function.

2.   The y-intercept is –27.

3.   There are at most four x-intercepts.

4.   From Example 4, the function has a relative minimum at (0, –27), which implies the curve is decreasing on the left and increasing on the right.

5.   Also from Example 4, the function has a stationary inflection point at (3, 0) with the curve increasing on the right.

6.   And because the graph of a polynomial function is known to be a smooth, continuous curve without any breaks or sharp corners, we can connect the critical points with a nice smooth curve as shown as the following page.

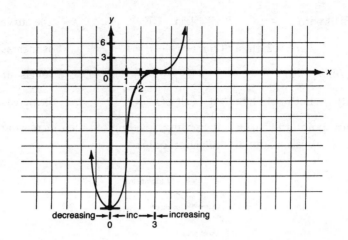

decreasing→|←inc→|←increasing
        0   3

Answers: a.   From the drawing, the absolute minimum value of $f(x)$ is -27 and occurs when $x = 0$. The function $f(x)$ does not have an absolute maximum.

           b.   The function $f(x)$ is increasing over the intervals $0 < x < 3$ and $3 < x$.

           c.   The function $f(x)$ is decreasing over the interval $x < 0$.

Here are three related problems for you to try before we continue with some more complicated functions.

Problem 1

Use the first-derivative test to determine any relative extrema for $y = f(x) = -x^5 + 1$.

Solution: $y = f(x) =$

            $f'(x) =$

            Find all critical points. Critical points occur where $f'(x) = 0$.

            Use the first-derivative test.

            For point   the curve:

            On the left: Let $x =$

            At point (,):   levels off

            On the right: Let $x =$

            Visualize the graph.

                                        Answer: The critical point $(0, 1)$ is a stationary inflection point.

Problem 2

Sketch the graph of $y = f(x) = -x^5 + 1$.

Solution: Answer:

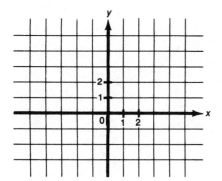   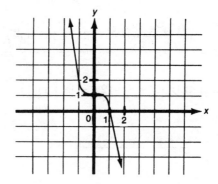

Problem 3

Use the information from Problems 1 and 2 to do the following:

a. Identify any absolute extrema of $f(x)$.

b. State the interval(s) over which $f(x)$ is increasing.

c. State the interval(s) over which $f(x)$ is decreasing.

Solution:

Answers: a. $f(x)$ has neither an absolute maximum nor an absolute minimum.

b. $f(x)$ is never increasing.

c. $f(x)$ is decreasing over the intervals $x < 0$ and $0 < x$. Note: At $x = 0$, the curve is neither increasing nor decreasing. At $x = 0$, the slope of the tangent line is 0; hence there is a critical point.

USING THE FIRST-DERIVATIVE TEST FOR MORE COMPLICATED FUNCTIONS

The procedure for using the first-derivative test is always the same. The only thing that remains for us to do is to consider functions that are more complicated. The examples that follow no longer involve polynomial functions, and we will not attempt to sketch the graphs.

EXAMPLE 6

Find all critical points for $y = f(x) = \dfrac{2x^2 + 18 + x}{x}$ with domain $x > 0$.

Use the first-derivative test to classify the critical points.

Solution: $y = f(x) = \dfrac{2x^2 + 18 + x}{x}$

$$f'(x) = \dfrac{x\dfrac{d(2x^2 + 18 + x)}{dx} - (2x^2 + 18 + x)\dfrac{d(x)}{dx}}{x^2}$$

$$= \dfrac{x(4x + 0 + 1) - (2x^2 + 18 + x)(1)}{x^2}$$

$$= \dfrac{4x^2 + x - 2x^2 - 18 - x}{x^2}$$

$$= \dfrac{2x^2 - 18}{x^2}$$

Find all critical points. Critical points occur where $f'(x) = 0$.

$$0 = \dfrac{2x^2 - 18}{x^2}$$

$0 = 2x^2 - 18$ (Multiplied entire equation by x^2.)

Notice the new equation is simply the numerator set equal to zero.

$0 = x^2 - 9$ (Divided equation by 2.)

$0 = (x - 3)(x + 3)$

$x - 3 = 0$ $x + 3 = 0$

$x^* = 3$ $x^* = -3$, but not in the domain.

If $x^* = 3$,

$$y = f(3) = \dfrac{2(9) + 3 + 18}{3} = 13$$

and (3, 13) is the only critical point.

Use the first-derivative test with $f'(x) = \dfrac{2x^2 - 18}{x^2}$.

For point (3, 13)   the curve:

On the left: Let $x = 2$,   then $f'(2) = (8 - 18)/4\ \ = -$ is decreasing

At point (3, 13):   levels off

On the right: Let $x = 4$,   then $f'(4) = (32 - 18)/4 = +$ is increasing

At this critical point, the curve is decreasing, leveling off, then increasing. Therefore, the critical point (3, 13) is a relative minimum.

EXAMPLE 7

Find all critical points for $y = f(x) = \dfrac{x}{x+1}$. Use the first-derivative test.

Solution: $y = f(x) = \dfrac{x}{x+1}$

$$f'(x) = \frac{(x+1)\dfrac{d(x)}{dx} - x\dfrac{d(x+1)}{dx}}{(x+1)^2}$$

$$= \frac{(x+1)(1) - x(1)}{(x+1)^2}$$

$$= \frac{x+1-x}{(x+1)^2}$$

$$= \frac{1}{(x+1)^2}$$

Find all critical points.

To find all values of x such that $f'(x) = 0$, set the **numerator** of $f'(x)$ equal to 0.

Since a variable does not appear in the numerator, there is no value of x that makes $f'(x) = 0$.

To find all values of x such that $f'(x)$ is undefined, set the **denominator** of $f'(x)$ equal to 0.

Since a variable appears in the denominator, there is a value of x that makes $f'(x)$ undefined, namely, $x = -1$. However, this value is not in the domain of $f(x)$, so it is not considered. If $f(-1)$ was defined, then $x = -1$ would be a critical point.

Answer: The function has no critical points.

You now should be able to use the first-derivative test to classify critical points. And further, for polynomial functions you should be able to state the intervals over which the function is increasing and decreasing.

To classify critical points for functions using the first-derivative test, the signs of the first derivative are examined on either side of the critical point; then, briefly:

1. If the derivative is negative on the left and positive on the right, the point is a relative minimum.

2. If the derivative is positive on the left and negative on the right, the point is a relative maximum.

3. If neither situation exists, the critical point is a stationary inflection point.

Here are some exercises to do before going on to the next unit. Optional—Confirm your answers using the various features on a graphing calculator before checking the results in the back of the book.

EXERCISES

Find all relative extrema using the first-derivative test.

1. $y = f(x) = 3 + 12x - x^2$

2. $y = f(x) = x^4 - 4x^3 + 10$

3. Sketch the graph of $y = f(x) = x^4 - 4x^3 + 10$.

4. For the graph in Exercise 3:

    a. Identify any absolute extrema.

    b. State the interval(s) over which $f(x)$ is increasing.

    c. State the interval(s) over which $f(x)$ is decreasing.

5. $y = f(x) = 7x^5$

6. $y = f(x) = \dfrac{1}{4}x^4 + x$

7. $y = f(x) = x^{10} + 2$

8. $y = f(x) = x^5 - 5x$

9. Sketch the graph of $y = f(x) = x^5 - 5x$.

10. For the graph in Exercise 9:

    a. Identify any absolute extrema.

    b. State the interval(s) over which $f(x)$ is increasing.

    c. State the interval(s) over which $f(x)$ is decreasing.

11. $y = f(x) = x^4 - 8x^2$

12. $y = f(x) = \dfrac{1}{x - 2}$

13. $y = f(x) = \dfrac{x^2 + 9}{x}$

14. $y = f(x) = \dfrac{x}{x^2 + 1}$

15. $y = f(x) = \dfrac{x^2}{x - 3}$

C16. Given $y = f(x) = \dfrac{x^2 - x + 1}{x^2 + 1}$.

    a. Graph the function. Hint: Both numerator and denominator must be enclosed in parentheses.

    b. Locate all relative extrema. Hint: Use the TRACE and ZOOM In features.

    c. Identify any absolute extrema.

    d. Rework the problem analytically to verify your results.

C17. Given $y = f(x) = \dfrac{12}{x^2 + 2}$.

    a.   Graph the function.

    b.   Locate all relative extrema.

    c.   Identify any absolute extrema.

18.   Ken is the General Operations Manager for an exclusive resort club with golf, tennis, marina, and food and beverage facilities on the island of Cat Cay in the Bahamas. Since taking the job as manager in 1995, Ken has determined that the number of guests per year staying at the club can be estimated by the following function:

$$f(x) = 2x^3 - 21x^2 + 60x + 10$$

where x = the number of years since 1995.

For example, $f(1) = 51$ means that in 1996 there were 51 guests at the club.

    a.   Graph $f(x)$. Be sure to include critical points.

    b.   What is the restricted domain for $f(x)$?

    c.   How many guests were there in 2001?

    d.   At what year between 1995 and the present did the club have the greatest number of guests?

    e.   Which year after the first had the fewest number of guests?

    f.   What was the fewest number of guests after the first year?

    g.   Describe the number of guests between 1998 and 1999.

    h.   How would you describe the future prospects for the club?

If additional practice is needed:

Barnett, Ziegler, and Byleen, pages 253–255, problems 59–85
Harshbarger and Reynolds, pages 747–750, problems 1–63
Hoffmann and Bradley, pages 206–210, problems 1–57
Waner and Costenoble, page 269, problems 21–47

UNIT 17

Classifying Critical Points— A Third Procedure

In this unit you will learn yet a third method, the second-derivative test, for determining where a function achieves maximum or minimum values. The unit will conclude with the introduction of an economic order quantity (EOQ) application problem.

Before introducing the test, we need to define some terms. There are numerous definitions for concavity; these are the ones I happen to like.

Definition:	A curve is **concave up** over an interval if every chord in that interval lies above the curve.

Definition:	A curve is **concave down** over an interval if every chord in that interval lies below the curve

Think of a chord as being a line segment connecting any two points on the curve. If, no matter which two points you select, the line segment connecting them is above the curve, the curve is said to be concave up.

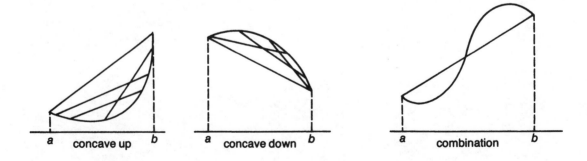

concave up   concave down   combination

Definition: The point at which the curve changes from concave up to down or vice versa is called an **inflection point**.

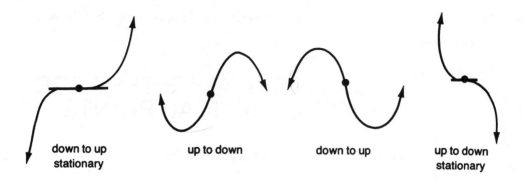

down to up
stationary

up to down

down to up

up to down
stationary

All of the above are examples of inflection points. However, two of them, the first and the last, are stationary inflection points, meaning that they are critical points as well.

CONCEPT OF DERIVATIVES

Recall that the first derivative at a point gives the slope of the tangent line at that point or the instantaneous rate of change of y with respect to x.

Since the **second derivative,** denoted by $f''(x)$, is defined as the derivative of the first derivative, the second derivative gives the rate of change of the slope of the tangent.

Consider a relative minimum point:

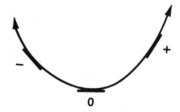

- At points to the left of the critical point, the tangent lines have negative slopes.

- At the minimum point, the slope is zero.

- At points to the right of the critical point, the tangent lines have positive slopes.

Thus as you move from left to right, the slope of the tangent line is increasing from negative values, to zero, to positive values. Hence, the rate of change of the slope (the second derivative) should be positive, and the curve is concave up.

By a similar argument, if the second derivative is negative on an interval, then the first derivative is decreasing, and the curve is concave down. In mathematical terms, we have the following theorems:

THEOREM: If $f''(x) > 0$ over an interval,

then $f(x)$ is concave up over the interval.

> THEOREM: If $f''(x) < 0$ over an interval,
>
> then $f(x)$ is concave down over the interval.

And now we have the basis for a new way of testing critical points; it is called the second-derivative test.

THE SECOND-DERIVATIVE TEST FOR CLASSIFYING CRITICAL POINTS

Let $f(x)$ be a function with a critical point at $x^* = a$; determine the sign of $f''(x)$.

1. If $f''(x^*) > 0$, then $f(x)$ is concave up, ◡, and the critical point is a relative minimum.

2. If $f''(x^*) < 0$, then $f(x)$ is concave down, ◠, and the critical point is a relative maximum.

3. If $f''(x^*) = 0$, the test fails.

To say that the test fails means this procedure gives no information that would allow you to classify the critical point. You must rework the problem using either of the other two methods.

EXAMPLE 1

Use the second-derivative test to determine all relative extrema for $f(x) = x^4 - 32x + 56$.

Solution: $f(x) = x^4 - 32x + 56$

$f'(x) = 4x^3 - 32$

$f''(x) = 12x^2$

Find all critical points. Critical points occur where $f'(x) = 0$.

$0 = 4x^3 - 32$

$4x^3 = 32$

$x^3 = 8$

$x^* = 2$

If $x^* = 2$,

$f(2) = (2)^4 - 32(2) + 56 = 16 - 64 + 56 = 8$

and $(2, 8)$ is a critical point.

Use the second-derivative test with $f''(x) = 12x^2$.

At point $(2, 8)$,

$f''(2) = 12(2)^2 = +$; hence the curve is concave up, ◡, and the critical point is a relative minimum.

Be sure to notice the following two items regarding the use of the second-derivative test.

1. The x-coordinate of the critical point is used to evaluate the second derivative.

2. The actual numerical value of the second derivative is unimportant; only the sign, positive or negative, need be determined.

EXAMPLE 2

Sketch the graph of $y = f(x) = x^4 - 32x + 56$.

Solution: $f(x)$ is a fourth-degree polynomial function.

The y-intercept is 56.

From Example 1, the function has a relative minimum at $(2, 8)$.

The graph will be a smooth, continuous curve without any breaks or sharp corners.

Answer:

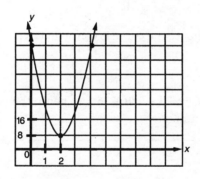

Comment: Although this is a smooth U-shaped curve, it is *not* a parabola. Only second-degree polynomial functions (or quadratic functions) yield parabolic curves, and this is a fourth-degree polynomial.

EXAMPLE 3

Use the second-derivative test to determine all relative extrema for $y = f(x) = 3x^4 - 16x^3 + 18x^2$.

Solution: $y = f(x) = 3x^4 - 16x^3 + 18x^2$.

$f'(x) = 12x^3 - 48x^2 + 36x$

$f''(x) = 36x^2 - 96x + 36$

Find all critical points. Critical points occur where $f'(x) = 0$.

$0 = 12x^3 - 48x^2 + 36x$

$0 = x^3 - 4x^2 + 3x$

$0 = x(x^2 - 4x + 3)$

$$0 = x(x - 1)(x - 3)$$

$$x^* = 0 \quad x^* = 1 \quad x^* = 3$$

If $x^* = 0$,

$y = f(0) = 0 - 0 + 0 = 0$, and $(0, 0)$ is a critical point.

If $x^* = 1$,

$y = f(1) = 3 - 16 + 18 = 5$, and $(1, 5)$ is another critical point.

If $x^* = 3$,

$y = f(3) = 3(81) - 16(27) + 18(9) = 243 - 432 + 162 = -27$, and $(3, -27)$ is still another critical point.

Use the second-derivative test with $f''(x) = 36x^2 - 96x + 36$.

Answer: At point $(0, 0)$:

$f''(0) = 0 - 0 + 36 = +$; hence the curve is concave up, $\smile$, and the critical point is a relative minimum.

At point $(1, 5)$:

$f''(1) = 36 - 96 + 36 = -$; hence the curve is concave down, $\frown$, and the critical point is a relative maximum.

At point $(3, -27)$:

$f''(3) = 36(9) - 96(3) + 36 = 324 - 288 + 36 = +$; hence the curve is concave up, and the critical point is a relative minimum.

Since this is the first example we have had with three critical points, let's sketch it as well.

EXAMPLE 4

Sketch the graph of $y = f(x) = 3x^4 - 16x^3 + 18x^2$.

Solution: The function is a fourth-degree polynomial function.

The y-intercept is 0.

There are three critical points:

$(0, 0)$ is a relative minimum.

$(1, 5)$ is a relative maximum.

$(3, -27)$ is a relative minimum.

Because the graph of a polynomial function is a continuous curve, connect the critical points with a smooth curve as shown.

Answer:

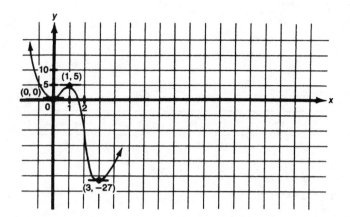

It's time for a review problem on increasing and decreasing intervals.

Problem 1

Use the graph from Example 4 to do the following:

a.   Identify any absolute extrema values of $f(x)$.

b.   State the interval(s) over which $f(x)$ is increasing.

c.   State the interval(s) over which $f(x)$ is decreasing.

Solution:

Answers: a.   The absolute minimum value is $f(x) = -27$ and it occurs when $x = 3$.
The function has no absolute maximum.

b.   The function is increasing over the intervals $0 < x < 1$ and $3 < x$.

c.   The function is decreasing over the intervals $x < 0$ and $1 < x < 3$.

Since you need the practice, I am going to have you work the first part of Example 5, that of finding the critical points, as a problem. Then I will classify them.

Problem 2

Find all critical points for $y = f(x) = 3x^5 - 5x^3$. Hint: There are three.

Solution:

Answer: $(1, -2); (-1, 2); (0, 0)$

EXAMPLE 5

Classify all critical points for $y = f(x) = 3x^5 - 5x^3$.

Solution: $y = f(x) = 3x^5 - 5x^3$

$f'(x) = 15x^4 - 15x^2$

$f''(x) = 60x^3 - 30x$

Use the second-derivative test with $f''(x) = 60x^3 - 30x$.

At point $(1, -2)$:

$f''(1) = 60 - 30 = +$; hence the curve is concave up, $\smile$, and the critical point is a relative minimum.

At point $(-1, 2)$:

$f''(-1) = 60(-1) - 30(-1) = -60 + 30 = -$; hence the curve is concave down, $\frown$, and the critical point is a relative maximum.

At point $(0, 0)$:

$f''(0) = 0 - 0 = 0$ and the test fails.

To classify the critical point, we must rework the problem using either the original-function test or the first-derivative test.

I will use the first-derivative test with $f'(x) = 15x^4 - 15x^2$.

To simplify the work, factor the first derivative.

$f'(x) = 15x^4 - 15x^2$

$= 15x^2(x^2 - 1)$

$= 15x^2(x + 1)(x - 1)$

Recall that, to use the first-derivative test, we select a value on either side of the critical point and determine the sign of the first derivative. Typically we would use -1 on the left and 1 on the right. But in this example those two values are too far away; they are critical points themselves. Therefore we need to use some values closer to 0; I will use 0.5 and -0.5. Remember that we need only the signs of the derivative, not the numerical values.

At point $(0, 0)$ with $f'(x) = 15x^2(x + 1)(x - 1)$ the curve:

On the left: Let $x = -0.5$, $f'(-0.5) = 15(-.5)^2(-0.5 + 1)(-0.5 - 1)$

$= (+)(+)(+)(-)$

$= -$ is decreasing

At point $(0, 0)$ levels off

On the right: Let $x = .5, f'(.5)$ $= 15(.5)^2(.5 + 1)(.5 - 1)$

$= (+)(+)(+)(-)$

$= -$ is decreasing

According to the first-derivative test, the curve is decreasing, leveling off, then decreasing. The critical point $(0, 0)$ is a stationary inflection point.

EXAMPLE 6

Sketch the graph of $y = f(x) = 3x^5 - 5x^3$.

a.   Identify any absolute maximum and absolute minimum values of $f(x)$.

b.   State the interval(s) over which $f(x)$ is increasing.

c.   State the interval(s) over which $f(x)$ is decreasing.

Solution: The function is a fifth-degree polynomial function.

The y-intercept is 0.

The x-intercepts are 0, $\sqrt{5/3}$, and $-\sqrt{5/3}$ because

$$0 = 3x^5 - 5x^3$$

$$0 = x^3(3x^2 - 5)$$

$$x^3 = 0 \quad 3x^2 - 5 = 0$$

$$x = 0 \qquad x^2 = 5/3$$

$$x = \pm\sqrt{\frac{5}{3}}$$

From Example 5, $(-1, 2)$ is a relative maximum,

$(0, \quad 0)$ is a stationary inflection point, and

$(1, -2)$ is a relative minimum.

Answers:

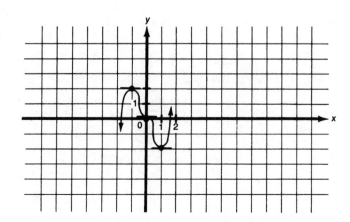

a. The function has neither an absolute maximum nor an absolute minimum value.

b. The function is increasing over the intervals $x < -1$ and $1 < x$.

c. The function is decreasing over the intervals $-1 < x < 0$ and $0 < x < 1$.

AN ECONOMIC ORDER QUANTITY PROBLEM

We will conclude this unit with an inventory cost application, which also illustrates the usefulness of the second-derivative test. It is a bit long, but interesting. First, from business and economics the formula for the cost of inventory is as follows:

$$C(x) = \left(\frac{Q}{x}\right)c + \left(\frac{x}{2}\right)s$$

where Q = the number of units needed per year,

c = the cost of placing an order,

s = the cost of storing one unit for 1 year,

and x = the number of units per order.

I will leave the actual explanation of the formula to your studies in economics, but we will use it to solve the next problem.

EXAMPLE 7

The Alan Kirby Electrical Company buys 151,200 on-off switches each year for use in the manufacture of its cordless drills. The ordering fee is $35 per order. Once the switches arrive at the plant, the storage cost is known to be $0.125 per switch per year. The switches are used at a constant rate throughout the year. How many switches should the company order each time to minimize inventory costs, and what will the costs be?

Solution: We are given the following information:

$Q = 151,200$, switches needed per year,

$c = \$35$, the cost of placing an order,

$s = \$0.125$, the cost of storing one switch for 1 year, and

x = the number of switches per order.

$$C(x) = \left(\frac{Q}{x}\right)c + \left(\frac{x}{2}\right)s$$

$$= \left(\frac{151,200}{x}\right)35 + \left(\frac{x}{2}\right)0.125$$

$$= \left(\frac{5,292,000}{x}\right) + 0.0625x$$

Before continuing let's look at one possibility. Suppose the Accounting Department decides it will place an order just twice a year. What will the company's inventory costs be? Since 151,200 drills are needed per year, ordered twice a year, each order will consist of 75,600 switches, or $x = 75,600$.

$$C(x) = \left(\frac{5{,}292{,}000}{x}\right) + 0.0625x$$

$$= \left(\frac{5{,}292{,}000}{75{,}600}\right) + 0.0625(75{,}600)$$

$$= 70 + 4{,}725$$

$$= 4{,}795$$

If the switches are ordered twice a year, the inventory costs will be $4,795.

Now let's return to the original problem and determine what the order size should be so as to minimize inventory costs. We need to find critical points and classify in our attempt to find an absolute minimum.

$$C(x) = \left(\frac{5{,}292{,}000}{x}\right) + 0.0625x$$

$$= 5{,}292{,}000x^{-1} + 0.0625x$$

$$C'(x) = -5{,}292{,}000x^{-1-1} + 0.0625$$

$$= -5{,}292{,}000x^{-2} + 0.0625$$

$$= \left(\frac{-5{,}292{,}000}{x^2}\right) + 0.0625$$

$$0 = \left(\frac{-5{,}292{,}000}{x^2}\right) + 0.0625$$

$$0 = -5{,}292{,}000 + 0.0625x^2$$

$$0.0625x^2 = 5{,}292{,}000$$

$$x^2 = 84{,}672{,}000$$

$$x^* = 9{,}201.74$$

$$x^* = \cancel{-9{,}201.74}$$

(−9,201.74 is not within the domain and should not be considered further.)

In this application it would be reasonable to round and let $x = 9{,}202$.

Since the first derivative is given by $C'(x) = -5{,}292{,}000x^{-2} + 0.0625$, the second-derivative test is reasonable. It takes one step and minimal calculations.

Test: Second-derivative test with $C''(x) = \dfrac{10{,}584{,}000}{x^3}$

At $x = 9{,}201.74$, $C''(9{,}201.74) = +$, concave up, minimum point.

In fact, the critical point is an absolute minimum. Do you understand why? The restricted domain is the set $0 < x \le 151{,}200$, and the curve has no breaks. Since there is one and only one relative minimum on the interval, it must be the absolute minimum.

$151{,}200 / 9{,}202 = 16.43$ number of orders per year

$$C(9{,}202) = \left(\frac{5{,}292{,}000}{9{,}202}\right) + 0.0625(9{,}202)$$

$$= 575.09 + 575.125$$

$$= 1{,}150.215$$

Answer: To minimize inventory costs, the Alan Kirkland Electric Company should order 9,202 on-off switches each time (approximately 16 orders per year). Its inventory costs (ordering costs and storage costs) will be $1,150.215.

Notice that our solution represents a savings of well over $3,500 to the company. Could we have found it by trial and error? It's doubtful; besides, the calculus wasn't all that difficult.

You should now be able to classify critical points using the second-derivative test.

To classify critical points for functions using the second-derivative test, the sign of the second derivative is determined for the critical point then, briefly:

1. If the second derivative is positive, the curve is concave up, and the point is a relative minimum.

2. If the second derivative is negative, the curve is concave down, and the point is a relative maximum.

3. If the second derivative is zero, the test fails, and the point must be reworked using either of the other two tests.

A COMPARISON OF PROCEDURES

Let us conclude this entire section with a brief summary of the four methods for classifying critical points.

Knowledge of the graph:

If the graph of $f(x)$ is available, it can be used to visually classify the critical point. If the function is a first-, second-, or third-degree polynomial function, you should use your knowledge of the graph.

The original-function test:

Use $f(x)$ and determine a point on either side of the critical point. Two complete numerical calculations are required.

The first-derivative test:

Use $f'(x)$ and determine the sign of a point on either side of the critical point. Two calculations are required, but only the sign has to be computed, not the numerical value.

The second-derivative test:

Use $f''(x)$ and determine the sign of the critical point. Only one calculation is required, that of the sign only. The disadvantage is that you must find the second derivative, a task that is sometimes more work than it is worth.

Clearly there are advantages and disadvantages to each method. So that you will have a better appreciation of each procedure, in each of Exercises 14, 15, and 16 you are directed to work the problem using a different one of the three tests. After this unit, you may select the method of your choice.

Now try the following exercises.

EXERCISES

Find all relative extrema using the second-derivative test whenever it applies. If the second-derivative test fails, use the first-derivative test. Optional—Now that you are proficient in doing graphs on the calculator, you might find it helpful to do a quick graph of the function just to confirm that your manual calculations are reasonable. That is frequently what I do.

1. $y = f(x) = \dfrac{-6}{x^2 + 2}$

2. $y = f(x) = x^4 - 18x^2$

3. $f(x) = x^{16}$

4. $y = f(x) = 2 - x^5$

5. $f(x) = \dfrac{x}{1 - x^2}$

6. $y = f(x) = 10x^6 + 24x^5 + 15x^4 + 2$

7. $y = f(x) = \dfrac{1}{4}x^4 - x^3 + x^2$

8. $y = f(x) = (x - 5)^3$

9. Use the critical points from Exercise 2 to sketch the graph of $y = f(x) = x^4 - 18x^2$.

10. For the graph of Exercise 9:

    a. Identify any absolute extrema.

    b. State the interval(s) over which $f(x)$ is increasing.

    c. State the interval(s) over which $f(x)$ is decreasing.

11. Use the critical points in Exercise 6 to sketch the graph of $y = f(x) = 10x^6 + 24x^5 + 15x^4 + 2$.

12. For the graph of Exercise 11:

    a. Identify any absolute maximum and absolute minimum values.

    b. State the interval(s) over which $f(x)$ is increasing.

    c. State the interval(s) over which $f(x)$ is decreasing.

13. Find all critical points for $y = f(x) = \dfrac{x^2 + 1}{x}$.

14.   Use the original-function test to classify the critical points for the function in Exercise 13.

15.   Use the first-derivative test to classify the critical points for the function in Exercise 13.

16.   Use the second-derivative test to classify the critical points for the function in Exercise 13.

17.   Compare the three methods used in Exercises 14, 15, and 16. Which test do you prefer?

18.   Sketch the graph of $y = f(x) = 2x^3 - 15x^2 + 300$.

C19.   Graph $y = \dfrac{x^2 - 2x + 4}{x - 2}$ and identify all relative extrema.

C20.   Graph $y = f(x) = \dfrac{2(x^2 - 9)}{x^2 - 4}$.

   a.   Identify all relative extrema.

   b.   Identify any absolute extrema.

21.   An exclusive restaurant opened in 1995. The manager has determined that profit, in thousands of dollars, per year at the restaurant can be estimated by the following function:

$$P(x) = x^3 - 6x^2$$

where x is the number of years since 1995. For example, $f(1) = -5$ means that in 1996, the restaurant's first year of operation, profit (or in this case loss) was $5,000.

   a.   How much profit (or loss) was made in the second year of operation?

   b.   How much profit (or loss) was made in 1998?

   c.   Which year was the most profitable?

   d.   Which year was the least profitable?

   e.   What was the lowest profit (or greatest loss) ever earned?

   f.   How would you describe the restaurant's situation for 2001?

   g.   How would you describe the future prospects for the restaurant?

22.   The ABC Company plans to introduce a new type of soft drink. The Marketing Department forecasts that the daily total revenue function, which is price times quantity, will be

$$R(x) = x(40 - x) = 40 - x^2$$

where x is the price, in dollars. The CEO is considering pricing the new soft drink, which tastes great and has no calories, at about $30 a case. What would you recommend to the CEO?

If additional practice is needed:

   Barnett, Ziegler, and Byleen, pages 266–270, problems 1–91
   Harshbarger and Reynolds, pages 761–764, problems 1–44
   Hoffmann and Bradley, pages 222–226, problems 1–53
   Waner and Costenoble, pages 290–294, problems 1–80

UNIT 18

Absolute Maxima and Minima

In this unit you will learn how to locate the absolute maximum and absolute minimum for a continuous function on a closed interval.

USING THE GRAPH OF A FUNCTION TO LOCATE ITS ABSOLUTE EXTREMA— A REVIEW

Recall that the term *absolute maximum* is used to denote the largest value of the function, if it exists. The absolute maximum is located at the highest point on the graph. Similarly, *absolute minimum* refers to the smallest value of the function, if it exists, and the absolute minimum is located at the lowest point on the graph. The absolute maxima and minima need not be at a critical point.

We need the graph of a function to use as an example. I am confident you can do Problem 1 with ease.

Problem 1

Graph: $y = f(x) = x^3 + 3x^2 + 1$.

Solution:

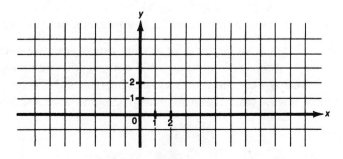

Answer:

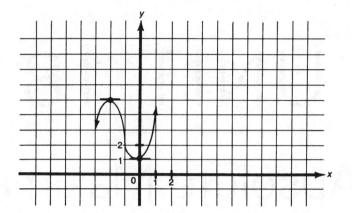

As we already know and can observe from our graph, a cubic has neither an absolute maximum nor an absolute minimum. On the right the curve is continuing upward, and on the left the curve is continuing downward. But if we restrict our attention to just a portion of the curve, say from $-1 \leq x \leq 2$, all of that changes.

A solid line in the next drawing indicates the portion of the curve from $-1 \leq x \leq 2$, with the rest being shown with a dashed line. From the drawing it should be obvious that the solid portion of the curve has an absolute maximum and an absolute minimum.

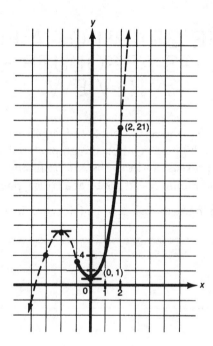

The absolute maximum occurs at an endpoint and the absolute minimum occurs at a critical point.

In preceding units we used the graph to determine the absolute extrema for various examples. In this unit we will introduce a procedure for locating absolute maxima and minima without having to draw a graph, provided that the function is continuous and defined over a closed interval. Such a continuous function, meaning there are no breaks in the curve, always will have an absolute maximum and an absolute minimum.

> THEOREM: If $y = f(x)$ is a continuous function over the interval $a \leq x \leq b$, its absolute maximum and absolute minimum will occur either at a critical point or at an endpoint.

Two criteria must be met before the theorem applies. First, the function must be defined over a closed interval, meaning both the lower and upper limits of the interval must be included. Second, the function must be continuous over the entire interval, meaning there are no breaks in the curve. In other words, the function must be defined for all x in the interval. In still other words, the function has a restricted domain.

If all the conditions stated in the theorem are met, it is actually less work to find absolute maxima and minima than to classify critical points. It is a two-step procedure.

PROCEDURE FOR LOCATING ABSOLUTE MAXIMUM AND MINIMUM ON A CLOSED INTERVAL

1. Find the x-coordinates of all criteria points *within* the interval.

2. Determine the y-coordinate, using $f(x)$, for each of the critical points within the interval, and for the lower and upper limits of the interval.
    The absolute maximum is the largest of these y-coordinates.
    The absolute minimum is the smallest of these y-coordinates.

Notice that in this short procedure there is no test to perform and no graph is necessary. The values must exist. The absolute maximum will be the largest, and the absolute minimum will be the smallest, of the y-coordinates found in step 2.

EXAMPLE 1

Find the absolute maximum and absolute minimum for $y = f(x) = x^3 + 3x^2 + 1 \quad -1 \leq x \leq 2$.

Solution: 1.   Find the x-coordinates of critical points within the interval.

$$y = f(x) = x^3 + 3x^2 + 1 \qquad -1 \leq x \leq 2$$

$$f'(x) = 3x^2 + 6x$$

$$0 = 3x^2 + 6x$$

$$= 3x(x + 2)$$

$$x = 0 \qquad x = -2$$

(Note: -2 is not within the interval and therefore not used in step 2.)

2.   Determine the y-coordinate for each of the critical points within the interval and for the lower and upper limits of the interval.

critical: $y = f(0) = 0 + 0 + 1$ $= 1$

lower: $y = f(-1) = (-1)^3 + 3(-1)^2 + 1 = -1 + 3 + 1 = 3$

upper: $y = f(2) = (2)^3 + 3(2)^2 + 1 = 8 + 12 + 1$ $= 21$

The absolute maximum is the largest of the y-coordinates and the absolute minimum is the smallest.

Answers: The absolute maximum is 21 and occurs when $x = 2$.

The absolute minimum is 1 and occurs when $x = 0$.

Note that our answers are consistent with those of Problem 1.

EXAMPLE 2

Given: $y = f(x) = x^2 - 12x - 64$ $-4 \le x \le 4$.

Find the absolute maximum and absolute minimum for the function over the interval.

Solution: Before starting, what do we know and what do we really want to find? We know that the function is a quadratic, opening up, with an absolute minimum at its vertex and no absolute maximum. But in this example we are considering only the portion of the curve from $-4 \le x \le 4$. We want to locate its absolute maximum and absolute minimum values just over the given interval.

1.   Find the x-coordinates of critical points within the interval.

$y = f(x) = x^2 - 12x - 64$ $-4 \le x \le 4$

$f'(x) = 2x - 12$

$0 = 2x - 12$

$2x = 12$

$\cancel{x = 6}$

(6 is not within the interval and therefore not used in step 2.)

2.   Determine the y-coordinate for each of the critical points within the interval and for the lower and upper limits of the interval.

lower: $y = f(-4) = (-4)^2 - 12(-4) - 64 = 0$

upper: $y = f(4) = (4)^2 - 12(4) - 64 = -96$

The absolute maximum is the largest of the y-coordinates and the absolute minimum is the smallest.

Answers: The absolute maximum is 0 and occurs when $x = -4$.

The absolute minimum is -96 and occurs when $x = 4$.

For reference, the graph of the function is shown below. The section of the curve under consideration is shown with a solid line; the rest of the parabola, with a dashed line.

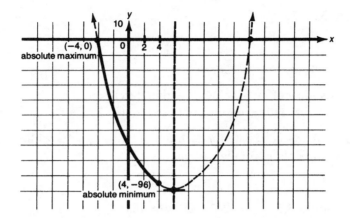

Try a problem yourself.

Problem 2

Given: $y = f(x) = x^2 - 8x + 6$ $0 \le x \le 6$.

Find the absolute extrema for the function over the given interval.

Solution: 1.   Find the x-coordinates of critical points within the interval.

2.   Determine the y-coordinate for each of the critical points within the interval and for the lower and upper limits of the interval.

critical:

lower:

upper:

The absolute maximum is the largest of the y-coordinates and the absolute minimum is the smallest.

Answers: The absolute maximum is 6 and occurs when $x = 0$.

The absolute minimum is −10 and occurs when $x = 4$.

I know you got the right answer, but, just to be on the safe side, we will work one more example for practice.

EXAMPLE 3

Given: $y = f(x) = -x^5 + 5x^4 + 1$ $1 \le x \le 3$.

Find the absolute extrema for the function over the given interval.

Solution: 1.   Find the x-coordinates of critical points within the interval.

$$y = f(x) = -x^5 + 5x^4 + 1 \quad 1 \leq x \leq 3$$

$$f'(x) = -5x^4 - 20x^3$$

$$0 = -5x^3(x + 4)$$

$$\cancel{x = 0} \quad \cancel{x = -4}$$

(Neither of these values is within the closed
interval and therefore is not used in step 2.)

2.   Determine the y-coordinate for each of the critical points within the interval and for the
lower and upper limits of the interval.

$$y = f(1) = -1 + 5 + 1 \qquad\qquad\qquad = 5$$

$$y = f(3) = -(3)^5 + 5(3)^4 + 1 = -243 + 405 + 1 = 163$$

The absolute maximum is the largest of the y-coordinates and the absolute minimum is the
smallest.

Answers: The absolute maximum is 163 and occurs when $x = 3$.

The absolute minimum is 1 and occurs when $x = 1$.

You should now be able to locate both the absolute maximum and the absolute minimum for a
continuous function defined on a closed interval. This is a two-step procedure and does not require
drawing the graph.

Briefly the procedure is to find the x-coordinates of all critical points within the interval, then determine the
corresponding y-coordinates for the critical points as well as for the endpoints of the interval. The absolute
maximum will be the largest, and the absolute minimum will be the smallest, value of these y-coordinates.

Fourteen exercises are provided below, which you should try before going on to the next unit.

EXERCISES

In each case, determine the location and values of the absolute maximum and absolute minimum for $y = f(x)$
over the given interval:

1. $y = f(x) = x^2 - 12x + 1$ $1 \leq x \leq 10$

2. $y = f(x) = 5x - 1$ $0 \leq x \leq 5$

3. $y = f(x) = \dfrac{1}{3}x^3 - x^2 - 3x + 1$ $0 \leq x \leq 3$

4. $y = f(x) = -x^3 - 3x^2 + 1$ $-2 \leq x \leq 2$

5. $y = f(x) = x^3 - 3x^2$ $-1 \leq x \leq 4$

6. $y = f(x) = x^4 - 8x^3 + 18x^2 - 27$ $1 \leq x \leq 2$

7. $y = g(x) = x^{1/3} + 1$ $-1 \leq x \leq 8$

8. $y = f(x) = 3x^4 - 16x^3 + 18x^2$ $0 \leq x \leq 2$

9. $y = h(x) = 3x^4 - 8x^3 + 6x^2$ $-1 \leq x \leq 1$

10. $y = f(x) = 3x^5 - 50x^3 + 135x + 20$ $-2 \leq x \leq 1$

11. $y = F(x) = \dfrac{x - 1}{(x + 1)^2}$ $0 \leq x \leq 3$

12. $y = g(x) = \sqrt{x} + x$ $0 \leq x \leq 4$

C13. $y = f(x) = 7.25x^2 - 6.5x + \sqrt{7}$ $2 \leq x \leq 4$

C14. $y = F(x) = 2x^3 + x^2 - x - 5$ $3 \leq x \leq 5$

If additional practice is needed:

    Barnett, Ziegler, and Byleen, pages 299–300, problems 1–26
    Harshbarger and Reynolds, page 773, problems 1–4
    Hoffmann and Bradley, pages 254–260, problems 1–42

UNIT 19

Optimization Problems

In Units 19 and 20 we will consider a variety of application problems, each of which has the objective of maximizing (or minimizing) a particular quantity. In this unit we will concentrate on familiar concepts, such as revenue, cost, profit, and price. In the next unit we will expand the concepts to include a wider range of application problems. The emphasis will be not only on the mathematical answer, but also on the interpretation of the answer in terms of the problem.

If you are tempted to skip this unit, please don't. It is meant to be interesting and fun. We are now able to put together the technical information and the skills you have been mastering to solve practical problems. Hopefully you will start to appreciate the power of calculus and the usefulness of what we have been doing. Although the problems here are of my choosing, I am optimistic that you will be able to relate the concepts to problems that are of interest to you.

Word problems or application problems sometimes are terrifying to students. Having to write the equations needed to solve a problem is much harder than simply solving a given equation. What I will do to help is to set down a general technique that is applicable to the majority of optimization problems. After that, success takes practice and a bit of self-confidence on your part.

GENERAL TECHNIQUE FOR OPTIMIZATION PROBLEMS

1. Decide what quantity is to be maximized or minimized and write the equation for this quantity—in words first, if necessary.

2. Use the constraints of the problem to write the equation in terms of only one independent variable, and simplify.

3. Find the first derivative, set it equal to zero, and solve. Test to determine whether critical points are maxima or minima. In other words, find all critical points and classify.

4. Answer the question posed in the problem.

Some of the above should sound very familiar. Step 3 is basically what we have spent the last four units doing; by now it should be second nature to you. From here on the unit will involve a variety of application problems, all of which will be worked using this four-step general technique.

Reminder: application problems often require restricted domains.

EXAMPLE 1

$y = f(x) = -\frac{2}{3}x^3 + \frac{19}{2}x^2 + 10x$ with $x \geq 0$ is the profit, in dollars, when x units are produced.

a.   How many units should be produced to make profit as large as possible?

b.   What is profit at this level?

Solution: 1.   What is to be maximized in the problem? Profit

Write the equation for profit.

The equation for profit was given as:

$$y = f(x) = -\frac{2}{3}x^3 + \frac{19}{2}x^2 + 10x \qquad x \geq 0$$

2.   Write the function in terms of a single independent variable.

The function was given in terms of a single variable.

$$y = f(x) = -\frac{2}{3}x^3 + \frac{19}{2}x^2 + 10x \qquad x \geq 0$$

3.   Find critical points and classify.

$$y = f(x) = -\frac{2}{3}x^3 + \frac{19}{2}x^2 + 10x \qquad x \geq 0$$

$$f'(x) = -\frac{2}{3}(3x^2) + \frac{19}{2}(2x) + 10$$

$$= -2x^2 + 19x + 10$$

$$0 = 2x^2 - 19x - 10$$

$$0 = (2x + 1)(x - 10)$$

$$x^* = -\frac{1}{2} \qquad x^* = 10$$

$\left(-\frac{1}{2} \text{ is not within the domain and should not be considered further.}\right)$

If $x^* = 10$,

$$f(10) = -\frac{2}{3}(10)^3 + \frac{19}{2}(10)^2 + 10(10)$$

$$= -\frac{2}{3}(1{,}000) + \frac{19}{2}(100) + 100$$

$$= -666.67 + 950 + 100$$

$$= 383.33$$

Test: The function is a cubic with

$$a = -\frac{2}{3} \text{ and opening } \curvearrowright .$$

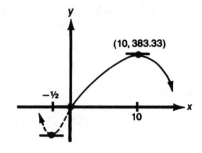

The y-intercept is at 0.

Point (10, 383.33) is a relative maximum.

Point (10, 383.33) is an absolute maximum on the restricted domain.

4.   Answer questions posed in problem.

Answer: a.   Maximum profit will be achieved when 10 units are produced.

b.   Profit at this level of production will be $383.33.

Some preliminary explanations will be helpful before the next example.

Recall that demand functions describe the relationship between the number of units sold (the demand for a product) and its selling price.

Let the demand function for a particular product be

$$q = f(p) = 120{,}000 - 3p$$

where q = the quantity sold at price p

and p = the price, in dollars.

Suppose we price our product at $1; how many units would we sell? And what should be our total revenue? Recall that total revenue equals (selling price) (quantity sold).

Let $p = \$1$;

then $q = f(1) = 120{,}000 - 3(1) = 120{,}000 - 3 = 119{,}997$ units.

Revenue = price · quantity = (1)(119,997) = $119,997

Suppose we increase our price to $10; what happens?

Let $p = \$10$;

then $q = f(10) = 120{,}000 - 30 = 119{,}970$ units,

Revenue = price · quantity = (10)(119,970) = $1,199,700.

We certainly increased our revenue with a small increase in price.

Business must be great. If we increase the price to $100, what happens?

Let $p = \$100$;

then $q = f(100) = 120{,}000 - 300 = = 119{,}700$ units.

Revenue = price · quantity = (100)(119,700) = $11,970,000.

Notice that each time we increase the price we sell fewer units, but total revenue increases. Let's do a few more. You do the calculations.

Let $p = \$1,000$;

then $q = f(1,000)$

Revenue =

Let $p = \$10,000$;

then $q = f(10,000)$

Revenue =

Let $p = \$100,000$;

then $q = f(100,000)$

Revenue =

What happened? We were doing fine until the last price increase. At a price of $10,000 per unit, we would have sold 90,000 units with total revenue of $900,000,000. But at a price of $100,000 per unit, no one was willing to buy any of the product. We had priced ourselves out of the market.

Where is all of this leading? The real question to be answered is, What price should we charge for the product so as to maximize total revenue? This problem is what is typically called a classical optimization problem. The next example will illustrate how to apply the concepts of calculus to solve it, rather than using trial-and-error as we were doing above.

EXAMPLE 2

Let the demand function for a particular product be

$$q = f(p) = 120{,}000 - 3p$$

where q = the quantity sold at price p

and p = the price, in dollars.

a.   What price will result in maximum total revenue?

b.   What is the maximum revenue that can be expected?

c.   How many units will be sold when revenue is maximized?

Solution: 1.   What is to be maximized in the problem? Revenue

Write the equation for revenue, in words first if necessary.

$$\text{Revenue} = \text{price} \cdot \text{quantity}$$

$$= p \cdot q$$

2.   Use the constraints of the problem to write the function in terms of a single variable and simplify.

Use $q = 120{,}000 - 3p$ and substitute.

$$\text{Revenue} = p \cdot q$$

$$R(p) = p(120{,}000 - 3p)$$

$$= 120{,}000p - 3p^2$$

3. Find critical points and classify.

$$R(p) = 120{,}000p - 3p^2$$

$$R'(p) = 120{,}000 - 6p$$

$$0 = 120{,}000 - 6p$$

$$6p = 120{,}000$$

$$p^* = 20{,}000$$

If $p^* = 20{,}000$,

$$R(20{,}000) = 20{,}000[120{,}000 - 3(20{,}000)]$$

$$= 20{,}000(120{,}000 - 60{,}000)$$

$$= 20{,}000(60{,}000)$$

$$= 1{,}200{,}000{,}000$$

Test: $R(p) = -3p^2 + 120{,}000p$

The function is a quadratic with $a = -3$.

The parabolic curve opens down.

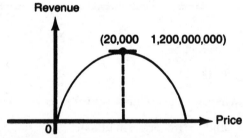

The point (20,000, $1,200,000,000) is an absolute maximum.

4. Answer questions posed in problem.

Answers: a. A price of $20,000 per unit will result in maximum total revenue.

b. At a price of $20,000 per unit, maximum revenue of $1,200,000,000 can be expected.

c. At a price of $20,000 per unit, when revenue is maximized, 60,000 units will be sold.

$$q = f(p) = 120{,}000 - 3p$$

$$q = f(20{,}000) = 120{,}000 - 3(20{,}000)$$

$$= 120{,}000 - 60{,}000$$

$$= 60{,}000$$

Several observations need to be made about Example 2. Could we have found the answer by trial and error? Probably, but using calculus was much faster. How do we know that, if we increased the price to, say, $20,025, total revenue would not increase? Because the revenue function is a quadratic with an absolute maximum, which is the value we are using.

EXAMPLE 3

a.   Sketch the revenue function from Example 2.

b.   Determine the restricted domain.

Solution: a.   The revenue function is $R(p) = p(120{,}000 - 3p)$

$$= 120{,}000p - 3p^2$$

$$= -3p^2 + 120{,}000$$

The function is a quadratic with $a = -3$.

The parabolic curve opens down.

The y-intercept is 0. (Recall this is actually the q-axis here.)

The x-intercepts are 0 and 40,000 because

$$0 = p(120{,}000 - 3p)$$

$$p = 0 \qquad 120{,}000 - 3p = 0$$

$$120{,}000 = 3p$$

$$40{,}000 = p$$

(As before, these are really the p-intercepts.)

The vertex is located at (20,000, 1,200,000,000) from Example 2.

Answer:

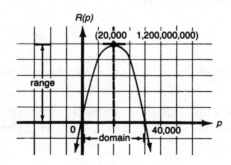

b.   Regarding the restricted domain, the question to be answered is, Are there any limitations on the values that can be used for p?

Because p represents price, a negative number would be meaningless. And because $R(p)$ represents total revenue, again a negative number would be meaningless. Therefore the relevant portion of the graph must be in quadrant I. From the graph, that means the restricted domain for p is $0 \le p \le 40{,}000$. If $p > 40{,}000$, the total revenue would be negative and that would imply a negative number for quantity, which would be meaningless.

Answer: Restricted domain: $0 \le p \le 40{,}000$

Try Problem 1. Use Example 2 as your guide.

Problem 1

Suppose that the number of cookbooks sold in a small community is given by

$$q = 100 - 25p$$

where p = the price, in dollars, of the cookbooks

and q = the quantity sold at price p.

Use calculus to find the price that will lead to maximum revenue.

Solution: 1.   What is to be maximized in the problem?

Write the equation, in words first if necessary.

2.   Use the constraints of the problem to write the function in terms of a single variable and simplify.

3.   Find critical points and classify.

Test:

4.   Answer questions posed in problem.

Answer: A price of $2 per cookbook would lead to maximum revenue.

You might have found the following additional information from problem 1, even though it was not asked for. At a price of $2 per cookbook, the maximum revenue would be $100, and 50 cookbooks would be sold.

Examples 4 and 5 will look at maximizing profit.

EXAMPLE 4

Paul is a freelance TV cameraman. To supplement his income between jobs, he started the Spaniel Manufacturing Company, a small firm making sunglasses. The sunglasses are priced at $8 and the company can sell all it produces. Paul has determined that the firm's total cost is given by

$$C(x) = 20 + 2x + 0.01x^2$$

where x = the number of sunglasses

and $C(x)$ = the total cost, in dollars, of producing x sunglasses.

a.   How many sunglasses should the Spaniel Manufacturing Company produce to maximize profit?

b.   What is the maximum profit?

Solution: Before actually solving the problem, I am going to do some preliminary computations again so that I am sure you understand the various bits of information that are given and what we want to eventually find.

$$\text{Total revenue} = \text{selling price} \cdot \text{quantity sold}$$

$$\text{Total cost} \quad = 20 + 2x + 0.01x^2$$

$$\text{Profit} \qquad\quad = \text{total revenue} - \text{total cost}$$

Suppose $x = 10$:

$$\text{Total revenue} = 8(10) \qquad\qquad\qquad\qquad = 80$$

$$\text{Total cost} \quad = 20 + 2(10) + 0.01(100) = 20 + 20 + 1 = 41$$

$$\text{Profit} \qquad = \qquad\qquad\qquad\qquad\qquad = 39$$

The interpretation of the above is that, if Paul produces 10 sunglasses and sells them for $8 each, he will take in as revenue $80. At the same time it will cost him $41 to produce the glasses, leaving the company with a profit of $39.

Suppose $x = 100$:

$$\text{Total revenue} = 8(100) \qquad\qquad\qquad\qquad = 800$$

$$\text{Total cost} \quad = 20 + 2(100) + 0.01(10{,}000) = 20 + 200 + 100 = 320$$

$$\text{Profit} \qquad = \qquad\qquad\qquad\qquad\qquad = 480$$

If Paul decides instead to produce 100 sunglasses, his revenue will be $800. The total cost to produce the 100 pairs will be $320, leaving the company with a profit of $480.

Now that you see how the various pieces of information fit together, we can proceed with solving the problem.

1. What is to be maximized in the problem? Profit

    Write the equation for profit, in words first if necessary.

    $$\text{Profit} = \text{total revenue} \quad - \text{total cost}$$

    $$= (\text{price})(\text{quantity}) - \text{total cost}$$

2. Use the constraints of the problem to write the function in terms of a single variable and simplify.

    Substitute $20 + 2x + 0.01x^2$ for total cost, $8 for price, and x for quantity.

    $$\text{Profit} = (\text{price})(\text{quantity}) - \text{total cost}$$

    $$P(x) = 8x - (20 + 2x + 0.01x^2)$$

    $$= 8x - 20 - 2x - 0.01x^2$$

    $$= -0.01x^2 + 6x - 20$$

3. Find critical points and classify.

    $$P(x) = -0.01x^2 + 6x - 20$$

    $$P'(x) = -0.02x + 6$$

    $$0 = -0.02x + 6$$

$$0.02x = 6$$

$$2x = 600$$

$$x^* = 300$$

If $x^* = 300$,

$$P(300) = -0.01(300)^2 + 6(300) - 20$$

$$= -0.01(90{,}000) + 1{,}800 - 20$$

$$= -900 + 1{,}800 - 20$$

$$= 880$$

Test: $P(x) = -0.01x^2 + 6x - 20$

The function is a quadratic with $a = -0.01$.

The parabolic curve opens down.

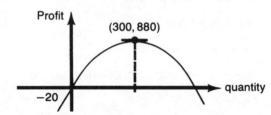

Point (300, 880) is an absolute maximum.

4.   Answer questions posed in problem.

Answers: a.   The Spaniel Manufacturing Company should produce 300 pairs of sunglasses to achieve maximum profit.

b.   Maximum profit is expected to be \$880.

___

Example 4 was okay, but the answer was somewhat predictable. I prefer examples for which the answers are not so obvious.

EXAMPLE 5

During a week, a manufacturer can sell q units of a product at a price of p cents per unit, where $p = 300 - 0.05q$. The total cost of making q units is c cents, where $c = 200 + q$.

a.   How many units should be manufactured per week so as to maximize profit?

b.   What price per unit will maximize profit?

c.   What will the maximum profit be?

___

Solution: This time we will do an even briefer analysis of the information before starting to work the problem.

Suppose $q = 100$;

then $p = 300 - 0.05(100) = 300 - 5 = 295$

and $c = 200 + 100$ $= 300$

The interpretation thus far is that, if 100 units were manufactured, the selling price would be 295 cents or \$2.95 per unit, and the total cost of making the 100 units would be 300 cents or \$3.00.

Continuing with the analysis,

$$\text{Total revenue} = \text{price} \cdot \text{quantity} = 2.95(100) = \$295$$

$$\text{Total cost} \quad = (\text{calculated above}) \qquad = \$3$$

$$\text{Profit} \qquad = \text{total revenue} - \text{total cost} \quad = \$292$$

In this problem we are being asked to determine the quantity to produce so as to maximize profit. The analysis above merely showed the profit if 100 units were produced.

1. What is to be maximized in the problem? Profit.

    Write the equation for profit, in words first if necessary.

$$\text{Profit} = \text{total revenue} \qquad - \text{total cost}$$

$$= (\text{price})(\text{quantity}) - \text{total cost}$$

2. Use the constraints of the problem to write the function in terms of a single variable and simplify.

    Substitute $200 + q$ for total cost, $300 - 0.05q$ for price, and q for quantity.

$$\text{Profit} = (\text{price})(\text{quantity}) - \text{total cost}$$

$$P(q) = (300 - 0.5q)q - (200 + q)$$

$$= 300q - 0.5q^2 - 200 - q$$

$$= -0.5q^2 + 299q - 200$$

3. Find the critical points and classify.

$$P(q) = -0.5q^2 + 299q - 200$$

$$P'(q) = -1.0q + 299$$

$$0 = -q + 299$$

$$q^* = 299$$

$$\text{If } q^* = 299,$$

$$P(299) = -0.5(299)^2 + 299(299) - 200$$

$$= -0.5(89{,}401) + 89{,}401 - 200$$

$$= -44{,}700.5 + 89{,}401 - 200$$

$$= 44{,}500.5 \text{ cents or } \$445.005$$

Test: $P(q) = -0.5q^2 + 299q - 200$

The function is a quadratic with $a = -0.5$.

The parabolic curve opens down.

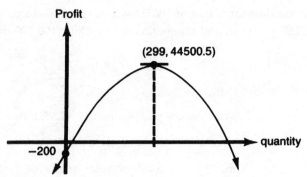

Point (299, 44500.5) is an absolute maximum.

4. Answer questions posed in problem.

Answers: a. The company should produce 299 units per week so as to maximum profit.

b. Price $= p = 300 - 0.5q$

$= 300 - 0.5(299)$

$= 300 - 149.5$

$= 150.5$

To maximize profit, the manufacturer should sell the product at a price of 150.5 cents or $1.505 per unit.

c. Maximum profit is expected to be $445.005.

Ready to try a problem similar to Example 5?

Problem 2

The total cost, in dollars, to produce q units is given by $C = 3q^2 + 15q + 45$. The company sells all it produces at $87 per unit.

a. How many units should be produced so as to maximize profit?

b. What is the maximum profit?

Solution: 1. What is to be maximized in the problem? Profit

Write the equation, in words first if necessary.

Profit = total revenue   − total cost

= (price)(quantity) − total cost

2. Use the constraints of the problem to write the equation in terms of a single variable and simplify.

3. Find the critical points and classify.

    Test:

4. Answer questions posed in problem.

Answers: a.   The company should produce 12 units to achieve maximum profit.

b.   Maximum profit is expected to be $387.

MARGINAL ANALYSIS REVISITED

Recall from Unit 11 that in economics, the term **marginal analysis** is used to denote the approximation procedure of using the derivative of a function to estimate the change in the function produced by a one-unit increase in the size of the variable.

First we will do a review example followed by an example using marginal analysis as an alternative method for maximizing profit. Recall the following definitions:

Marginal cost $= MC = C'(x)$, where $C(x) =$ total cost of producing x units, and

Marginal revenue $= MR = R'(x)$, where $R(x) =$ total revenue from selling x units.

EXAMPLE 6

Susan Garriques, a talented artist, lives on a boat at OceanSide Marina in Key West. To earn a bit of extra money for the holidays, she is planning to sell small watercolor paintings of various boats in the marina. Calculating her expenses (paint, paper, etc.) she determines that

$$C(x) = 40 + 5x + 0.25x^2$$

where $x =$ the number of paintings produced

and $C(x) =$ the total cost, in dollars, of producing x paintings.

Further, she believes the demand function for the paintings, that is, the relationship between price and quantity, is given by

$$p = 60 - x$$

where $x =$ the number of paintings sold

and $p =$ price in dollars.

a.   Determine the revenue function, $R(x)$.

b.   Determine the profit function, $P(x)$.

c.   Find the equation for marginal cost.

d.   Find $C'(40)$ and interpret.

e. Find the equation for marginal revenue.

f. Find $R'(40)$ and interpret.

Solution: a. Total revenue = (selling price per unit)(number of units sold)

$$R(x) = p \cdot x$$

$$= (60 - x)x$$

$$= 60x - x^2$$

b. Profit = total revenue − total cost

$$P(x) = R(x) - C(x)$$

$$= (60x - x^2) - (40 + 5x + 0.25x^2)$$

$$= 60x - x^2 - 40 - 5x - 0.25x^2$$

$$= -1.25x^2 + 55x - 40$$

c. By definition, marginal cost equals the derivative of cost.

$$C(x) = 40 + 5x + 0.25x^2$$

$$MC = C'(x) = 5 + 0.5x$$

d. $MC = C'(40) = 5 + 0.5(40) = 5 + 20 = 25$

At a production level $x = 40$, one more painting costs approximately $25 to produce. Or, the marginal cost of the 41st painting is $25.

e. By definition, marginal revenue equals the derivative of revenue.

$$R(x) = 60x - x^2$$

$$MR = R'(x) = 60 - 2x$$

f. $MR = R'(40) = 60 - 2(40) = 60 - 80 = -20$

At a production level $x = 40$, approximately $20 less in sales revenue would be derived from selling one more painting. Or, the marginal revenue of the 41st painting is −$20. In this instance total revenue actually decreases with the sale of an additional painting. This happens, for example, when a market becomes "flooded" with a product so that to sell additional units requires that the price be lowered far enough to more than offset the effect of additional sales.

For the moment we are going to leave Example 6 and look at maximizing profit in a more general sense by using the technique for optimization as introduced in this unit.

1. What is to be maximized? Profit.

Write the equation for profit, in words first if necessary.

$$\text{Profit} = \text{total revenue} - \text{total cost}$$

2. Use the constraints of the problem to write the function in terms of a single variable and simplify.

$$P(x) = R(x) - C(x)$$

3. Find the first derivative, set it equal to zero, and solve.

$$P(x) = R(x) - C(x)$$

$$P'(x) = R'(x) - C'(x)$$

$$0 = R'(x) - C'(x)$$

$$C'(x) = R'(x)$$

$$MC = MR$$

Believe it or not, we have just proven a theorem from economics, namely, that profit is maximized when marginal revenue equals marginal cost. The theorem provides a second, although very similar, method for maximizing profit when total cost and total revenue functions are given.

THEOREM: Profit is maximized when marginal revenue equals marginal cost.

EXAMPLE 7

Use the information provided in Example 7 to answer the following:

a.   Use marginal analysis to determine the number of paintings for which profit is maximized.

b.   What will the maximum profit be?

c.   What price should Susan charge for her paintings so as to maximize profit?

Solution: a.   Profit is maximized when marginal revenue equals marginal cost.

From Example 6, $MR = 60 - 2x$ and $MC = 5 + 0.5x$

$$MR = MC$$

$$60 - 2x = 5 + 0.5x$$

$$55 = 2.5x$$

$$x = 22$$

To maximize her profits, Susan should plan to produce and sell 22 paintings.

b. $$P(x) = -1.25x^2 + 55x - 40$$

$$P(22) = -1.25(22)^2 + 55(22) - 40$$

$$= -605 + 1210 - 40$$

$$= 565$$

Maximum profit is expected to be $565.

c.   From Example 6, the demand function was given where p was price in dollars:

$$p = 60 - x$$

$$= 60 - 22$$

$$= 38$$

To maximize profit, Susan should sell her paintings for $38.

I hope you saw the similarity in part a to solving the problem by using marginal revenue and marginal cost versus solving the problem by taking the derivative of the profit function. If you did not, I suggest you work the next problem and compare the methods.

Problem 3

Given: $P(x) = -1.25x^2 + 55x - 40$.

Take the derivative, set it equal to zero, and solve for x.

Solution:

Answer: $x = 22$

Next is a review problem for you.

Problem 4

Suppose the total cost, in dollars, of manufacturing x items is given by the function $C(x) = 3x^2 + x + 48$. Find $C(2)$, $C(5)$, $C(10)$, $C(0)$ and interpret.

Solution:

Answers: The total cost of manufacturing 2 items is $62.

The total cost of manufacturing 5 items is $128.

The total cost of manufacturing 10 items is $358.

Fixed costs are $48.

MINIMIZING AVERAGE COSTS

In Problem 4 total costs for manufacturing various numbers of units were computed. It is not possible to minimize total cost because costs continue to increase when more and more units are manufactured. However, it is possible to minimize the average cost per unit, as the next example shows. We will use the same total cost function as in Problem 4.

EXAMPLE 8

The total cost, in dollars, of manufacturing x items is given by the function $C(x) = 3x^2 + x + 48$.

a. For what value of x is the average cost per unit a minimum?

b. What is the minimum average cost?

Solution: Again let us first analyze what we are attempting to find.

Suppose $x = 2$.

From Problem 4, the total cost of producing 2 units was found to be $62. Therefore the average cost for the 2 units is $62/2 = \$31$ per unit.

Suppose $x = 5$.

The total cost of producing 5 units was found to be $128. Now the average cost becomes $128/5 = \$25.60$ per unit.

Suppose $x = 10$.

The total cost of producing 10 units was found to be $358 and the average cost per unit becomes $358/10 = \$35.80$ per unit.

The question being asked in this problem is, How many units should be produced so that the average cost per unit is lowest?

1. What is to be minimized in the problem? Average cost

   Write the equation for average cost, in words first.

   $$\text{Average cost} = \frac{\text{total cost}}{\text{quantity}}$$

2. Use the constraints of the problem to write the equation in terms of a single variable and simplify.

   Substitute $3x^2 + x + 48$ for total cost and x for quantity.

   $$A(x) = \frac{3x^2 + x + 48}{x}$$

3. Find critical points and classify.

   $$A(x) = \frac{3x^2 + x + 48}{x}$$

   $$A'(x) = \frac{x \dfrac{d(3x^2 + x + 48)}{dx} - (3x^2 + x + 48)\dfrac{d(x)}{dx}}{x^2}$$

   $$= \frac{x(6x + 1) - (3x^2 + x + 48)(1)}{x^2}$$

   $$= \frac{6x^2 + x - 3x^2 - x - 48}{x^2}$$

   $$= \frac{3x^2 - 48}{x^2}$$

$$0 = \frac{3x^2 - 48}{x^2}$$

$$0 = 3x^2 - 48$$

$$3x^2 = 48$$

$$x^2 = 16$$

$$x^* = 4 \qquad \cancel{x^* = -4}$$

(–4 is not within the domain and should not be considered further.)

If $x^* = 4$,

$$A(4) = \frac{3(4)^2 + 4 + 48}{4}$$

$$= \frac{48 + 4 + 48}{4}$$

$$= 25$$

Test: First-derivative test with $A'(x) = \frac{3x^2 - 48}{x^2}$

For point (4, 25)		the curve:
On the left:	let $x = 3$, $A'(3) = -$	is decreasing
At point (4, 25):		levels off
On the right:	let $x = 5$, $A'(5) = +$	is increasing

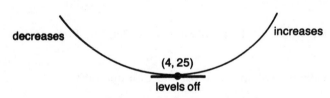

decreases   increases

(4, 25)

levels off

Point (4, 25) is a relative minimum.

4.   Answer questions posed in problem.

Answers: a.   The average cost per unit is minimized if 4 units are produced.

b.   At this level the average cost is $25 per unit.

Thus far we have looked at maximizing total revenue, maximizing profit, and minimizing average cost per unit. We will consider a greater variety of application problems in the next unit. By now you should be familiar with the general technique for optimization problems and be able to apply it to revenue, profit, and average cost functions.

Briefly stated, the general technique is as follows:

1.   Decide what quantity is to be maximized or minimized, and write the equation for it.

2.   Use the constraints of the problem to write the equation in terms of a single variable.

3. Find all critical points and classify.

4. Answer the question posed in the problem.

Before beginning the next unit, you should do the following exercises. Do not get discouraged if you are unable to solve all of them. Remember that complete solutions for all even-numbered problems are provided at the back of the book if you need help.

EXERCISES

1. The profit, in dollars, when x units are produced is given by

$$P(x) = 1{,}850x - 0.05x^2 - 123{,}000.$$

   How many units should be produced to maximize profit?

2. $$P(x) = 50x - 0.01x^2 - 1{,}000 \quad x \geq 0$$

   is the profit, in dollars, when x units are produced. How many units should be produced to maximize profit? What would the maximum profit be?

3. Given the following revenue and cost functions:

$$R(x) = 5{,}000x - 0.1x^2 \quad \text{and} \quad C(x) = 80{,}000 + 200x,$$

   find the number of items, x, that produces a maximum profit. Both functions are expressed in dollars per unit of $x \geq 0$. What would the maximum profit be?

4. Let the demand function for a particular product be

$$q = f(p) = 1{,}152 - 5p$$

   where q = the quantity sold at price p

   and p = the price, in dollars.

   a. What price will result in maximum total revenue?

   b. What is the maximum revenue that can be expected?

   c. How many units will be sold when revenue is maximized?

   d. Sketch the revenue function.

   e. Determine the restricted domain for the revenue function.

5. The demand function for a product is

$$p = 10 - 0.25q$$

   where p = the price, in dollars,

   and q = the quantity sold at price p.

   a. Determine the total revenue function.

   b. Find the value of q that results in maximum revenue.

   c. What is the maximum revenue that can be expected?

   d. When revenue is maximized, how many units will be sold?

   e. When revenue is maximized, what will be the price per unit?

6. A manufacturing company can sell all it produces of a certain product at a price of $50. The company's total cost is given by

$$C(x) = 110 - 2.5x^2 + x^3 \quad x \geq 0$$

where $x =$ the number of units produced

and $C(x) =$ the total cost, in dollars, of producing x units.

How many units should the company produce to maximize profit? What is the maximum profit? Verify that marginal revenue equals marginal cost when profit is maximized.

7. The demand for a commodity is $p = 200 - q$. The cost of producing the commodity consists of a fixed cost of $5,000 and a variable cost of $0.50 per unit. Determine the profit function for the commodity and the quantity of output that leads to maximum profit.

8. The Far Western Manufacturing Company is engaged in the business of manufacturing souvenirs that are sold in various shops throughout the western states. The company has found that its total cost, in dollars, for manufacturing x souvenirs is given by

$$C(x) = 0.01x^2 + 0.2x + 40$$

and the total revenue, in dollars, for selling x souvenirs is given by

$$R(x) = 20x.$$

Determine each of the following:

a. The number of units for which profit is maximized

b. The maximum profit

c. The unit price at which profit is maximized

d. The total cost at the production level that maximizes profit

e. The average cost per unit at the production level that maximizes profit

9. John Whitaker, my piano teacher, is the founder of the Southern Park Music School. Recently he decided to do a survey of the school's students. From past years he knew that the total cost of surveying x students is given by

$$C(x) = 20(x - 100)^2 + 1,000 \quad x \geq 90.$$

a. How many students should be surveyed in order to minimize the total cost of the survey? Is this number a relative minimum? Is this number an absolute minimum?

b. Find the minimum total cost of the survey.

10. The total cost, in dollars, of manufacturing x items is given by the function

$$C(x) = x^2 + 4x + 16 \quad x \geq 0.$$

a. Determine $C(10)$ and interpret.

b. Find the average cost per unit if 10 units are produced.

c. For what value of x is the average cost per unit a minimum?

d. What is the minimum average cost?

11. A product has a total cost function given by

$$C(x) = 0.125x^2 + 20{,}000$$

where x is the number of units produced.

Find the value of x at which the average cost per unit is a minimum. What is the minimum average cost per unit?

12. Pen, the leading stylist at the top beauty salon in the city, is paid primarily on commission. His most profitable service is a haircut because it does not require a great deal of time and he is allowed to set his own price within reason. Pen, being a bright person, knows that his profit is a function of the price he charges. In fact, he has determined that his profit per day on haircuts is

$$y = P(x) = 25x - x^2 - 100$$

where $P(x) = $ the profit, in dollars,

and $x = $ the price, in dollars, for a haircut.

The owner, however, has stipulated that Pen may not charge more than $14 nor less than $6 for a haircut. What price should Pen charge so as to maximize his profit and still stay within the owner's limitations?

13. The ABC Bus Tour Company conducts tours of Boston. At a price of $50 per person, ABC experienced an average demand of 1,000 customers per week. After a price reduction to $40 per person, the average demand increased to 1,200 customers per week. Assuming a linear demand function, determine the total revenue function. Also, find the tour fare and the number of customers that result in maximum revenue, and determine the total maximum revenue.

14. The profit gained from producing and selling x units of some product is given by

$$P(x) = 3x - \frac{x^3}{1{,}000{,}000} \qquad 0 \le x \le 1{,}500.$$

a. How many units should be produced and sold in order to maximize profit?

b. What is the maximum profit?

If additional practice is needed:

Barnett, Ziegler, and Byleen, pages 300–301, problems 37–40
Harshbarger and Reynolds, pages 773–777, problems 5–44
Waner and Costenoble, pages 278–279, problems 1–24

UNIT 20

Optimization Problems (Continued)

In Unit 19 we looked at maximizing total revenue, maximizing profit, and minimizing average cost per unit. Actually there are a whole host of optimization problems covering such topics as ordering costs, inventory expenses, material requirements, and surfaces.

In this unit we will consider a variety of application problems, each of which again has the objective of maximizing or minimizing a particular quantity.

"FENCE" PROBLEMS

We will start with what I call fence problems for lack of a better title. I like fence problems because they cover the concept of optimization, they can be easily visualized by almost everyone, and best of all they require only two familiar formulas from geometry: the formulas for the area and the perimeter of a rectangle.

> **Area of a rectangle** = (length)(width) = lw

> **Perimeter of a rectangle** = 2(length) + 2(width) = $2l + 2w$

Remember that the area of a rectangle is the number of square units contained within the rectangle, whereas the perimeter is the distance measured around the outside.

Keep the two formulas in mind and the basic technique stated at the beginning of Unit 19. We are about to tackle some fence problems.

EXAMPLE 1

Ben, a farmer with a field adjacent to a straight river, wishes to fence a rectangular region for grazing. He has available 1,600 feet of fencing with none needed along the river. Being a bit tight with his money, Ben does not want to buy any more fencing, but he does want to design the region so that he achieves the maximum possible area.

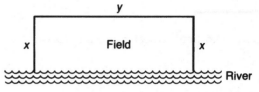

What should be the dimensions of the field in order that it have maximum area?

Solution: Again let's look at some of the possibilities before starting right in with the solution. Ben has 1,600 feet of fencing and is going to fence a rectangular region on three sides. Below are some extreme possibilities.

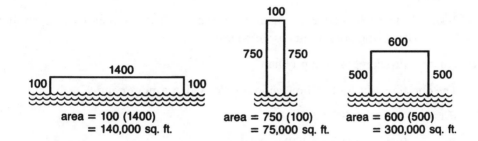

The problem asks us to determine the dimensions so that the field will have maximum area, given the constraint of 1,600 feet of fencing for the three sides.

1. What is to be minimized in the problem? Area of rectangle

    Write the equation for area of rectangle, in words first.

    $$\text{Area} = (\text{length})(\text{width})$$

    $$= y \cdot x$$

2. Use the constraints of the problem to write the equation in terms of a single variable and simplify.

    What are the constraints of this problem?

    There is only 1,600 feet of fencing and it must go around three sides, or, in symbols:

    $$x + y + x = 1,600$$

    $$y + 2x = 1,600$$

    You have a choice; you may solve the above equation for either x or y. Which variable is easier to solve for? I would have to say y.

    $$y = 1,600 - 2x$$

Now use substitution to rewrite the equation in step 1 in terms of x and simplify.

$$\text{Area} = y \cdot x$$

$$A(x) = (1{,}600 - 2x)x$$

$$= 1{,}600x - 2x^2$$

$$= -2x^2 + 1{,}600x$$

3. Find critical points and classify.

$$A(x) = -2x^2 + 1{,}600x$$

$$A'(x) = -4x + 1{,}600$$

$$0 = -4x + 1{,}600$$

$$4x = 1{,}600$$

$$x^* = 400$$

Test: $A(x) = -2x^2 + 1{,}600x$ is a quadratic with $a = -2$. The parabolic curve opens downward. The critical point is an absolute maximum.

4. Answer question posed in problem.

The question asks for the dimensions of the field, meaning x and y.

If $x = 400$ and $y = 1{,}600 - 2x$

then $y = 1{,}600 - 2(400)$

$= 800$

Answer: To maximize the area, given 1,600 feet of fencing for three sides, Ben should lay out the field to be 400×800 feet with the longest side parallel to the river as shown in the sketch. The area of the field will be 320,000 square feet.

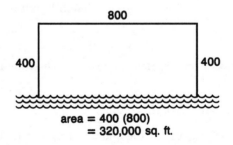

The next example is probably the most difficult we have attempted thus far.

EXAMPLE 2

A rectangular playground must contain 1,050 square feet and is to be enclosed by a fence, then divided down the middle by another piece of fence. A sketch is provided on the next page. The cost of the interior fencing for the middle section is $6.164 per foot. The exterior fencing, which must be heavier, costs $7.00 per foot.

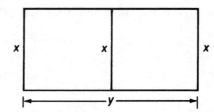

a. How should the playground be designed so as to minimize the cost of the fencing?

b. What is the minimum cost?

Solution: Let's consider some of the possibilities again so that we are sure we understand the problem. The rectangular area is to be 1,050 square feet with fencing as described above.

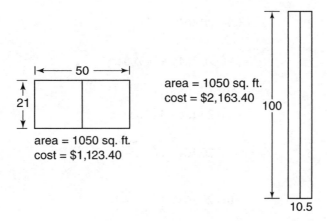

The problem asks us to determine the dimensions so that the playground's area will be 1,050 square feet, while at the same time the cost of the fencing will be minimized.

1. What is to be minimized in the problem? Cost of fencing.

Write the equation for the cost of fencing, in words first.

Total cost = cost of interior fence + cost of exterior fence

$$= 6.164 \left(\begin{array}{c} \text{length of interior} \\ \text{fencing required} \end{array} \right) + 7.00 \left(\begin{array}{c} \text{length of exterior} \\ \text{fencing required} \end{array} \right)$$

$$= 6.164x + 7.00(2x + 2y)$$

$$= 6.164x + 14x + 14y$$

$$= 20.164x + 14y$$

2. Use the constraints of the problem to write the equation in terms of a single variable and simplify.

What are the constraints of this problem?

The area of the rectangle must be 1,050 square feet, or, in symbols:

$$xy = 1,050$$

Again you have a choice; you may solve the above equation for either x or y. Which variable is easier to solve for it? It doesn't really matter, but I will solve for y.

$$y = \frac{1,050}{x}$$

Now use substitution to rewrite the equation in step 1 in terms of x and simplify.

$$\text{Total cost} = 20.164x + 14y$$

$$= 20.164x + 14\left(\frac{1,050}{x}\right)$$

$$= 20.164x + \left(\frac{14,700}{x}\right)$$

3. Find critical points and classify.

$$C(x) = 20.164x + \left(\frac{14,700}{x}\right)$$

$$= 20.164x + 14,700x^{-1}$$

$$C'(x) = 20.164 + 14,700(-1x^{-1-1})$$

$$= 20.164 - 14,700x^{-2}$$

$$= 20.164 - \left(\frac{14,700}{x^2}\right)$$

$$0 = 20.164 - \left(\frac{14,700}{x^2}\right)$$

$$0 = 20.164x^2 - 14,700$$

$$20.164x^2 = 14,700$$

$$x^2 = 729$$

$$x^* = 27 \qquad \cancel{x^* = -27}$$

(−27 is not within the domain and should not be considered further.)

Test: Second-derivative test with $C''(x) = \dfrac{29,400}{x^3}$

At $x = 27$, $C''(27) = +$, concave up, minimum point

4. Answer questions posed in problem.

The questions ask for the dimensions of the playground, meaning x and y, that minimize cost.

$$\text{If } x = 27 \text{ and } y = 1,050/x,$$

$$\text{then } y = 1,050/27$$

$$= 38.9$$

Answers: a.   To minimize the cost for the fencing, the area should be 27×38.9 feet with the fence in the middle being parallel to the 27-foot side, as shown in the sketch.

$$\text{Total cost} = 20.164x + 14y$$

$$= 20.164(27) + 14(38.9)$$

$$= 544.428 + 544.6$$

$$= 1{,}089.03$$

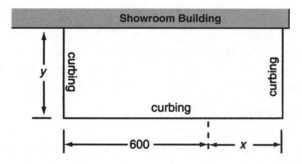

b.   The minimum cost for the fencing would be $1,089.03.

We will do one more example of this type and then move on. This one I will start and you will finish.

EXAMPLE 3

The manager of a car dealership wants to enclose a rectangular area in front of the showroom. There must be at least 600 feet of frontage along the street side. Find the dimensions that will maximize the enclosed area if the manager has 2,000 feet of curbing.

Solution: 1.   What is to be maximized in the problem? Area

Write the equation for area in words first.

$$\text{Area} = (\text{length})(\text{width})$$

Notice that in this example we have something new. The length must be at least 600 feet. To incorporate that condition into the problem, the length is described as $600 + x$. Thus the area becomes

$$\text{Area} = (600 + x)y$$

2.   Use the constraints to write in terms of a single variable, and simplify.

What are the constraints?

There is only 2,000 feet of curbing and it must go around three sides, or, in symbols:

$$y + 600 + x + y = 2{,}000$$

$$x + 2y = 1{,}400$$

$$x = 1{,}400 - 2y$$

Now use substitution to rewrite equation in step 1 in terms of y and simplify.

From here on you can finish step 2 yourself.

3. Find critical points and classify.

4. Answer question posed in problem.

Answer: To maximize the area, the manager should use the 2,000 feet of curbing for three sides by setting the dimensions to be $500 \times 1,000$ feet, with the longest side parallel to the building.

OTHER TYPES OF OPTIMIZATION PROBLEMS

It's time to leave fence problems and look at other types of optimization problems.

EXAMPLE 4

An apartment building has 200 apartments. If the rent is $750 per month, the rental agent is able to rent all 200 apartments. However, the owner knows from past experience that, for each additional $25 increase in monthly rent, 2 additional apartments will become vacant. The rental agent receives a bonus when the building is fully occupied and is urging that the rent remain at $750 per month.

How much would you advise the owner to charge for rent?

Solution: The owner's objective should be to determine how much rent to charge so as to maximize the total monthly rental income.

Again let us first analyze what we are attempting to find.

If rent is $750 per month, total monthly income will be

$$150,000 = (750 \quad)(200 \quad)$$

If rent is increased by $25, the owner will rent 2 fewer apartments, and total monthly income will be

$$153,450 = (750 + 25)(200 - 2)$$

If rent is increased by a second $25, occupancy will decrease by a second 2 apartments, and total monthly income will be

$$156,800 = [750 + 25(2)][200 - 2(2)]$$

If a third rent increase takes place, total monthly income will be

$$160,050 = [750 + 25(3)][200 - 2(3)]$$

Is the pattern apparent? I hope so. Therefore, if x rent increases take place, total monthly income will be

$$(750 + 25x)(200 - 2x)$$

1. What is to be maximized? Total monthly income.

Write the equation.

$$M(x) = (750 + 25x)(200 - 2x)$$

2. Simplify.

$$M(x) = 150{,}000 + 3{,}500x - 50x^2$$

3. Find critical points and classify.

$$M'(x) = 3{,}500 - 100x$$

$$0 = 3{,}500 - 100x$$

$$100x = 3{,}500$$

$$x^* = 35$$

$M(x)$ is a quadratic opening down, which has an absolute maximum.

4. Answer question posed in problem.

Rent owner should charge $= \$750 + \$25(35)$

$$= \$750 + \$875$$

$$= \$1{,}625$$

Answer: At a monthly rent of \$1,625, a total of 130 apartments would be rented $[200 - 2(35) = 130]$, and monthly rental income would be maximized at \$211,250.

We will look at one more example like Example 4.

EXAMPLE 5

An orange grower knows from past experience that, if she harvests the oranges now, each tree will yield, on average, 130 pounds, and she can sell the oranges for \$0.73 per pound. However, for each additional week that she waits before harvesting, the yield per tree will increase by 5 pounds, while the price per pound will decrease by 2¢.

How many weeks should the grower wait before harvesting the oranges in order to maximize the sales revenue per tree?

Solution: Let's consider some of the possibilities.

$$\text{Price} \cdot \text{quantity} = \text{Revenue}$$

If harvest now, $(0.73)(130)$ $= 94.90$ sales revenue per tree

If wait 1 week, $(0.73 - 0.02)(130 + 5)$ $= 95.85$ sales revenue per tree

If wait 2 weeks, $[0.73 - 0.02(2)][130 + 5(2)] = 96.60$ sales revenue per tree

If wait 3 weeks, $[0.73 - 0.02(3)][130 + 5(3)] = 97.15$ sales revenue per tree

If wait x weeks, $(0.73 - 0.02x)(130 + 5x)$ $=$ sales revenue per tree

With the pattern established, we can now go on to solve the problem.

1. $$S(x) = (0.73 - 0.02x)(130 + 5x)$$

2. $$= 94.9 - 2.6x + 3.65x - 0.1x^2$$

$$= 94.9 + 1.05x - 0.1x^2$$

3. $S'(x) = 1.05 - 0.2x$

$$0 = 1.05 - 0.2x$$

$$0.2x = 1.05$$

$$x^* = 5.25$$

$S(x)$ is a quadratic opening down, so the critical point is an absolute maximum.

Answer: The owner should wait $5\frac{1}{4}$ weeks before harvesting the oranges in order to maximize her sales revenue per tree.

Our final example will deal with maximizing volume.

EXAMPLE 6

A manufacturer needs to package a product in a closed rectangular box. The sides of the box cost $4.00 per square foot, and the base and the top (which are squares) cost $8.00 per square foot. The manufacturer has $96 to spend on construction materials.

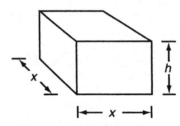

Find the dimensions of the box with the greatest volume that can be constructed for $96.

Solution: 1.   Volume = (length)(width)(height)

$$= x \cdot x \cdot h$$

$$= x^2 h$$

In this example the base and the top are squares, so the length and the width are equal.

       2.   Constraints given:

Cost of top = 8(area of top)

$$= 8(x^2)$$

$$= 8x^2$$

Cost of bottom = 8(area of bottom)

$$= 8x^2$$

Cost of one side = 4(area of side)

$$= 4(x \cdot h)$$

$$= 4xh$$

Manufacturer has $96 for materials

$96 = cost of top + cost of bottom + cost of four sides

$$= 8x^2 + 8x^2 + 4(4xh)$$

$$= 8x^2 + 8x^2 + 16xh$$

$$= 16x^2 + 16xh$$

In this example I would certainly solve for h.

$$-16xh = 16x^2 - 96$$

$$h = -x + \frac{6}{x}$$

Now use substitution to rewrite the equation in step 1 in terms of x, and simplify.

$$\text{Volume} = x^2h$$

$$V(x) = x^2(-x + \frac{6}{x})$$

$$= -x^3 + 6x$$

3. Find critical points and classify.

$$V'(x) = -3x^2 + 6$$

$$0 = -3x^2 + 6$$

$$0 = x^2 - 2$$

$$0 = (x + \sqrt{2})(x - \sqrt{2})$$

(About now I wish I hadn't started this example, but it's too late to quit.)

$V(x)$ is a cubic, with $a = -1$ and opens. $x^* = \sqrt{2}$ would be where the relative maximum occurs.

4. Answer question posed in the problem.

The question asks for the dimensions of the box, meaning x and h.

If $x = -\sqrt{2}$ and $h = -x + \frac{6}{x}$,

$$\text{then } h = -\sqrt{2} + \frac{6}{\sqrt{2}}$$

$$= -1.414 + \frac{6}{1.414}$$

$$= -1.414 + 4.243$$

$$= 2.829$$

Answer: To maximize volume, given the constraint of $96 for materials, the dimensions of the box
           should be $1.4' \times 1.4' \times 2.8'$.

We've had enough examples. What is needed now is for you to try some problems on your own. You should now be familiar with the general technique for optimization problems.

Before beginning the next unit, you should do the following exercises. I have provided a variety of applications. If you need help, remember that complete solutions to the even-numbered problems are provided at the back of the book as well as the answers to the odd-numbered problems.

EXERCISES

1. A small company has determined that its profit, P, depends on the amount of money spent on advertising, x. This relationship is given by the equation

$$P(x) = \frac{8x}{x^2 + 16} + 11,$$

where P and x are measured in thousands of dollars. What amount should the company spend on advertising to ensure maximum profit?

2. A machine shop has determined that its cost is a function of the number of machines in operation and is given by

$$C(x) = 20x + \frac{1,280}{x}$$

where x = the number of machines in operation

and $C(x)$ = the total cost, in dollars.

How many machines should the company operate so as to minimize its cost?

3. A retailer has determined that the cost C for ordering and storing x units of a product is

$$C(x) = 5x + \frac{50,000}{x} \qquad 1 \le x \le 125$$

Find the order size that will minimize the cost if the delivery truck can transport a maximum of 125 units per order.

4. A rectangle is to be constructed so that the perimeter will be 60 feet. What is the maximum possible area for the rectangle?

5. A farmer has 1,000 feet of fence and wishes to enclose a rectangular plot of land. The land borders a river, and no fence is required on that side. What should the length of the side parallel to the river be in order to enclose the largest possible area?

6. Richard McHenry has 800 feet of new fencing. He wishes to fence a rectangular plot of ground and to divide it into three pens by placing two fences parallel to one side of the rectangle. Determine the largest total area he can enclose. (See the accompanying drawing.)

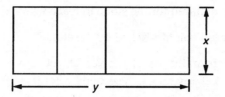

7. A farmer wishes to enclose an area of 20,000 square feet into three pens. Fencing is required on all four sides, as well as two sections in the interior, as illustrated below. Find the dimensions, x and y, that minimize the length of fence required.

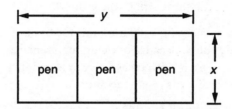

8. A park manager has $100 to buy fencing to enclose a rectangular garden along the side of a building. No fence is needed along the building. The fence along the ends costs $1.25 per foot, and the fence along the front costs $2 per foot. Use calculus to determine the dimensions, x and y, that will maximize the enclosed area. What will the maximum area be?

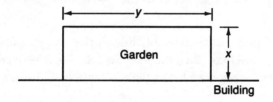

9. From economics, the formula for the cost of inventory is as follows:

$$C(x) = \left(\frac{Q}{x}\right)c + \left(\frac{x}{2}\right)s$$

where Q = the number of units needed per year and sales occur at a constant rate,

  c = the cost of placing an order,

  s = the cost of storing one unit for 1 year,

and x = the number of units per order.

Determine the order size, x, that will minimize cost.

10. A fence must be built around a rectangular area of 1,600 square feet. The fence along three sides is to be made of a material that costs $12 per foot. The material for the front side costs $6 per foot. Find the dimensions (rounded to the nearest foot) of the rectangle that would be the least expensive to build, provided that no side is less than 10 feet long.

11. A stone is thrown vertically upward from the top of a tower 400 feet high. The height h, in feet, of the stone from the ground at any time t, in seconds, after it is thrown until it strikes the ground is given by

$$h = 400 + 120t - 16t^2.$$

a.   How much time will be required for the stone to reach its maximum height?

b.   What will be the maximum height reached by the stone?

12.   A stone is thrown vertically from the top of a tower 80 feet above ground level. The height h, in feet, of the stone from the ground at any time t, in seconds, after it is thrown until it strikes the ground is

$$h = -16t^2 + 64t + 80.$$

a.   How much time will be required for the stone to reach its maximum height?

b.   What is the maximum height reached by the stone?

c.   Determine the time when the stone strikes the ground.

13.   A new apartment complex is charging $400 a month for each of its 200 two-bedroom apartments, and the complex is currently filled. Because of rising maintenance costs, the manager needs to raise the rent, but he knows that, for each $20 increase, five apartments will become vacant. What monthly rent should he charge to maximize his rental income?

14.   A tour agency is booking a tour and has 100 people signed up. The price of a ticket is $2,000 per person. The agency has booked a plane seating 150 people at a cost of $125,000. Additional costs to the agency are incidental fees of $500 per person. For each $5 that the price is lowered, a new person will sign up. By how much should the price be lowered to maximize the profit for the tour agency?

15.   Suppose the tour agency described in Exercise 14 is able to obtain a booking for a plane that will seat 225 people at the same price for the charter as the first airline. Answer the same question for these conditions.

16.   An electronics firm has found that an assembly line worker can produce an average of 600 units per week when there are no more than 12 workers on the line. Because of crowded conditions and other factors, for each additional worker added to the assembly line, the output per worker decreases by 30 units per week.

a.   How many workers should be on the line to maximize the number of units produced per week?

b.   What is the maximum number of units that can be produced?

17.   The Great Sound Company is ready to introduce its new stereo. The price equation for the new product is estimated to be

$$p = 1,600 - 4x$$

where p = the price, in dollars,

and x = the number of stereos produced.

The cost function, also in dollars, is known to be

$$C(x) = 40x + 2,000.$$

a.   Find the maximum profit, and determine the price that should be charged to make the maximum profit.

b.   Suppose that a tax of $8 per stereo is imposed on the product. This new tax modifies the cost equation so that

$$C(x) = \text{original cost} + \text{tax}$$

$$= 40x + 2,000 + 8x$$

Find the new maximum profit, and determine the price that should be charged to make the maximum profit.

c.   How has the tax changed the price that the Great Sound Company charges for the product?

d.   What portion of the tax is passed on to the consumer?

18.   A sheet of tin, 10 by 20 inches, has equal squares cut from the corners, and the edges are turned up to form a box with open top. If x is the length of the side of the squares removed, then the volume of the box is

$$V(x) = x(20 - 2x)(10 - 2x).$$

a.   Determine the volume of the largest box that can be so formed.

b.   What should be the dimensions of the squares cut from each corner if the volume is to be maximized?

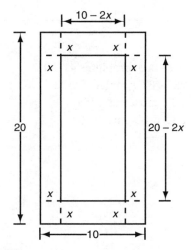

19.   A fast-food service supplier is constructing open boxes for takeout orders. The boxes are to be constructed from rectangular pieces of cardboard by cutting a square from each corner and bending up the sides. If the pieces of cardboard are 12 by 18 inches, what size squares should be cut from each corner to maximize the volume of the boxes? See the accompanying figure.

Hint: See the solution to Exercise 18 at the back of the book.

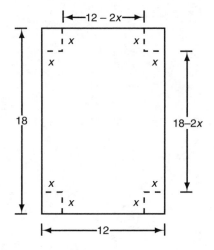

20. A company needs to build a small warehouse, half of which will be a refrigeration area. The floor plan of the warehouse is shown in the accompanying figure. The walls will all be 8 feet high. The four interior walls for the refrigerated area are to be covered with material that costs $6 per square foot, and the four interior walls for the remaining part are to be covered with material that costs $2 per square foot. If the total floor space needed is 3,200 square feet, what dimensions should the refrigerated area have to minimize the total cost?

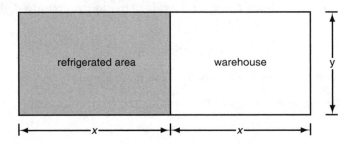

If additional practice is needed:

Barnett, Ziegler, and Byleen, pages 300–303, problems 41–60
Harshbarger and Reynolds, pages 785–787, problems 1–32
Hoffmann and Bradley, pages 274–283, problems 1–53
Waner and Costenoble, pages 279–282, problems 25–60

UNIT 21

Implicit Differentiation

In this unit implicit differentiation will be explained. When you have completed the unit, you will be able to differentiate an implicit equation without first having to rewrite the equation explicitly.

All of our work with derivatives so far has been with explicit functions or equations. Each equation has been written as $y = f(x)$, where the dependent variable is a function of the independent variable. In other words, an explicit equation has one variable expressed in terms of the other variables.

Examples of explicit functions:

$$y = f(x) = x^2 - 2x - 3$$

$$y = f(x) = 2x^5 - 7$$

$$y = f(x) = x^3 + x - 6$$

$$y = f(x) = \frac{x^2 + 2}{x - 1}$$

An implicit equation is one that does *not* directly express one variable in terms of the others.

Examples of implicit equations:

$$x^2 + x + y = 5$$

$$x^2 + y^2 = 5$$

$$xy = 1$$

$$y^3 + y^2 = 3x - x^7$$

The last explicit function requires the use of the Quotient Rule, but I am confident you could differentiate the first three equations without using paper and pencil. But how do we differentiate implicit equations like the four examples above? The answer to this question is the topic of this unit.

DIFFERENTIATING IMPLICIT EQUATIONS

To find the derivative, $\dfrac{dy}{dx}$, the implicit equation may be solved for the dependent variable and the derivative determined as usual. This method is illustrated by Example 1.

EXAMPLE 1

Find $\dfrac{dy}{dx}$ for $x^2 + x + y = 5$.

Solution: Given: $x^2 + x + y = 5$

Solve for the dependent variable. $y = 5 - x^2 - x$

Differentiate as usual. $\dfrac{dy}{dx} = -2x - 1$

In the above example, rewriting the equation explicitly for y in terms of x was easy. In other situations, however, it may be difficult or impossible to solve for y. I doubt that any of us could solve $y^3 + y^2 = 3x - x^7$ for y.

Implicit differentiation is a technique for finding the derivative of an implicit equation without first having to rewrite the equation explicitly.

PROCEDURE FOR IMPLICIT DIFFERENTIATION

To find the derivative, $\dfrac{dy}{dx}$, of an implicit equation:

1. Differentiate the equation term by term, regarding y as a function of x.

2. Solve the resulting equation for $\dfrac{dy}{dx}$. If $\dfrac{dy}{dx}$ appears in more than one term, you will need to factor it out as a common factor.

We will redo Example 1 using implicit differentiation so that you can contrast the methods.

EXAMPLE 2

Find $\dfrac{dy}{dx}$ for $x^2 + x + y = 5$.

Solution: 1.   Differentiate the equation term by term, regarding y as a function of x.

As I have done in the past, I will include an extra step to show you what is intended before actually taking the derivative.

$$x^2 + x + y = 5$$

$$\frac{d(x^2)}{dx} + \frac{d(x)}{dx} + \frac{d(y)}{dx} = \frac{d(5)}{dx}$$

$$2x + 1 + \frac{dy}{dx} = 0$$

2. Solve resulting equation for $\frac{dy}{dx}$.

$$\frac{dy}{dx} = -1 - 2x$$

Don't let the first step confuse you. We are differentiating and regarding y as a function of x. Here is what you need to keep in mind:

a. $\frac{d(x)}{dx} = \frac{dx}{dx} = 1$, because it is the derivative of x with respect to x;

b. but $\frac{d(y)}{dx} = \frac{dy}{dx}$, because it is the derivative of y with respect to x. There is nothing more that can be done to simplify the expression.

EXAMPLE 3

Find the derivative of $2x^2 + 3y = 10$ using implicit differentiation.

Solution:

$$2x^2 + 3y = 10$$

1. Differentiate term by term, regarding y as a function of x.

$$\frac{d(2x^2)}{dx} + \frac{d(3y)}{dx} = \frac{d(10)}{dx}$$

$$4x + 3\frac{dy}{dx} = 0$$

2. Solve for $\frac{dy}{dx}$.

$$3\frac{dy}{dx} = -4x$$

$$\frac{dy}{dx} = \frac{-4x}{3}$$

THE CHAIN RULE REVISITED

From Unit 12, the Chain Rule is repeated here:

Chain Rule

If $f(x) = [u(x)]^n$, then $f'(x) = n[u(x)]^{n-1}u'(x)$.

In words, the derivative of a function raised to a power is equal to the power times the function raised to the power minus 1, and all of that is multiplied times the derivative of the function.

With implicit differentiation we regard y as a function of x and use the Chain Rule to differentiate term by term. To make the method clearer, think of replacing $u(x)$ in the Chain Rule with the letter y.

For example;   If $f(x) = [u(x)]^3$, then $f'(x) = 3[u(x)]^2u'(x)$.

Replace with y: If $f(x) = y^3$, then $f'(x) = 3y^2\dfrac{dy}{dx}$.

We will use the above result in the next example.

EXAMPLE 4

Find the derivative of $y^3 = 2x^2 + 1$ using implicit differentiation.

Solution: $$y^3 = 2x^2 + 1$$

1. Differentiate term by term. $\dfrac{d(y^3)}{dx} = \dfrac{d(2x^2)}{dx} + \dfrac{d(1)}{dx}$

$$3y^2\,\dfrac{dy}{dx} = 4x + 0$$

2. Solve for $\dfrac{dy}{dx}$. $3y^2\,\dfrac{dy}{dx} = 4x$

$$\dfrac{dy}{dx} = \dfrac{4x}{3y^2}$$

Do you notice something different about this example? Previously all derivatives were in terms of x, the independent variable. With this example the derivative is in terms of both variables, x and y. However, the interpretation remains the same. The derivative can be thought of as a formula for the slope of the tangent line at any point on the curve. Except with implicit differentiation, to evaluate the formula often requires both the x- and y-values.

Are you feeling somewhat comfortable with the technique? Try two problems.

Problem 1

Find the derivative for $2y + 3x + 15 = 0$ using implicit differentiation.

Solution:

Answer: $\dfrac{dy}{dx} = \dfrac{-3}{2}$

You just used implicit differentiation to find the slope of the line.

Problem 2

Differentiate $x^2 + y^2 = 5$ using implicit differentiation.

Solution:

Answer: $\dfrac{dy}{dx} = \dfrac{-x}{y}$

EXAMPLE 5

Find the slope of the tangent line to the curve $y^3 = x^2$ at point (8, 4).

Solution: The first derivative is a formula for the slope of the tangent line at any given point on the curve. Use implicit differentiation to find $\dfrac{dy}{dx}$.

$$y^3 = x^2$$

$$\frac{d(y^3)}{dx} = \frac{d(x^2)}{dx}$$

$$3y^2 \frac{dy}{dx} = 2x$$

$$\frac{dy}{dx} = \frac{2x}{3y^2}$$

At point (8, 4), $\dfrac{dy}{dx} = \dfrac{2(8)}{3(4^2)} = \dfrac{16}{48} = \dfrac{1}{3}$

Answer: At point (8, 4) the slope of the tangent line to the curve is $\dfrac{1}{3}$.

Observe that, to evaluate the derivative in this example, both the x- and the y-coordinates were needed for the calculations.

The next example looks easier than it is, which is why I am doing it.

EXAMPLE 6

Differentiate $xy = 1$ using implicit differentiation.

Solution: The xy term is a product. It is x times y. Both are factors, and as such must be differentiated using the Product Rule.

$$xy = 1$$

Differentiate term by term.

$$\frac{d(xy)}{dx} = \frac{d(1)}{dx}$$

Use the Product Rule on left side. $$x\frac{d(y)}{dx} + y\frac{d(x)}{dx} = 0$$

$$x\frac{dy}{dx} + y(1) = 0$$

Simplify.

$$x\frac{dy}{dx} = -y$$

Solve for $\frac{dy}{dx}$.

$$\frac{dy}{dx} = \frac{-y}{x}$$

Don't worry; we will do a few more with factors and the Product Rule.

EXAMPLE 7

Differentiate the equation $x^3y^2 = 3$.

Solution: Again the left side of the equation is a product and must be differentiated using the Product Rule.

$$x^3y^2 = 3$$

$$\frac{d(x^3y^2)}{dx} = \frac{d(3)}{dx}$$

$$x^3\frac{d(y^2)}{dx} + y^2\frac{d(x^3)}{dx} = 0$$

$$x^3(2y)\frac{dy}{dx} + y^2(3x^2) = 0$$

$$2x^3y\frac{dy}{dx} + 3x^2y^2 = 0$$

$$2x^3y\frac{dy}{dx} = -3x^2y^2$$

$$\frac{dy}{dx} = \frac{-3x^2y^2}{2x^3y}$$

$$= \frac{-3y}{2x}$$

Thus far all of the examples could have been solved for y and differentiated as usual. That is not the case with the next example, which we could not work without implicit differentiation.

The major difference from the earlier examples will occur in line 3, where $\frac{dy}{dx}$ will appear in more than one term. When $\frac{dy}{dx}$ appears in more than one term, it must be factored out as a common factor. Then to solve for $\frac{dy}{dx}$, we divide the entire equation by the coefficient of $\frac{dy}{dx}$.

EXAMPLE 8

Differentiate $y^3 + y^2 = 3x - x^7$.

Solution:
$$y^3 + y^2 = 3x - x^7.$$

Differentiate term by term.
$$\frac{d(y^3)}{dx} + \frac{d(y^2)}{dx} = \frac{d(3x)}{dx} - \frac{d(x^7)}{dx}$$

$$3y^2 \frac{dy}{dx} + 2y \frac{dy}{dx} = 3 - 7x^6$$

Factor out $\frac{dy}{dx}$.
$$\frac{dy}{dx}(3y^2 + 2y) = 3 - 7x^6$$

Divide by the coefficient.
$$\frac{dy}{dx} \frac{(3y^2 + 2y)}{(3y^2 + 2y)} = \frac{3 - 7x^6}{3y^2 + 2y}$$

$$\frac{dy}{dx} = \frac{3 - 7x^6}{3y^2 + 2y}$$

That wasn't too bad, was it?

We will conclude this unit with an example (Example 10) to illustrate another advantage of implicit differentiation. First, however, the problem will be worked in Example 9 as an explicit function using the Chain Rule.

EXAMPLE 9

If $y = \sqrt{1 - x^2}$, find $\frac{dy}{dx}$ using the Chain Rule.

Solution:
$$y = \sqrt{1 - x^2}$$

Rewrite as a power of x.
$$y = (1 - x^2)^{1/2}$$

Use the Chain Rule.
$$\frac{dy}{dx} = \frac{1}{2}(1 - x^2)^{(1/2)-1} \frac{d(1 - x^2)}{dx}$$

Simplify.	$= \dfrac{1}{2}(1-x^2)^{-1/2}(-2x)$
Rewrite with positive exponents.	$= \dfrac{1}{2} \cdot \dfrac{1}{(1-x^2)^{1/2}} \cdot (-2x)$
Rewrite without fractional exponents.	$= \dfrac{-x}{\sqrt{1-x^2}}$

Now compare how much shorter, and hence easier, the problem becomes when implicit differentiation is used. The solution changes at the first step. Instead of rewriting the equation as a power of x as we did in Example 9, we will square both sides of the equation to remove the radical sign, and then differentiate implicitly.

EXAMPLE 10

If $y = \sqrt{1-x^2}$, find $\dfrac{dy}{dx}$ using implicit differentiation.

Solution:	Given:	$y = \sqrt{1-x^2}$
	Square both sides of the equation.	$y^2 = (\sqrt{1-x^2})^2$
		$y^2 = 1 - x^2$
	Use implicit differentiation.	$\dfrac{d(y^2)}{dx} = \dfrac{d(1)}{dx} - \dfrac{d(x^2)}{dx}$
		$2y\dfrac{dy}{dx} = 0 - 2x$
	Solve for $\dfrac{dy}{dx}$.	$\dfrac{dy}{dx} = \dfrac{-2x}{2y}$
	Simplify	$= \dfrac{-x}{y}$

Are the answers equivalent? Yes. If you were to substitute $\sqrt{1-x^2}$ for y into the second answer, the two derivatives would be identical.

To my way of thinking, the second way was much easier than having to deal with fractional exponents and radicals. But as always, the final choice is yours. Use whichever method gives you better results.

You should now be able to differentiate an implicit equation without having to first rewrite the equation explicitly.

Remember: to differentiate an implicit equation, the equation is differentiated term by term, regarding y as a function of x. Then the resulting equation is solved for $\dfrac{dy}{dx}$. If $\dfrac{dy}{dx}$ appears in more than one term, you will need to factor it out as a common factor.

Use implicit differentiation to find the derivatives of the equations in the exercises before continuing to the next unit.

EXERCISES

Differentiate implicitly.

1. $7x^3 + y = 5$

2. $x^2 + 5y = 10$

3. $6x^2 + 8x + 1 - 2y = 0$

4. $y^2 = x^3$

5. $y^3 = 60 + x^2 - x$

6. $x^2 + y^2 = 9$

7. $x^3 + y^2 = 20$

8. $x^2 - y^2 + 3y = 7$

9. $3x + 4y^2 - y = 28$

10. $x^3 y = 2$

11. $xy^3 + x^4 + y = 2$

12. Find the slope of the tangent line to the curve $x^2 - y^2 - x = 1$ at point (2, 1).

13. Find the slope of the tangent line to the curve $4x - 2y^2 + 5y = -7$ at point (–1, 3).

14. Evaluate the derivative of $xy = 15$ at (3, 5).

15. Given $x^3 - 2xy + y^2 = 4$:

    a. Find $\dfrac{dy}{dx}$ by implicit differentiation.

    b. Find the slope of the tangent line at (1, 3).

    c. Find the equation of the tangent line in part b.

16. Differentiate implicitly: $y = \sqrt{x}$.

17. Use implicit differentiation to find $\dfrac{dy}{dx}$ for $y = \sqrt{3x^2 + 7x - 1}$.

18. For a certain product, the demand and unit price are related by

$$q = p^2 - 40p + 2{,}150 \qquad 0 \le p \le 20$$

$$\text{where } p = \text{the price, in dollars,}$$

$$\text{and} \quad q = \text{the quantity demanded at price } p.$$

a. Find $\dfrac{dp}{dq}$.

b. Evaluate q at $p = 10$.

c. Evaluate $\dfrac{dp}{dq}$ at $p = 10$.

If additional practice is needed:

Barnett, Ziegler, and Byleen, pages 354–356, problems 1–30
Harshbarger and Reynolds, pages 836–837, problems 1–62
Hoffmann and Bradley, pages 177–178, problems 1–28
Waner and Costenoble, pages 245–246, problems 1–56

UNIT 22

Exponential Functions

Exponential functions will be introduced and defined in this unit. When you have finished it you will be able to identify an exponential function, differentiate it, and find and classify all its critical points.

We've gone this far together, so it is only fair to warn you. Typically this is the least favorite topic in calculus, probably because the material is brand new. Many of you will have never seen an exponential function before, much less the ones we will be dealing with. Add the fact that we use an irrational number as the base, and it all becomes a bit much for some students. My advice is don't give up. It is only one unit. You can do it.

Thus far in the book we have been working with power functions. In a **power function** the independent variable appears only in the base and the exponent is a real number. In contrast, an **exponential function** has the independent variable appearing in the exponent.

Power Function:	$f(x) = x^n$, n a real number
Exponential Function:	$f(x) = b^x$, b a real number, $b \neq 0$

We have covered many units learning how to graph, to differentiate, and to locate and classify critical points for power functions. In this unit you will learn how to do all of the same things for exponential functions. Hopefully it will not take us as long.

THE NUMBER e

Although any number may be used as the base in an exponential function, one of the most common bases is the number e. From geometry you most likely remember the irrational number π, which is approximately equal to 3.14 and is used in formulas for the area and circumference of a circle. Like π, e is an irrational number, meaning that it is a nonrepeating and nonterminating decimal, with a value approximately equal to 2.718.

This book will deal exclusively with exponential functions to the base e.

THE GRAPH OF $f(x) = e^x$

The simplest exponential function to the base e for us to examine is $f(x) = e^x$. Again we will start by graphing the function in Example 1.

EXAMPLE 1

Graph: $f(x) = e^x$.

Solution: Since this is a new type of function and its graph is probably unfamiliar to you, we will plot a reasonable number of points.

Most calculators today contain a key for e. If yours does, use it. Otherwise use 2.718 in place of e to obtain the various values for e^x. Your answers will be correct to three decimal places. Practice using either e or 2.718 on your calculator by verifying that the values listed below are correct.

Let $x = -3$,   then $f(-3) = e^{-3} = 0.050$ Let $x = 0.5$, then $f(.5) = e^{.5} = 1.649$

Let $x = -2$,   then $f(-2) = e^{-2} = 0.135$ Let $x = 1$,   then $f(1) = e^1 = 2.718$

Let $x = -1$,   then $f(-1) = e^{-1} = 0.368$ Let $x = 2$,   then $f(2) = e^2 = 7.389$

Let $x = -0.5$, then $f(-.5) = e^{-0.5} = 0.607$ Let $x = 3$,   then $f(3) = e^3 = 20.086$

Let $x = 0$,   then $f(0) = e^0 = 1.0$

Answer:

For most of our work after this graph, the answers will be left in terms of e.

The above nine points have been plotted on the grid at the right and connected with a smooth curve. The curve has no breaks because the domain for the function is the set of all real numbers. Why?

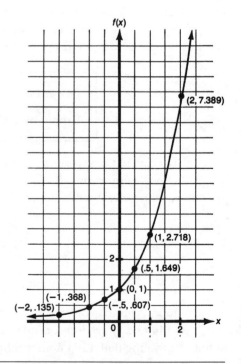

FACTS ABOUT e^x

Several observations can be made regarding the function $f(x) = e^x$. Refer to Example 1, especially the graph, to verify each of the following statements.

1. e^x is always positive.

    To generalize further, e raised to any power is always positive.

2. $e^x = 0$ has no solution. That is the same as saying the curve has no x-intercept, but we will use the no-solution form more frequently than we will refer to the x-intercept.

    To generalize further, e raised to any power can never equal 0.

3. $e^0 = 1$.

4. As x gets very large (say as $x = 4$, 100, or 1,000), e^x gets very, very large. (In Example 1, the point where $x = 3$ was already too large to be included on the graph.)

5. As x gets very small (say as $x = -4$, -100, or $-1,000$), e^x gets closer and closer to zero. Mathematically, that would be stated as "e^x is **asymptotic** to the x-axis on the left." This means that the value of e^x is approaching, but never reaches, the value of 0 as x is getting smaller and smaller. (In Example 1, the point where $x = -3$ was already too close to zero to be shown on the graph.)

6. The domain of $f(x) = e^x$ is the set of real numbers. Remember: the only time values need to be excluded from the domain occurs when there are functions with even-rooted radicals or variables in the denominator, neither of which exist here.

7. The function is increasing over its entire domain.

While all of the above statements are important, for our purposes the first two are the ones you need to remember.

APPLICATIONS

Exponential functions to the base e are found throughout applied mathematics. For example, if interest is compounded continuously, the resulting bank balance is determined by an exponential function. In the absence of any environmental constraints, population increases exponentially. Sales of many products decrease exponentially when advertising is discontinued. Learning curves, which are exponential functions, describe the relationship between the efficiency with which an individual performs a task and the amount of training he or she has had. And logistic curves, which are exponential functions, describe the spread of epidemics or rumors through a community.

We will do several examples and let it go at that.

EXAMPLE 2

The total number of hamburgers sold by a national fast-food chain is growing exponentially according to the function

$$Q(t) = 4e^{2t-3}$$

where t = the number of years since 1995 ($t = 0$ corresponds to 1995)

and $Q(t)$ = the number, in hundreds, of hamburgers sold in t years.

The company is considering using the number of hamburgers sold in a future advertising campaign, somewhat as McDonald's once did.

Just to see how the number would look in the ads, determine how many hamburgers had been sold by 2001.

Solution: Since $t =$ the years since 1995, 2001 would correspond to $t = 6$.

$$Q(t) = 4e^{2t-3}$$

$$Q(6) = 4e^{2(6)-3}$$

$$= 4e^9$$

$$= 4(8,103.1)$$

$$= 32,412.4$$

Answer: By 2001, the fast-food chain had sold 32,412.4 hundred or 3,241,240 hamburgers.

As mentioned earlier, if interest is compounded continuously, the resulting bank balance is determined by an exponential function. That function is

$$A(t) = Pe^{rt}$$

where $P =$ amount of money invested,

$r =$ annual interest rate expressed as a decimal,

$t =$ number of years P is invested,

and $A(t) =$ amount after t years

The next example illustrates this formula.

EXAMPLE 3

Suppose $1,000 is invested in a money market account that pays interest at a rate of 6.5% per year compounded continuously. Determine the balance in the account after 2 years.

Solution: Use the formula $A(t) = Pe^{rt}$,

with $P = 1,000$

$r = 0.065$

$t = 2$

$$A(t) = Pe^{rt}$$

$$A(2) = 1,000e^{0.065(2)}$$

$$= 1,000e^{0.13}$$

$$= 1,000(1.13883)$$

$$= 1,138.83$$

Answer: The balance in the account after 2 years will be $1,138.83.

DIFFERENTIATION

By now I know you are able to take the derivative of any polynomial function, as well as of the product of two polynomials, of the quotient of two polynomials, and of a polynomial raised to a power.

Here is a brief summary of the rules thus far:

Power: If $f(x) = x^n$, then $f'(x) = nx^{n-1}$

Constant: If $f(x) = c$, then $f'(x) = 0$

Constant · Function: If $f(x) = c \cdot u(x)$, then $f'(x) = c \cdot u'(x)$

Special Case: If $f(x) = c \cdot x^n$, then $f'(x) = cnx^{n-1}$

Sum: If $f(x) = u(x) + v(x)$, then $f'(x) = u'(x) + v'(x)$

Difference: If $f(x) = u(x) - v(x)$, then $f'(x) = u'(x) - v'(x)$

Product Rule: If $f(x) = F(x) \cdot S(x)$, then $f'(x) = F(x)S'(x) + S(x)F'(x)$

Quotient Rule: If $f(x) = \dfrac{N(x)}{D(x)}$, then $f'(x) = \dfrac{D(x)N'(x) - N(x)D'(x)}{[D(x)]^2}$

Chain Rule: If $f(x) = [u(x)]^n$, then $f'(x) = n[u(x)]^{n-1}u'(x)$

The differentiation of exponential functions requires the addition of only one more rule.

Exponential Rule

If $f(x) = e^{u(x)}$, where $u(x)$ is a function of x, then $f'(x) = u'(x)e^{u(x)}$.

In words, the derivative of e raised to some exponent containing x equals the derivative of the exponent times the original function.

Notice that we are working with a function of x, so all the preceding rules still apply.

EXAMPLE 4

If $f(x) = e^{5x}$, find $f'(x)$.

Solution: To differentiate e raised to some exponent containing x, take the derivative of the exponent times the original function.

$$f(x) = e^{5x}$$

$$f'(x) = \frac{d(5x)}{dx}e^{5x}$$

$$= 5e^{5x}$$

Keep in mind that all that is happening here is that we are considering another group of functions, called exponentials. The derivative remains a formula for finding the slope of the tangent line at any point along the curve. Critical points are located and classified in the same way as before. And exponential functions are maximized or minimized using the same general technique presented in Units 15–18.

EXAMPLE 5

If $f(x) = e^{5x}$, find and interpret: a. $f(0), f'(0)$, b. $f(1), f'(1)$, and c. $f(-1), f'(-1)$.

Solution: a.

$$f(x) = e^{5x}$$

$$f(0) = e^0 = 1, \text{ which is point } (0, 1) \text{ on the curve.}$$

$$f'(x) = 5e^{5x}$$

$$f'(0) = 5e^0 = 5(1) = 5$$

Answer: The slope of the tangent line to the curve at point (0, 1) is 5.

b.

$$f(x) = e^{5x}$$

$$f(1) = e^{5(1)} = e^5, \text{ which is point } (1, e^5) \text{ on the curve.}$$

> Note: Leaving the answer as e^5 is perfectly acceptable. This is not an engineering course. We do not need a decimal approximation for any further use in the problem. Mentally I think of e as being almost 3 and let it go at that. If more precision is required, using a calculator, $e^5 = 148.4$ and the point would be about (1, 148.4).

$$f'(x) = 5e^{5x}$$

$$f'(1) = 5e^{5(1)} = 5e^5$$

Answer: The slope of the tangent line to the curve at point $(1, e^5)$ is $5e^5$.

c.

$$f(x) = e^{5x}$$

$$f(-1) = e^{5(-1)} = e^{-5}, \text{ which is point } (-1, e^{-5}).$$

$$f'(x) = 5e^{5x}$$

$$f'(-1) = 5e^{5(-1)} = 5e^{-5}$$

Answer: The slope of the tangent line to the curve at point $(-1, e^{-5})$ is $5e^{-5}$.

If the answer is to be written with positive exponents only, the slope of the tangent line to the

curve at point $\left(-1, \dfrac{1}{e^5}\right)$ is $\dfrac{5}{e^5}$.

Again, as stated in part b, if you are going to do something more with the answer and need an approximate value, $e^{-5} = 0.00674$ and $5(0.00674) = 0.0337$. The answer would now read as follows: The slope of the tangent line to the curve at point (-1, 0.00674) is approximately 0.0337.

In case you are curious, the graph of $f(x) = e^{5x}$ is shaped about the same as the function in Example 1 except that the curve goes up faster on the right and drops down to the x-axis faster on the left.

Unfortunately, the graphs of exponential functions as a whole are not predictable. Although some generalizations can be made about specific types of exponential functions, the graphing of exponential functions will not be discussed in this book.

EXAMPLE 6

If $f(x) = e^{x^2+1}$, find $f'(x)$.

Solution: The derivative of e raised to some exponent containing the independent variable equals the derivative of the exponent times the original function.

$$f(x) = e^{x^2+1}$$

$$f'(x) = \frac{d(x^2+1)}{dx} e^{x^2+1}$$

$$= 2xe^{x^2+1}$$

derivative of exponent   original function

EXAMPLE 7

If $f(x) = e^x$, find $f'(x)$.

Solution: $f(x) = e^x$

$$f'(x) = \frac{d(x)}{dx} e^x$$

$$= 1 \cdot e^x$$

$$= e^x$$

Example 7 illustrates that the derivative of e^x is e^x. It is a special case of the Exponential Rule.

Special Case

If $f(x) = e^x$, then $f'(x) = e^x$.

By now you are probably anxious to try a few problems before they get harder.

Problem 1

Find $f'(x)$ if $f(x) = e^{6x+5}$.

Solution:

Answer: $f'(x) = 6e^{6x+5}$

Problem 2

Find $f'(x)$ if $f(x) = e^{x^2+3x-1}$.

Solution:

Answer: $f'(x) = (2x + 3)e^{x^2+3x-1}$

The parentheses are required to show that the entire quantity is a factor.

Before preceding with more complicated exponential functions we will do another example involving interest that is compounded continuously.

EXAMPLE 8

Continuing with Example 3, suppose $1,000 is invested in a money market account that pays interest at a rate of 6.5% per year compounded continuously. Use the formula $A(t) = Pe^{rt}$. Find and interpret: a. $A'(2)$, b. $A'(10)$, and c. $A'(20)$.

Solution: a.   From Example 3, we found the balance in the account after 2 years would be $1,138.83, and the formula was:

$$A(t) = 1,000e^{0.065t}$$

$$A'(t) = 1,000(0.065e^{0.065t})$$

$$= 65e^{0.065t}$$

$$A'(2) = 65e^{0.065(2)}$$

$$= 65e^{0.13}$$

$$= 65(1.1388)$$

$$= 74.02$$

Answer: The interest earned the next year (year 3) will be about $74.02.

Answers may vary because of rounding.

b. $A'(10) = 65e^{0.065(10)}$

$$= 65e^{0.65}$$

$$= 65(1.9155)$$

$$= 124.51$$

Answer: The interest earned the next year (year 11) will be about $124.51.

     c. $A'(20) = 65e^{0.065(20)}$

                 $= 65e^{1.3}$

                 $= 65(3.6692)$

                 $= 238.50$

Answer: The interest earned the next year (year 21) will be about $238.50.

Problem 3

Continuing with Examples 3 and 8, determine the balance in the account after 20 years.

Solution:

Answer: $3,669.30

Examples 9–11 require the use of more than one rule.

EXAMPLE 9

If $f(x) = \dfrac{e^x}{x}$, find $f'(x)$.

Solution: I hope you recognize that $f(x)$ is a quotient and therefore that we must use the Quotient Rule.

     Given: $f(x) = \dfrac{e^x}{x}$

     Use the Quotient Rule. $f'(x) = \dfrac{x\dfrac{d(e^x)}{dx} - e^x\dfrac{d(x)}{dx}}{x^2}$

     Differentiate. $= \dfrac{x(e^x) - e^x(1)}{x^2}$

     Simplify. $= \dfrac{xe^x - e^x}{x^2}$

     Or factor, if you prefer. $= \dfrac{e^x(x-1)}{x^2}$

That wasn't too bad, but the next one is not quite so obvious.

EXAMPLE 10

If $f(x) = 4xe^x$, find $f'(x)$.

Solution: This time $f(x)$ is a product. Do you see why?

Given:
$$f(x) = 4xe^x$$

first factor   second factor

Use the Product Rule.
$$f'(x) = (4x)\frac{d(e^x)}{dx} + (e^x)\frac{d(4x)}{dx}$$

Differentiate.
$$= 4x(e^x) + e^x(4)$$

Simplify.
$$= 4xe^x + 4e^x$$

Or factor, if you prefer.
$$= 4e^x(x + 1)$$

Here is another example with the Product Rule.

EXAMPLE 11

If $f(x) = (1 - 3x)e^{x^2}$, find $f'(x)$.

Solution:
$$f(x) = (1 - 3x)e^{x^2}$$

first factor   second factor

Use the Product Rule.
$$f'(x) = (1 - 3x)\frac{d(x^{x^2})}{dx} + (e^{x^2})\frac{d(1 - 3x)}{dx}$$

Differentiate.
$$= (1 - 3x)(2xe^{x^2}) + (e^{x^2})(-3)$$

Simplify.
$$= 2xe^{x^2} - 6x^2e^{x^2} - 3e^{x^2}$$

$$= -6x^2e^{x^2} + 2xe^{x^2} - 3e^{x^2}$$

Or factor, if you prefer.
$$= e^{x^2}(-6x^2 + 2x - 3)$$

LOCATING AND CLASSIFYING CRITICAL POINTS

Recall that critical points, if they exist, are located by taking the first derivative, setting it equal to zero, solving for x, and then determining the corresponding y-coordinate. Once located, a critical point may be classified using either the original-function test, the first-derivative test, or the second-derivative test.

The same procedures are used for exponential functions. Do not let that irrational number e confuse you. It is nothing more than a commonly accepted symbol used to represent a number approximately equal to 2.718. And exponential functions are just another group of functions with different characteristics and applications than polynomial functions, but the concepts and techniques remain the same.

To save ourselves some work, I will use functions for which we have already found the derivatives.

EXAMPLE 12

Find and classify all critical points for $f(x) = 4xe^x$.

Solution: $f(x) = 4xe^x$

$f'(x) = 4e^x(x + 1)$ from Example 10

Find all critical points. Critical points occur where $f'(x) = 0$.

$f'(x) = 4e^x(x + 1)$

$0 = 4e^x(x + 1)$

To solve an equation of this type, as before, we set each factor equal to zero and solve individually.

$x + 1 = 0$ $4e^x = 0$

$x^* = -1$ $e^x = 0,$

(Recall from Example 1 that
this equation has no solution.)

If $x^* = -1$,

$f(-1) = -4e^{-1} = \dfrac{-4}{e}$; thus $\left(-1, \dfrac{-4}{e}\right)$ is the only critical point.

If the e makes you uncomfortable, rewrite the coordinates as (1, −1.472), but this is actually more work and is unnecessary.

Classify the critical point.

My choice would be to use the first-derivative test with $f'(x) = 4e^x(x + 1)$.

To save ourselves time, remember we need determine only the sign, positive or negative, of the first derivative. Again recall from Example 1, that e^x is always positive, so $4e^x$ is always positive.

For point $\left(-1, \dfrac{-4}{e}\right)$ with $f'(x) = 4e^x(x + 1)$ the curve

On the left: Let $x = -2$; then $f'(-2) = (+)(-2 - 1) = (+)(-) = -$ is decreasing

At point $\left(-1, \dfrac{-4}{e}\right)$:   levels off

On the right: Let $x = 0$;   then $f'(0) = (+)(0 + 1) = (+)(+) = +$ is increasing

At the critical point, the curve is decreasing, leveling off, then increasing.

Answer: The critical point $\left(-1, \dfrac{-4}{e}\right)$ is a relative minimum.

In fact, the critical point $\left(-1, \frac{-4}{e}\right)$ is an absolute minimum. Do you understand why? The domain is the entire set of real numbers, meaning that the curve has no breaks. Since there is one and only one relative minimum, it must be the absolute minimum.

That one was a bit long; let's try a shorter one.

EXAMPLE 13

Find and classify all critical points for $f(x) = e^x$.

Solution:
$$f(x) = e^x$$
$$f'(x) = e^x$$

Find all critical points. Critical points occur where $f'(x) = 0$.

$$0 = e^x$$

but this equation has no solution.

Answer: $f(x) = e^x$ has no critical points.

Before we do a really long one, I want you to try a problem.

Problem 4

Given: $f(x) = e^{x^2+1}$.

a.   Find $f'(x)$

b.   Find all critical points for $f(x)$.

c.   Classify each critical point.

Solution:

Answers: a.   Refer to Example 6.

b. $(0, e)$

c.   Relative minimum

EXAMPLE 14

Find and classify all critical points for $f(x) = \frac{e^x}{x}$.

Solution: $f(x) = \dfrac{e^x}{x}$

$f'(x) = \dfrac{xe^x - e^x}{x^2}$ from Example 9.

Find all critical points. Critical points occur where $f'(x) = 0$.

$0 = \dfrac{xe^x - e^x}{x^2}$

$0 = xe^x - e^x$ Multiply the entire equation by x^2.

$0 = e^x(x - 1)$ Factor out the common factor e^x.

Set each factor equal to zero and solve.

$x - 1 = 0$ $e^x = 0$

$x^* = 1$ From Example 1, e^x has no solution.

If $x^* = 1$,

$f(1) = \dfrac{e^1}{1} = e$; thus $(1, e)$ is the only critical point.

Classify the critical point.

My choice would be to use the first-derivative test with

$$f'(x) = \frac{xe^x - e^x}{x^2} = \frac{e^x(x - 1)}{x^2}$$

To save ourselves time, we need determine only the sign. Also notice that the denominator will always be positive, because it is being squared and e^x is always positive.

For point $(1, e)$		the curve:
On the left:	Let $x = 0.5$, then $f'(0.5) = (+)(0.5 - 1) / + = -$	is decreasing
At point $(1, e)$:		levels off
On the right:	Let $x = 2$, then $\quad f'(2) = (+)(2 - 1) \;/ + = +$	is increasing

At the critical point, the curve is decreasing, leveling off, then increasing.

Answer: The critical point $(1, e)$ is a relative minimum.

OPTIMIZATION REVISITED

As you well know, the real purpose behind all of this is to be able to determine where an exponential function achieves its maximum and minimum values. The four-step general technique in condensed form is repeated below for reference.

The general technique is:

1. Decide what quantity is to be maximized or minimized and write the equation for it.

2. Use the constraints of the problem to write the equation in terms of a single variable.

3. Find all critical points and classify.

4. Answer the original question.

One example is all you are likely to want to do.

EXAMPLE 15

Edward, a publisher's rep, has observed that his sales of college textbooks are a function of the number of years the book has been on the market. Typically a new book reaches its peak in sales a few years after it is first published; then sales level off and remain fairly constant. This year the publishing company is introducing a new economics text. Edward estimates that his sales for the book will be given by

$$f(x) = x^2 e^{-x} \qquad x \geq 0$$

where $f(x)$ = the number, in thousands, of books sold in year x

and x = the number of years since the book was published.

a. In what year of publication should Edward expect to maximize his sales of the new economics book?

b. What is the maximum number of copies of the book that he estimates he will sell in any year?

Solution: 1. What is to be maximized in the problem? Number of books sold

Write the equation for number of books sold.

The equation for number of books sold was given as

$$f(x) = x^2 e^{-x} \text{ with } x \geq 0$$

2. Write function in terms of a single independent variable.

The function was given in terms of a single variable.

3. Find critical points and classify.

$f(x) = x^2 e^{-x}$

$$f'(x) = x^2 \frac{d(e^{-x})}{dx} + e^{-x} \frac{d(x^2)}{dx}$$

$$= x^2(-e^{-x}) + e^{-x}(2x)$$

$$= -x^2 e^{-x} + 2x e^{-x}$$

$$= -x e^{-x}(x - 2)$$

$$0 = -x e^{-x}(x - 2)$$

$-x = 0 \qquad x - 2 = 0 \qquad e^{-x} = 0$

$x^* = 0 \qquad\qquad x^* = 2 \qquad$ This has no solution.

If $x^* = 0$,

$f(0) = 0$ and is of little interest to the problem.

If $x^* = 2$,

$$f(2) = (2)^2 e^{-2}$$

$$= 4(0.135)$$

$$= 0.54$$

Test: First-derivative test with $f'(x) = -xe^{-x}(x - 2)$

Remember e^{-x} is always positive.

At point (2, 0.54)		the curve:
On the left:	Let $x = 1, f'(1) = (-)(+)(-)= +$	is increasing
At point (2, 0.54):		levels off
On the right:	Let $x = 3, f'(3) = (-)(+)(+) = -$	is decreasing

The curve is increasing, then leveling off, then decreasing. Point (2, 0.54) is a relative maximum. In other words, the maximum value for $f(x)$ is 0.54 and occurs when $x = 2$.

4. Answer the questions posed in the problem.

Answers: a.   Edward should expect to maximize the sale of the new economics textbook in its second year on the market.

b.   At that time he estimates he will sell 0.54 thousand or 540 books.

CALCULUS AND THE CALCULATOR (Optional)

By now I expect that you automatically view most graphs to verify the reasonableness of your analytical answers. And I hope you have already done so with all the examples thus far. If you haven't, for practice, quickly view the graphs of the functions in this unit. You should see why I said that as a whole the graphs of exponential functions are unpredictable.

In this last example we will use the graph, as well as the derivative, to forecast an outcome.

EXAMPLE 16

The market penetration for cell phones has followed a logistic growth model, given by

$$G(x) = \frac{77}{1 + 585e^{-0.68x}}$$

where $G(x) =$ the percentage of the market penetrated by cell phones

and $x =$ years since 1985.

a.   View the graph of $G(x)$ with a viewing window [0, 25] by [0, 100].

b.   Find $G'(x)$.

c.   Use $G(x)$ and $G'(x)$ and create a table of x-values of 10, 15, 20, and 25.

d.   Find and interpret $G(20)$ and $G'(20)$.

e.   What are the long-term prospects for the market penetration for cell phones?

Solution: a.   \Y₁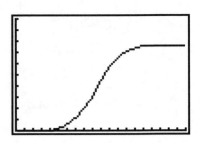

b. $G(x) = \dfrac{77}{1 + 585e^{-0.68x}}$

$$G'(x) = \frac{(1 + 585e^{-0.68x})\dfrac{d(77)}{dx} - 77\dfrac{d(1 + 585e^{-0.68x})}{dx}}{(1 + 585e^{-0.68x})^2}$$

$$= \frac{(1 + 585e^{-0.68x})(0) - 77\left[0 + \dfrac{d(-0.68x)}{dx}(585e^{-0.68x})\right]}{(1 + 585e^{-0.68x})^2}$$

$$= \frac{0 - 77[0 + (-0.68)(585e^{-0.68x})]}{(1 + 585e^{-0.68x})^2}$$

$$= \frac{-77(0 - 397.8e^{-0.68x})}{(1 + 585e^{-0.68x})^2}$$

$$= \frac{-0 + 77(397.8e^{-0.68x})}{(1 + 585e^{-0.68x})^2}$$

$$= \frac{30{,}630.6e^{-0.68x}}{(1 + 585e^{-0.68x})^2}$$

c.   To key in the derivative requires quite a few keystrokes and numerous parentheses. Try it on your own first, before looking at the following sequence.

\Y₂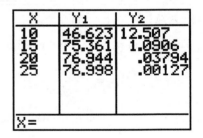

d. $G(20) = 76.944$ and $G'(20) = 0.03794$

Approximately 76.94% of all households are projected to have cell phones in the year 2005, with an increase of less than 0.04% for the following year.

e.   From the table of values and graph it can be seen that the function levels off at about 77%. In other words, eventually approximately 77% of all households are projected to have cell phones.

You should now be able to identify an exponential equation, to differentiate it, and to find and classify all of its critical points.

Before continuing to the next unit, practice with the following exercises.

EXERCISES

For each of the following, find and classify all critical points:

1. $f(x) = 3e^{5x}$

2. $f(x) = e^{x^2}$

3. $f(x) = e^{x^4+1}$

4. $f(x) = \dfrac{e^x}{x^2}$

5. Let the total cost, in dollars, of producing x units of a product be given by

    $$TC = f(x) = e^x \qquad x > 0$$

    a. Determine the function, $AC = q(x)$, for the average cost per unit of producing x units.

    b. Find $f(2)$ and interpret.

    c. Find $q(3)$ and interpret.

    d. How many units should be produced so as to minimize average cost per unit?

6. a. Find and classify the two critical points for the function $y = f(x) = x^2 e^x$.

    b. Optional—Identify any absolute extrema.

7. If \$1,000 is invested for t years at 8% compounded continuously, the total amount is given by

    $$M(t) = 1,000e^{0.08t}$$

    a. Find $M'(t)$

    b. Find $M(5)$ and interpret.

    c. Find $M'(5)$ and interpret.

8. Jude manages the concession stand at a local theater. She is planning the introduction and promotion of a new popcorn line. After testing the new line for several months, she has found that the demand is given approximately by

    $$p = 5e^{-x} \qquad 0 \le x \le 2$$

    > where $x =$ the number, in thousands, of boxes of popcorn
    > sold per week at a price of p dollars each.

    At what price will the weekly revenue be maximum?

9. The Wonder Company introduced a new product recently and has been spending a fixed budget on advertising. The company knows that, if the advertising is stopped, its monthly sales will decrease.

The Marketing Department has found that the monthly sales and advertising cost can be estimated by

$$y(t) = 2{,}000e^{-0.60t} \quad t \geq 0$$

where $y(t)$ = the monthly sales, in dollars,

and t = the number of months since the cancellation of any advertising.

a. Find $y(0)$ and interpret.

b. Find $y(2)$ and interpret.

c. Find $y'(t)$.

d. Find $y'(2)$ and interpret.

C10. Bill McKinley purchased a new automobile last month. He estimates its trade-in value after so many years is given by

$$V(t) = 19{,}000e^{-0.225t} \quad 0 \leq t \leq 10$$

where $V(t)$ = the trade-in value measured in dollars,

and t = the time in years.

a. What was the original purchase price of the car?

b. Find the rate of change of V with respect to t when $t = 0$ and $t = 5$, and interpret your answers.

c. If Bill keeps the car for 10 years, what will its trade-in value be?

C11. According to Newton's law of cooling, the rate at which an object cools is directly proportional to the difference in temperature between the object and the surrounding medium. A hot cup of coffee is poured at 8 A.M., set on the kitchen counter, and forgotten in the rush to get to work on time. Assuming that the room remains at a constant temperature, then the temperature of the coffee, $f(t)$, measured in degrees Fahrenheit after t hours of cooling is given by $f(t) = 72 + 45e^{-2t}$.

a. If $t = 0$ corresponds to 8 A.M., how hot was the coffee when first poured?

b. Approximate to the nearest tenth of a degree the temperature of the coffee at 8:30 A.M., 9:00 A.M., and noon.

c. Graph $f(t)$.

d. What was the temperature of the room?

If additional practice is needed:

Barnett, Ziegler, and Byleen, page 103, problems 7–18; pages 328–330, problems 1–74

Harshbarger and Reynolds, pages 389–390, problems 23–39; pages 415–418, problems 1–41; pages 825–828, problems 1–65

Hoffmann and Bradley, pages 308–314, problems 1–49; page 339, problems 1–12 and 21–23; pages 350–357, problems 1–51

UNIT 23

Logarithmic Functions

The purpose of this unit is to provide you with a brief overview of logarithms, especially natural logarithms. First the definition and notation used with natural logarithms will be introduced. Then you will learn to differentiate logarithmic functions and to find and classify all their critical points.

Keep in mind that all we are doing in this unit is considering still another group of functions called logarithmic functions. The derivative remains a formula for finding the slope of the tangent line at any point along the curve. Critical points are located and classified in the same way as before. And logarithmic functions will be maximized or minimized using the same general technique presented in Units 15–18.

Although some generalizations can be made about the graphs of specific types of logarithmic functions, the graphing of these functions will not be covered in this book.

The equation $\log_b N = x$ is read "the logarithm of N to the base b is x."

Definition: x is called the **logarithm of N to the base b** if $b^x = N$, where N and b are both positive numbers, $b \neq 1$.

In other words:

$$\log_b N = x \quad \text{if and only if } b^x = N.$$

Examples of logarithms are:

$$\log_3 9 = 2 \text{ because } 3^2 = 9$$

$$\log_2 8 = 3 \text{ because } 2^3 = 8$$

$$\log_7 7 = 1 \text{ because } 7^1 = 7$$

$$\log_2 16 = 4 \text{ because } 2^4 = 16$$

$$\log_{15} 1 = 0 \text{ because } 15^0 = 1$$

$$\log_2 \left(\frac{1}{2}\right) = -1 \text{ because } 2^{-1} = \left(\frac{1}{2}\right)$$

Notice that the logarithm of a positive number N is the exponent to which the base must be raised to produce the number N.

THE NATURAL LOGARITHM

Although any number may be used as the base in a logarithmic function, one of the most common bases is the number e. You remember from Unit 22 that e is an irrational number with a value approximately equal to 2.718. When e is used as the base of a logarithmic function, the function is referred to as a natural logarithm. Typically ln, rather than log, is used to denote a natural logarithm, and e is omitted.

The equation $\ln x = y$ is read "the natural logarithm of x is y." Often "the natural logarithm of x" is shortened and read as "natural log of x."

Definition: y is called the **natural logarithm** of x if $e^y = x$, where x is a positive number.

In words:

$$\ln x = y \quad \text{if and only if } e^y = x.$$

Here are examples of natural logarithms:

$$\ln e = 1 \text{ because } e^1 = e$$

$$\ln 1 = 0 \text{ because } e^0 = 1$$

$$\ln 7.3891 = 2 \text{ because } e^2 = 7.3891$$

$$\ln 20.086 = 3 \text{ because } e^3 = 20.086$$

You need to remember the first two examples. For the last two examples and any other values, most calculators today contain a key for a natural logarithm, which you will need to use throughout this unit. Practice using the ln key by first verifying that the values listed in the examples are correct and then doing the problems on your own.

EXAMPLE 1 $\ln 4 = 1.38629$ and $e^{1.38629} = 4$

EXAMPLE 2 $\ln 1.2 = 0.18232$ and $e^{0.18232} = 1.2$

EXAMPLE 3 $\ln 0.1 = -2.30259$ and $e^{-2.30259} = 0.1$

Several of the above points have been plotted on the grid on the following page and connected with a smooth curve. The curve has no breaks because the domain for the function was stated as the set of all positive real numbers.

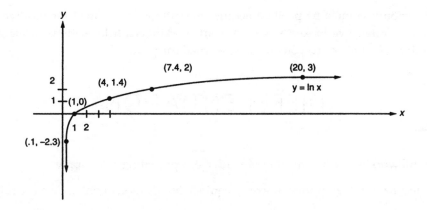

Cover the answers to Problems 1–4 and find the natural logs of the given numbers by using a calculator.

Problem 1	$\ln 1.49 =$
Problem 2	$\ln 0.50 =$
Problem 3	$\ln 0.3 =$
Problem 4	$\ln 0.61 =$

Answers: 0.39878; –0.69315; –1.20397; –0.49430

As in the earlier definition, the natural logarithm of a positive number x is the exponent to which the base e must be raised to produce the number x.

$$\ln x = y \text{ is called the logarithmic form,}$$

$$b^y = x \text{ is called the exponential form,}$$

and the two statements are equivalent.

This book will deal exclusively with natural logarithms.

The tendency at this point is to be a bit confused about what logarithmic functions are and what this unit is really about. Recall that logarithmic functions are merely another group of functions with properties and applications unique to their group. We are not going to graph logarithmic functions, but you will learn how to differentiate them and how to locate and classify their critical points.

FACTS ABOUT $y = \ln u(x)$

There are several observations to be remembered concerning the function $y = f(x) = \ln u(x)$. Refer to preceding material to verify each of the following statements.

1. $\ln 1 = 0$ because $e^0 = 1$.

2. $\ln e = 1$ because $e^1 = e$.

3. The function $u(x)$ must be positive because logarithms are defined for positive numbers only. This fact is often overlooked when we start working with logarithms of algebraic functions rather than of numbers, so you are being warned early on.

DIFFERENTIATION

Because we are still working with a function of x, all of our preceding rules apply.

To differentiate a natural logarithmic function requires the addition of only one more rule.

> **Logarithmic Rule**
>
> If $f(x) = \ln u(x)$, then $f'(x) = \dfrac{u'x}{u(x)}$.
>
> Verbally, the derivative of $\ln u(x)$ is the derivative of $u(x)$ divided by $u(x)$ itself.

EXAMPLE 4

If $f(x) = \ln x$, find $f'(x)$.

Solution: The derivative of $\ln u(x)$ is simply the derivative of $u(x)$ divided by $u(x)$ itself.

$$f(x) = \ln x$$

$$f''(x) = \frac{\dfrac{d(x)}{dx}}{x}$$

$$= \frac{1}{x}$$

Example 4 illustrates that the derivative of $\ln x$ is $\dfrac{1}{x}$. It is a frequently encountered function and can be considered a special case of the logarithmic rule.

> **Special Case**
>
> If $f(x) = \ln x$, then $f'(x) = \dfrac{1}{x}$.

EXAMPLE 5

If $f(x) = \ln (x^2 + 5x)$, find $f'(x)$.

Solution: The derivative of ln $u(x)$ is the derivative of $u(x)$ divided by $u(x)$.

$$f(x) = \ln (x^2 + 5x)$$

$$f'(x) = \frac{\frac{d(x^2 + 5x)}{dx}}{(x^2 + 5x)}$$

$$= \frac{2x + 5}{x^2 + 5x}$$

EXAMPLE 6

If $f(x) = \ln x^3$, find $f'(x)$.

Solution: The derivative of ln $u(x)$ is the derivative of $u(x)$ divided by $u(x)$.

$$f(x) = \ln x^3$$

$$f'(x) = \frac{\frac{d(x^3)}{dx}}{x^3}$$

$$= \frac{3x^2}{x^3}$$

$$= \frac{3}{x}$$

Aren't these examples easy compared to some of the functions we have been doing? Try Problems 5–7 on your own. Be sure to simplify your answers.

Problem 5

Find $f'(x)$ if $f(x) = \ln x^2$.

Solution:

Answer: $f'(x) = 2/x$

Problem 6

Find $f'(x)$ if $f(x) = \ln (x^2 + 7x + 1)$.

Solution:

Answer: $f'(x) = \dfrac{2x+7}{x^2 + 7x + 1}$

Problem 7

Find $f'(x)$ if $f(x) = \ln x^2 + 7x + 1$. Be careful.

Solution:

Answer: $f'(x) = \dfrac{2}{x} + 7$

Did I catch you on that last one? I really was not trying to trick you, but merely wanted to show you the importance of parentheses. In Problem 6 the ln is of the entire quantity in parentheses, whereas, in Problem 7 the ln is of only x^2 with $7x$ and 1 being separate terms. In Problem 7 you had to take the derivative term by term. The following example also illustrates having to take the derivative term by term.

EXAMPLE 7

Find $f'(x)$ if $f(x) = \ln x + \ln 3x$.

Solution: $f(x)$ is a function of two separate terms, much as in Problem 7.

$$f(x) = \ln x + \ln 3x$$

$$f'(x) = \frac{1}{x} + \frac{\dfrac{d(3x)}{dx}}{3x}$$

$$= \frac{1}{x} + \frac{3}{3x}$$

$$= \frac{1}{x} + \frac{1}{x}$$

$$= \frac{2}{x}$$

EXAMPLE 8

Find $f'(x)$ if $f(x) = \ln 3$.

Solution: What kind of function is $f(x)$? It is a constant. And the derivative of a constant is zero.

$$f(x) = \ln 3$$

$$f'(x) = 0$$

Having trouble with that? Let's rewrite $f(x)$ and try it again.

If $f(x) = \ln 3$,

then $f(x) = 1.09861$

and $f'(x) = 0$

There is no way to avoid it; Examples 9 and 10 require the Quotient Rule and the Product Rule.

EXAMPLE 9

Find $f'(x)$ if $f(x) = \dfrac{\ln x}{x}$.

Solution: $f(x)$ is a quotient, and therefore we must use the Quotient Rule.

$$f(x) = \frac{\ln x}{x}$$

Use the Quotient Rule. $f'(x) = \dfrac{x\dfrac{d(\ln x)}{dx} - (\ln x)\dfrac{d(x)}{dx}}{x^2}$

Special Case with $\ln x$: $= \dfrac{x\left(\dfrac{1}{x}\right) - (\ln x)(1)}{x^2}$

Simplify. $= \dfrac{1 - \ln x}{x^2}$

EXAMPLE 10

Find $f'(x)$ if $f(x) = x^2 \ln x$.

Solution: This time $f(x)$ is a product.

$$f(x) = x^2 \underset{\text{first factor \quad second factor}}{\ln x}$$

first factor   second factor

Use the Product Rule. $f'(x) = (x^2)\dfrac{d(\ln x)}{dx} + (\ln x)\dfrac{d(x^2)}{dx}$

Special Case with ln x:	$= (x^2)\left(\dfrac{1}{x}\right) + (\ln x)(2x)$
Simplify.	$= x + 2x \ln x$
Or, if you like it factored:	$= x(1 + 2 \ln x)$

LOCATING AND CLASSIFYING CRITICAL POINTS

Critical points, as you well know by now, occur where the first derivative equals zero. They are located by setting the first derivative equal to zero, solving the resulting equation, and then determining the corresponding y-coordinate. Once located, a critical point may be classified as either a relative maximum, a relative minimum, or a stationary inflection point. One of three tests is used to make the determination: the original-function test, the first-derivative test, or the second-derivative test.

The same procedures are used for logarithmic functions.

EXAMPLE 11

Find and classify all critical points for $f(x) = x - \ln x$ for $x > 0$.

Solution:
$$f(x) = x - \ln x$$

$$f'(x) = 1 - \frac{1}{x} \text{ from the Special Case.}$$

Find all critical points. Critical points occur where $f'(x) = 0$.

$$0 = 1 - \frac{1}{x}$$

$$0 = x - 1 \text{ (Cleared of fractions)}$$

$$x^* = 1$$

If $x^* = 1$,

$$f(1) = 1 - \ln 1 = 1 - 0 = 1;$$

thus $(1, 1)$ is the only critical point.

Classify the critical point.

For variety, we will use the second-derivative test.

$$f'(x) = 1 - \frac{1}{x}$$

$$= 1 - x^{-1}$$

$$f''(x) = 0 - (-1x^{-2})$$

$$= 0 + x^{-2}$$

$$= \frac{1}{x^2}$$

At point (1, 1)

$$f''(x) = +, \text{ curve is concave up, } \smile, \text{ point is minimum.}$$

Answer: The critical point (1, 1) is a relative minimum. In other words, the minimum value of $f(x)$ is 1 and occurs when $x = 1$.

Here is a short problem for you to do; then I will do one more example. Don't look at the answer until you have finished the problem.

Problem 8

Find and classify all critical points for $f(x) = \ln x$.

Solution:

Answer: $f(x)$ has no critical points.

EXAMPLE 12

Find and classify all critical points for $f(x) = \ln x - \dfrac{1}{2} x^2$ for $x > 0$.

Solution: $$f(x) = \ln x - \frac{1}{2} x^2$$

$$f'(x) = \left(\frac{1}{x}\right) - \frac{1}{2}(2x) \text{ from the Special Case.}$$

$$= \frac{1}{x} - x$$

Find all critical points. Critical points occur where $f'(x) = 0$.

$$0 = \frac{1}{x} - x$$

$$0 = 1 - x^2 \text{ (Cleared of fractions)}$$

$$x^2 = 1$$

$$x^* = 1 \quad \cancel{x^* = -1}$$

(−1 is not within the domain and should not be considered further.)

If $x^* = 1$,

$$f(1) = \ln 1 - \frac{1}{2} (1)^2 = 0 - \frac{1}{2} = -\frac{1}{2};$$

thus $\left(1, -\dfrac{1}{2}\right)$ is the only critical point.

Classify the critical point.

This time we will use the first-derivative test for practice.

For point $\left(1, -\dfrac{1}{2}\right)$ with $f'(x) = \dfrac{1}{x} - x$ the curve:

On the left: Let $x = \dfrac{1}{2}$, then $f'\left(\dfrac{1}{2}\right) = 2 - \dfrac{1}{2} = +$ is increasing

At point $\left(1, -\dfrac{1}{2}\right)$:   levels off

On the right: Let $x = 2$, then $f'(2) = \dfrac{1}{2} - 2 = -$ is decreasing

The curve is increasing, leveling off, then decreasing.

Answer: The critical point $\left(1, -\dfrac{1}{2}\right)$ is a relative maximum. In other words, the maximum value of $f(x)$ is

$-\dfrac{1}{2}$ and occurs when $x = 1$.

We will conclude the unit with an application example.

EXAMPLE 13

Suppose $1,000 is invested in a money market account that pays interest at a rate of 6.5% per year compounded continuously. How long will it take to amount to $5,000? Reminder: $A(t) = Pe^{rt}$ is the formula for the balance in the account, measured in dollars, after t years when P dollars are invested at rate r per year with interest compounded continuously.

Solution: In this problem we want to find how long it will take for the balance, $A(t)$, to equal $5,000 when $P = 1,000$ and $r = 0.065$. In other words, we are looking for t.

Use formula. $A(t) = Pe^{rt}$

Substitute in values. $5{,}000 = 1{,}000e^{0.065t}$

Simplify. $5 = e^{0.065t}$

Rewrite in logarithmic form. $0.065t = \ln 5$

Use a calculator to evaluate ln 5. $0.065t = 1.6094$

Divide by the coefficient. $t = 24.76$

Answer: It will take 24.76 years for $1,000 invested at 6.5% compounded continuously to reach $5,000.

> # CALCULUS AND THE CALCULATOR (Optional)
>
> The most difficult part of graphing is finding an appropriate viewing window. With natural log functions it is helpful to remember that they are defined for positive numbers only when setting the viewing window's dimensions. Hopefully you have been graphing the natural log functions as you worked your way through the unit. If you haven't, be sure to return to Examples 11 and 12, graph the function, and verify the reasonableness of the analytical answers.

You should now be able to identify a natural logarithmic function, to differentiate it, and to find and classify all of its critical points.

This concludes the section on differential calculus of a single variable. By now you should be able to take the derivative of all polynomial functions, rational functions, exponential functions to the base e, and natural logarithmic functions. A complete list of the rules for differentiation presented in this book is provided below.

A Brief Summary of All the Rules

Power:	If $f(x) = x^n$, then $f'(x) = nx^{n-1}$
Constant:	If $f(x) = c$, then $f'(x) = 0$
Constant · Function:	If $f(x) = c \cdot u(x)$, then $f'(x) = c \cdot u'(x)$
Special Case:	If $f(x) = c \cdot x^n$, then $f'(x) = cnx^{n-1}$
Sum:	If $f(x) = u(x) + v(x)$, then $f'(x) = u'(x) + v'(x)$
Difference:	If $f(x) = u(x) - v(x)$, then $f'(x) = u'(x) - v'(x)$
Product:	If $f(x) = F(x) \cdot S(x)$, then $f'(x) = F(x)S'(x) + S(x)F'(x)$
Quotient:	If $f(x) = \dfrac{N(x)}{D(x)}$, then $f'(x) = \dfrac{D(x)N'(x) - N(x)D'(x)}{[D(x)]^2}$
Chain:	If $f(x) = [u(x)]^n$, then $f'(x) = n[u(x)]^{n-1}u'(x)$
Exponential:	If $f(x) = e^{u(x)}$, then $f'(x) = u'(x)e^{u(x)}$
Special Case:	If $f(x) = e^x$, $f'(x) = e^x$
Logarithmic:	If $f(x) = \ln u(x)$, then $f'(x) = \dfrac{u'(x)}{u(x)}$
Special Case:	If $f(x) = \ln x$, then $f'(x) = \dfrac{1}{x}$

Before beginning the next unit, you should solve the following problems involving logarithmic functions. Reduce answers to lowest terms, but you need not change them to decimal approximations.

EXERCISES

1.   Given $f(x) = \ln(7x - 1)$, find $f'(x)$.

2.   Differentiate $f(x) = \ln 3x$.

3. Differentiate $f(x) = 50 \ln (x + 5)$.

4. If $f(x) = \ln (x^2 - 5)$, find $f'(x)$

5. Find $f'(x)$ if $f(x) = 2x \ln x$.

6. Find $\dfrac{dy}{dx}$ if $y = f(x) = \ln x^2 - 5$.

7. The Wonder Company's sales, $S(x)$, in thousands of dollars, are related to advertising expenditures, x, in thousands of dollars, by the function

$$S(x) = 100 + 500 \ln x \qquad x > 0$$

    a. Find $S(5)$ and interpret.

    b. Find $S'(x)$.

    c. Find $S'(5)$ and interpret.

8. If $f(x) = \dfrac{2x}{\ln x}$, find $f'(x)$.

9. Differentiate $y = x \ln x^2$.

10. If $f(x) = e^x \ln x$, find $f'(x)$.

11. The total cost of producing x units of a product is given by

$$TC(x) = 5{,}000 + 10 \ln (x + 2).$$

    a. Find $TC(3)$ and interpret.

    b. Find $TC'(x)$.

    c. Find $TC'(3)$ and interpret.

C12. a. Find all critical points for $f(x) = \ln (x^2 - 6x + 25)$.

    b. Graph in a standard viewing window.

    c. Classify all critical points.

13. The function

$$N(x) = 900 \ln (0.5x)$$

    denotes the number of items produced by a new worker during the xth day after a training period.

    a. Find $N(3)$ and interpret.

    b. Find $N'(3)$ and interpret.

14. Find and classify all critical points for $f(x) = 4 \ln x - 2x$.

If additional practice is needed:

Barnett, Ziegler, and Byleen, page 116, problems 1–12; pages 328–330, problems 1–74
Harshbarger and Reynolds, page 818, problems 1–47
Hoffmann and Bradley, pages 339–341, problems 13–20 and 24–45
Waner and Costenoble, pages 228–231, problems 1–96

UNIT 24

Integration—Indefinite Integrals

This unit starts with the beginning of what is traditionally called integral calculus. The purpose of the unit is to provide you with an understanding of an indefinite integral. When you have finished the unit, you will be able to integrate a variety of functions.

In this unit we perform an operation that is the reverse of differentiation and is called antidifferentiation or integration. Typically we will be given a function, $f(x)$, and be asked to find a function, $F(x)$, such that $F'(x), = f(x)$. Such a function is called an antiderivative of $f(x)$.

NOTATION AND TERMINOLOGY

Like differentiation, integration has its own notation and terminology.

$$\int f(x)\,dx \ = \ F(x) + C \qquad \text{where } F'(x) = f(x)$$

Here is the terminology for this equation:

$\displaystyle\int f(x)\,dx$ is called the **indefinite integral** of $f(x)$.

$\displaystyle\int$ is called an integral sign.

$f(x)$ is called the integrand and is the quantity to be integrated.

dx indicates that x is the variable with respect to which the integration is to take place.

$F(x)$ is the antiderivative.

C is called the constant of integration.

$\int f(x)\ dx$ is read "the indefinite integral" of f of $x\ dx$ and the process for finding $\int f(x)\ dx$ is called **indefinite integration.** Basically we will be starting with the derivative of a function and working to find the function itself.

We will start with a review differentiation problem from Unit 23, which I can then use as an example to illustrate the above notation.

Problem 1

If $f(x) = x \ln x - x$, find $f'(x)$. Hint: The first term is a product.

Solution:

Answer: $f'(x) = \ln x$

From Problem 1 you found that, if $f(x) = x \ln x - x$, then $f'(x) = \ln x$. In terms of our new notation, we could write

$$\int \ln\ x\ dx = x \ln x - x + C.$$

In other words, if we are asked to find a function whose derivative is $\ln x$, it is $x \ln x - x + C$. I will explain more about C, the constant of integration, after a few more examples. For the time being, accept it as a separate term added on at the end.

In effect, we have just derived our first integration formula.

Indefinite Integral of ln x

$\int \ln x\ dx = x \ln x - x + C$ where C is an arbitrary constant

In words, the indefinite integral of $\ln x$ is $x \ln x - x$, plus the constant of integration.

INTEGRATION OF POWER FUNCTIONS

Next we will work with one of the major rules, that for integrating power functions, and do a few examples to get you started with this topic.

Indefinite Integral of x^n

$$\int x^n\, dx = \frac{n^{n+1}}{n+1} + C \quad \text{for } n \neq -1 \text{ and } C \text{ an arbitrary constant}$$

In words, the indefinite integral of x to the nth power equals x to the $n+1$ power divided by $n+1$, plus the constant of integration.

In still other words, to integrate x to the nth power, we increase the exponent by 1, then divide by the new exponent, and finally add C, the arbitrary constant term, as a separate term.

EXAMPLE 1

Find: $\int x^5\, dx.$

Solution: We are being asked to find a function whose derivative is x^5.

According to the above rule, we increase the exponent by 1, divide by the new exponent, and then add C as a separate term.

$$\int x^5 dx = \frac{x^6}{6} + C$$

The nice part about integration is that we can always check our answer by differentiating.

Check: Let $F(x) = \dfrac{x^6}{6} + C$

$$= \frac{1}{6} x^6 + C$$

then $F'(x) = \dfrac{1}{6} (6x^5) + 0$

$$= x^5$$

Therefore the answer to the integration problem above is correct.

EXAMPLE 2

Find: $\int x^3 dx.$

Solution: We are being asked to find a function whose derivative is x^3.

According to the rule for integrating powers of x, we increase the exponent by 1 and then divide the term by the new exponent.

$$\int x^3 \, dx \ = \ \frac{x^4}{4} + C$$

Check: Let $F(x) = \frac{1}{4} x^4 + C$

then $F'(x) = \frac{1}{4} (4x^3) + 0$

$= x^3$, and the answer checks.

EXAMPLE 3

Find: $\int x \, dx$.

Solution: We are being asked to find a function whose derivative is x.

According to the rule, we increase the exponent by 1 and then divide by the new exponent.

$$\int x \, dx \ = \ \frac{x^2}{2} + C$$

Check: Let $F(x) = \frac{1}{2} x^2 + C$

then $F'(x) = \frac{1}{2} (2x)$

$= x$, and the answer checks.

Let's talk a bit about the constant of integration, C. In Example 3 we were asked to find a function whose derivative was x. There are actually infinitely many such functions. Here are just a few:

$$F_1(x) = \frac{1}{2} x^2 + 1$$

$$F_0(x) = \frac{1}{2} x^2$$

$$F_3(x) = \frac{1}{2} x^2 - 3$$

$$F_5(x) = \frac{1}{2} x^2 + 5$$

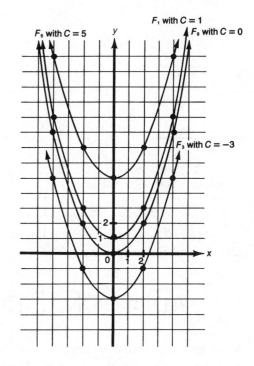

Verify for yourself that each of the functions on the previous page has x as its derivative. In summary, all antiderivatives of $f(x) = x$ are functions of the form

$$F(x) = \frac{1}{2}x^2 + C, \qquad \text{where } C \text{ is an arbitrary constant.}$$

If we were to graph the antiderivatives of $f(x)$, we would have a family of curves. Each is a parabola, differing from the others only in its y-intercept. A few of the curves are illustrated at the right.

EVALUATION OF THE CONSTANT OF INTEGRATION

If the problem states an additional property, often called an initial condition, it is then possible to find C. In other words, we are able to determine the one member of the family of curves that is an antiderivative, as well as satisfying the given initial condition.

EXAMPLE 4

Given $F'(x) = x$, find the function, $F(x)$, whose graph passes through point $(2, 1)$.

Solution: If $F'(x) = x$

then $F(x) = \displaystyle\int x \, dx$

$= \dfrac{x^2}{2} + C$

The problem asks for one curve that passes through point $(2, 1)$, or if $x = 2$, $y = 1$. Substitute and solve for C.

$$F(x) = \frac{x^2}{2} + C$$

$$1 = \frac{(2)^2}{2} + C$$

$$1 = 2 + C$$

$$C = -1$$

Answer: $F(x) = \frac{1}{2}x^2 - 1$

Notice that with an initial condition the family of antiderivatives is narrowed to only the one function that also satisfies the initial condition.

ADDITIONAL RULES FOR INTEGRATION

Remember that with derivatives, when there was more than one term, we took the derivative term by term. With integration the rule states that, if there is more than one term, we also integrate term by term.

Indefinite Integral of a Sum (or Difference)

$$\int [f(x) + g(x)] \, dx = \int f(x) \, dx + \int g(x) \, dx$$

In words, the indefinite integral of a sum (or difference) of two functions is the sum of (or difference) of their individual integrals.

EXAMPLE 5

Find: $\int (x^2 - x^3 + x) \, dx$.

Solution: To integrate more than one term, we integrate term by term.

$$\int (x^2 - x^3 + x) \, dx = \int x^2 \, dx - \int x^3 \, dx + \int x \, dx$$

$$= \left(\frac{x^3}{3} + c_1 \right) - \left(\frac{x^4}{4} + c_2 \right) + \left(\frac{x^2}{2} + c_3 \right)$$

$$= \frac{x^3}{3} - \frac{x^4}{4} + \frac{x^2}{2} + C \quad \text{where } C = c_1 - c_2 + c_3$$

Note: In the future it is not necessary to write the arbitrary constants associated with each integral as I have done above. Instead simply add a single constant of integration as the last term of the final answer.

Indefinite Integral of a Constant Times a Function

$$\int k\ f(x)\ dx = k \int f(x)\ dx$$

In words, the indefinite integral of a constant k times a function $f(x)$ is equal to the constant k times the indefinite integral of the function $f(x)$.

This rule states that, if the function is being multiplied by some constant, the constant can be pulled out in front of the integral sign.

EXAMPLE 6

Find: $\int 6x^2\ dx$.

Solution: $\int 6x^2\ dx = 6 \int x^2\ dx$

$$= 6\left(\frac{x^3}{3} + c_1\right)$$

$$= 2x^3 + 6c_1$$

$$= 2x^3 + C \quad \text{where } 6c_1 \text{ is rewritten as the arbitrary constant } C.$$

Note: As in Example 5, in the future simply add the constant of integrations as a term to the final answer.

Keep in mind that integration is the reverse of differentiation. We are given the derivative and asked to find the original function. Therefore a rule for integration can be obtained by stating in reverse the rule for differentiation, much as we did with Problem 1.

For the next three rules the differentiation rules will be stated, followed by the corresponding rules for integration. The $\frac{dy}{dx}$ notation will be used so as to make the similarity as obvious as possible. Also observe that each of the three rules is exactly the reverse of the differentiation rule. Therefore, if we read the differentiation rule from left to right, it follows that if we read the rule from right to left we have the corresponding integration rule.

EXPONENTIAL RULE: SPECIAL CASE

$$\frac{d(e^x)}{dx} = e^x$$

Indefinite Integral of e^x

$$\int e^x dx = e^x + C$$

In words, the indefinite integral of e^x is e^x.

LOGARITHMIC RULE: SPECIAL CASE

$$\frac{d(\ln x)}{dx} = \frac{1}{x}$$

Indefinite Integral of $\frac{1}{x}$

$$\int \frac{1}{x}\, dx = \ln |x| + C$$

or

$$\int x^{-1}\, dx = \ln |x| + C$$

In words, the indefinite integral of $\frac{1}{x}$ is $\ln |x|$.

Recall that the rule for the integration of powers of x was valid for all n, provided $n \neq -1$. This rule demonstrates how to integrate when n does equal -1.

The absolute value signs are required because, as you recall, logarithms are defined for nonnegative numbers only. The absolute value signs, therefore, guarantee that the number will be nonnegative.

POWER RULE: SPECIAL CASE

$$\frac{d(x)}{dx} = 1$$

Indefinite Integral of 1

$$\int dx = x + C$$

In words, the indefinite integral of 1 is x.

This one often causes trouble. Since the derivative of x is 1, it follows that the integral of 1 is x.

You might want to make a list of the various integration rules for handy reference as we do a variety of examples.

EXAMPLE 7

Find: $\displaystyle\int (3x^2 - 5x + 2)\, dx$.

Solution: $\displaystyle\int (3x^2 - 5x + 2)\, dx = \int 3x^2\, dx - \int 5x\, dx + \int 2\, dx$

$$= 3\int x^2\, dx - 5\int x\, dx + 2\int dx$$

$$= 3\left(\frac{x^3}{3}\right) - 5\left(\frac{x^2}{2}\right) + 2x + C$$

$$= x^3 - \frac{5}{2}x^2 + 2x + C$$

EXAMPLE 8

Find: $\displaystyle\int \frac{1}{x^2}\, dx$.

Solution: $\displaystyle\int \frac{1}{x^2}\, dx = \int x^{-2}\, dx$

$$= \frac{x^{-2+1}}{-1} + C$$

$$= -x^{-1} + C$$

$$= \frac{-1}{x} + C$$

Did you notice in Example 8 the similarity to differentiation? The function was written first as a power of x, then integrated, and finally rewritten with a positive exponent.

From now on I am going to shorten the process by omitting the first step.

EXAMPLE 9

Find: $\displaystyle\int \left(2e^x + \frac{6}{x} - 7x^2\right) dx$.

Solution: $\displaystyle\int \left(2e^x + \frac{6}{x} - 7x^2\right) dx = 2\int e^x\, dx + 6\int \frac{1}{x}\, dx - 7\int x^2\, dx$

$$= 2e^x + 6 \ln |x| - \frac{7x^3}{3} + C$$

Ready to try a few problems? Remember that you can always check your answer by differentiating.

Problem 2

Find: $\displaystyle\int (2x + 3)\, dx.$

Solution:

Answer: $x^2 + 3x + C$

Problem 3

Find: $\displaystyle\int (x^3 - 3x^2 + 5x - 3x^{-1})\, dx.$

Solution:

Answer: $\dfrac{x^4}{4} - x^3 + \dfrac{5x^2}{2} - 3 \ln |x| + C$

Thus far all of the examples of integration of powers have been whole numbers. As you well know, however, often we have to work with radicals or fractional exponents.

EXAMPLE 10

Find: $\displaystyle\int \sqrt{x}\, dx.$

Solution: $\displaystyle\int \sqrt{x}\, dx = \int x^{1/2}\, dx = \dfrac{x^{(1/2)+1}}{\dfrac{3}{2}} + C$

$$= \dfrac{2x^{3/2}}{3} + C$$

$$= \dfrac{2\sqrt{x^3}}{3} + C$$

We will conclude the unit with an example of an application problem.

EXAMPLE 11

A firm produces posters. At a production level of x posters, the marginal cost is $MC = TC'(x) = 5 - 0.001x$. Find the cost function, $TC(x)$, if the fixed cost is known to be $200.

Solution: $TC(x) = \displaystyle\int TC'(x)\, dx$

$$= \int (5 - 0.001x)\, dx$$

$$= 5 \int dx - 0.001 \int x\, dx$$

$$= 5x - 0.001\frac{x^2}{2} + C$$

$$= 5x - 0.0005x^2 + C$$

Since the fixed cost is $200, then $TC(0) = 200$,

and $TC(x) = 5x - 0.0005x^2 + C;$

therefore $200 = 5(0) - 0.0005(0)^2 + C$

$200 = C$

Answer: $TC(x) = 5x - 0.0005x^2 + 200$

You should now be able to identify an indefinite integral. In addition, you should be able to integrate a select group of functions such as e^x, ln x, and powers of x, using the seven rules of integration. And if an initial condition is specified, you should be able to evaluate the constant of integration.

Before continuing to the next unit, do the following exercises. Be sure to simplify your answers.

EXERCISES

Find the indefinite integral (if possible).

1. $\displaystyle\int (2x^4 + x^2)\, dx$

2. $\displaystyle\int \left(x + \frac{1}{x}\right) dx$

3. $\displaystyle\int x^7\, dx$

4. $\displaystyle\int x^{1/3}\, dx$

5. $\displaystyle\int \frac{5}{x}\,dx$

6. $\displaystyle\int 2x^{-5}\,dx$

7. $\displaystyle\int 3\,dx$

8. $\displaystyle\int (15x^2 - \sqrt{2})\,dx$

9. $\displaystyle\int 3e^x\,dx$

10. $\displaystyle\int \ln 3\,dx$

11. $\displaystyle\int 2\ln x\,dx$

12. $\displaystyle\int 5ax^4 q\,dq$

13. $\displaystyle\int \sqrt[3]{x}\,dx$

14. $\displaystyle\int \left(x^3 - \frac{2}{x^2} + 2\right)\,dx$

15. $\displaystyle\int (1 - \ln x)\,dx$

16. $\displaystyle\int \frac{e^x - 2x}{2}\,dx$

17. $\displaystyle\int \frac{50}{x}\,dx$

18. $\displaystyle\int (3x^{-2} + x^{-1})\,dx$

19. $\displaystyle\int (2x^7 + 6x^2 - 1)\,dx$

20. $\displaystyle\int \frac{\sqrt{x}}{2}\,dx$

Use algebra to rewrite each of the following so that the indefinite integral rules can be applied; then complete the integration.

21. $\displaystyle\int x\,(x^2 - 12x + 2)\,dx$

22. $\displaystyle\int x^3\!\left(\frac{2}{x} + 4x\right) dx$

23. $\displaystyle\int (x + 1)(x - 5)\,dx$

24. $\displaystyle\int \frac{1 + x}{x^2}\,dx$

25. $\displaystyle\int \frac{8x^5 + 4x}{x^2}\,dx$

26. Find the equation of the curve, given that $f'(x) = 2x^3 - x + 5$ and that point $(3, 38)$ satisfies $f(x)$.

27. Determine $f(x)$, given that $f'(x) = 2x - 5$ with an initial condition of $f(5) = 4$.

28. The marginal cost function is

$$MC = 6x + 1$$

where $x =$ the number of units produced

and $MC =$ the marginal cost, in dollars.

The fixed cost is \$50. What is the total cost of producing 10 units?

29. Find $\displaystyle\int (3xy^2 + 2x^2y + 7)\,dy$

30. The function describing the marginal profit from producing and selling a product is

$$MP = 100 - 2x$$

where $x =$ the number of units

and $MP =$ the marginal profit, in dollars.

When 10 units are produced and sold, total profit equals \$150. Determine the total profit function.

If additional practice is needed:

Barnett, Ziegler, and Byleen, pages 372–376, problems 1–106
Harshbarger and Reynolds, pages 870–872, problems 1–54; page 900, problems 1–9
Hoffmann and Bradley, pages 381–386, problems 1–49
Waner and Costenoble, pages 323–325, problems 1–64

UNIT 25

Integration by Substitution

This unit will continue our discussion of integration. Specifically, you will learn a technique called integration by substitution.

Below are listed the seven rules of integration from Unit 24;

1. $\int x^n \, dx = \dfrac{x^{n+1}}{n+1} + C \quad n \neq -1$

2. $\int dx = x + C$

3. $\int [\, f(x) + g(x)] \, dx = \int f(x) \, dx + \int g(x) \, dx$

4. $\int c \, f(x) \, dx = c \int f(x) \, dx$

5. $\int e^x \, dx = e^x + C$

6. $\int \dfrac{1}{x} \, dx = \ln |x| + C$

7. $\int \ln x \, dx = x \ln x - x + C$

As you probably guessed, the seven rules apply directly to only a limited number of integration problems. Fortunately, however, the use of the rules can be extended by a rather straightforward technique called integration by substitution. The technique is a procedure whereby certain functions can be integrated by substituting a new variable for a function and integrating with respect to the substituted variable.

The technique can be summarized by the following six steps.

1. Let u equal the quantity *inside* the parentheses and/or the most complicated algebraic expression.

2. Take the derivative $\dfrac{du}{dx}$.

3. "Pretend" to solve for dx by

    a. first cross-multiplying and

    b. then dividing by the coefficient of dx.

4. Substitute the u and du terms into the original problem and simplify.

5. If possible, integrate with respect to u.

6. Change back to the original variable.

EXAMPLE 1

Find: $\displaystyle\int (3x-5)^5 \; dx.$

Solution: If we had unlimited time and nothing better to do, we could multiply the quantity out and integrate term by term, using the rules from Unit 24. But lacking the time and inclination to do that, we use integration by substitution, an easier and certainly faster method.

Using the six steps:

1. Let u equal the quantity *inside* the parentheses.   Let $u = 3x - 5$.

2. Take the derivative $\dfrac{du}{dx}$. $\dfrac{du}{dx} = 3$

3. "Pretend" to solve for dx by

    a. first cross-multiplying and $du = 3 \, dx$

    b. then dividing by the coefficient of dx. $\dfrac{du}{3} = dx$

4. Substitute u and du terms into the original problem and simplify.

    Substitute: $\displaystyle\int (3x-5)^5 dx = \int u^5 \, \dfrac{du}{3}$

    Simplify: $= \dfrac{1}{3}\displaystyle\int u^5 \, du$

5. Integrate with respect to u. $= \dfrac{1}{3} \cdot \dfrac{u^6}{6} + C$

                                                                     $= \dfrac{u^6}{18} + C$

6. Change back to the original variable. $= \dfrac{(3x-5)^6}{18} + C$

EXAMPLE 2

Find: $\displaystyle\int (2-7x)^{2/3}\, dx$.

Solution: 1. Let u equal the quantity *inside* the parentheses.   Let $u = 2 - 7x$.

2. Take the derivative $\dfrac{du}{dx}$. $\dfrac{du}{dx} = -7$

3. "Pretend" to solve for dx by

    a. first cross-multiplying and $du = -7\, dx$

    b. then dividing by the coefficient of dx. $\dfrac{du}{-7} = dx$

4. Substitute u and du terms into the original problem and simplify.

    Substitute: $\displaystyle\int (2-7x)^{2/3}\, dx = \int u^{2/3}\, \dfrac{du}{-7}$

    Simplify: $= \dfrac{-1}{7}\displaystyle\int u^{2/3}\, du$

5. Integrate with respect to u. $= -\dfrac{1}{7}\cdot\dfrac{u^{5/3}}{\dfrac{5}{3}} + C$

                                                  $= \dfrac{-3}{35}u^{5/3} + C$

6. Change back to the original variable. $= \dfrac{-3}{35}(2-7x)^{5/3} + C$

By now I am sure you are ready to try a problem on your own.

Problem 1

Find: $\displaystyle\int (2x+3)^4\, dx$.

Solution: 1. Let u equal the quantity *inside* the parentheses.

2. Take the derivative $\dfrac{du}{dx}$.

3. "Pretend" to solve for dx by

    a. first cross-multiplying and

    b. then dividing by the coefficient of dx.

4. Substitute u and du terms into the original problem and simplify.

    Substitute:

    Simplify:

5. Integrate with respect to u.

6. Change back to the original variable.

$$\text{Answer:} \quad \frac{(2x+3)^5}{10} + C$$

Although the next example appears more difficult, you will soon see it really isn't.

EXAMPLE 3

Find: $\displaystyle\int (x^2 - x + 7)^3 \, (2x - 1) \, dx.$

Solution: Recall that the first step of this technique is to let u equal the quantity inside the parentheses and/or the most complicated algebraic expression. Since we have two quantities in parentheses, let u equal the more complicated, $x^2 - x + 7$.

1. Let u equal the more complicated quantity. $\qquad$ Let $u = x^2 - x + 7.$

2. Take the derivative $\dfrac{du}{dx}$. $\qquad\qquad\qquad\qquad \dfrac{du}{dx} = 2x - 1$

3. "Pretend" to solve for dx by

    a.   first cross-multiplying and $\qquad\qquad\qquad du = (2x - 1) \, dx$

    b.   then dividing by the coefficient of dx. $\qquad \dfrac{du}{(2x - 1)} = dx$

4. Substitute u and du terms into the original problem and simplify.

    Substitute: $\displaystyle\int (x^2 - x + 7)^3 \, (2x - 1) \, dx = \int u^3 (2x - 1) \, \frac{du}{(2x - 1)}$

    Simplify: $\displaystyle = \int u^3 (2x - 1) \, \frac{du}{(2x - 1)}$

    Note: the common factors
    $(2x - 1)$ cancel out. $\displaystyle = \int u^3 \, du$

5. Integrate with respect to u.

$$= \frac{u^4}{4} + C$$

6. Change back to the original variable.

$$= \frac{(x^2 - x + 7)^4}{4} + C$$

Integration by substitution works for a variety of problems. We will go through a few more examples and then let you try some problems.

From now on I will put the first three steps off to the left to provide a more condensed format for working the examples.

EXAMPLE 4

Find: $\displaystyle\int \frac{8x}{(4x^2 + 1)^2}\, dx.$

Solution:

$$\int \frac{8x}{(4x^2 + 1)^2}\, dx = \int \frac{8x}{u^2}\frac{du}{8x}$$

Let $u = 4x^2 + 1$.

$$= \int \frac{\cancel{8x}}{u^2}\frac{du}{\cancel{8x}}$$

$\dfrac{du}{dx} = 8x$

$$= \int \frac{du}{u^2}$$

$du = 8x\, dx$

$$= \int u^{-2}\, du$$

$\dfrac{du}{8x} = dx$

$$= \frac{u^{-2+1}}{-2+1} + C$$

$$= \frac{u^{-1}}{-1} + C$$

$$= \frac{-1}{u} + C$$

$$= \frac{-1}{4x^2 + 1} + C$$

The next example contains a quantity under a radical sign. As when we were taking derivatives, we will need to first rewrite the expression using a fractional exponent, then proceed as usual with integration by substitution.

EXAMPLE 5

Find: $\displaystyle\int x\sqrt{x^2+5}\ dx$.

Solution:

$$\int x\sqrt{x^2+5}\ dx = \int x(x^2+5)^{1/2}\ dx$$

Let $u = x^2 + 5$.

$$= \int x\ u^{1/2}\ \frac{du}{2x}$$

$$\frac{du}{dx} = 2x$$

$$= \int \cancel{x}\ u^{1/2}\ \frac{du}{2\cancel{x}}$$

$$du = 2x\ dx$$

$$= \frac{1}{2}\int u^{1/2}\ du$$

$$\frac{du}{2x} = dx$$

$$= \frac{1}{2}\cdot\frac{u^{3/2}}{\dfrac{3}{2}} + C$$

$$= \frac{1}{3}u^{3/2} + C$$

$$= \frac{(x^2+5)^{3/2}}{3} + C$$

Now it is time for you to try a few. We will start with some easy problems and work up to more difficult ones.

Problem 2

Find: $\displaystyle\int (10x-3)(5x^2-3x+17)^{1/7}\ dx$.

Solution:

Answer: $\dfrac{7}{8}(5x^2-3x-17)^{8/7} + C$

Problem 3

Find: $\int x(3x^2 + 2)^3 \, dx.$

Solution:

Answer: $\dfrac{(3x^2 + 2)^4}{24} + C$

Problem 4

Find: $\int \dfrac{dx}{(4x + 2)^2}.$

Solution:

Answer: $\dfrac{-1}{4(4x + 2)} + C$

Let's try two more; then I will explain some additional examples of the technique. The next problem looks harder than it really is. I am confident you will be able to solve it without any difficulty.

Problem 5

Find: $\int [(x - 1)^5 + 3\,(x - 1)^2 + 6(x - 1) + 5] \, dx.$

Solution:

Hint: Let $u = x - 1$.

Answer: $\dfrac{1}{6}(x - 1)^6 + (x - 1)^3 + 3(x - 1)^2 + 5(x - 1) + C$

Problem 6

Find: $\int x \sqrt[3]{3x^2 + 5}\, dx.$

Solution:

Answer: $\dfrac{1}{8}(3x^2 + 5)^{4/3} + C$

The next example involves the integration of a logarithm. While the technique remains the same, an additional step is required to simplify the final answer.

EXAMPLE 6

Find: $\int (6x - 1) \ln (3x^2 - x + 7)\, dx.$

Solution:

$$\int (6x - 1) \ln (3x^2 - x + 7)\, dx = \int \cancel{(6x-1)} \ln u \frac{du}{\cancel{(6x-1)}}$$

Let $u = 3x^2 - x + 7.$

$$= \int \ln u \, du$$

$$\frac{du}{dx} = 6x - 1$$

$$= u \ln u - u + C$$

$$du = (6x - 1)\, dx$$

$$= (3x^2 - x + 7) \ln (3x^2 - x + 7) - (3x^2 - x + 7) + C$$

$$\frac{du}{(6x + 1)} = dx$$

$$= (3x^2 - x + 7) \ln (3x^2 - x + 7) - 3x^2 + x - 7 + C$$

$$= (3x^2 - x + 7) \ln (3x^2 - x + 7) - 3x^2 + x + C$$

Be sure to notice what happened in the final steps of this problem. The parentheses were removed and the terms − 7 and + C were combined into a new constant of integration, still denoted simply by C.

Integration by substitution can be used with exponential functions as well, the only difference being that it is usually desirable to let u equal the exponent of e rather than the quantity in parentheses. As you will note, the exponent is the more complicated expression.

EXAMPLE 7

Find: $\displaystyle\int (2x+3)e^{x^2+3x+5}\ dx.$

Solution:

$$\int (2x+3)e^{x^2+3x+5}\ dx = \int \cancel{(2x+3)}e^u \frac{du}{\cancel{(2x+3)}}$$

Let $u = x^2 + 3x + 5$.

$$= \int e^u\ du$$

$$\frac{du}{dx} = 2x + 3$$

$$= e^u + C$$

$$du = (2x+3)\ dx$$

$$= e^{x^2+3x+5} + C$$

$$\frac{du}{2x+3} = dx$$

Problem 7

Find: $\displaystyle\int 2x\ e^{x^2-1}\ dx.$

Solution:

Answer: $e^{x^2-1} + C$

Problem 8

Find: $\displaystyle\int x\ e^{x^2}\ dx.$

Solution:

Answer: $\frac{1}{2}e^{x^2} + C$

Now we will consider one last situation, that of the integration of a rational function, which is the quotient of two polynomials. With a quotient no parentheses appear; but if we consider the numerator and denominator as separate functions, the procedure remains the same. In other words, let u be the more complicated expression, either the numerator or the denominator, and proceed as usual.

EXAMPLE 8

Find: $\displaystyle\int \frac{2}{x+1}\, dx$.

Solution:

$$\int \frac{2}{x+1}\, dx = \int \frac{2}{u}\, du$$

Let $u = x + 1$.

$$= 2 \int \frac{du}{u}$$

$$\frac{du}{dx} = 1$$

$$= 2 \ln |u| + C$$

$$du = dx$$

$$= 2 \ln |x + 1| + C$$

Problem 9

Find: $\displaystyle\int \frac{\ln 3x}{x}\, dx$.

Solution:

Answer: $\dfrac{1}{2}(\ln 3x)^2 + C$

By now you should be wondering what happens if the factors involving the original variable do *not* cancel out so conveniently after the substitution step. The next and final example should answer the question.

EXAMPLE 9

Find: $\displaystyle\int (x^3 - 1)^{1/2}\ dx$.

Solution:
$$\int (x^3 - 1)^{1/2}\ dx = \int u^{1/2}\ \frac{du}{3x^2}$$

$$= \frac{1}{3}\int u^{1/2}\ \frac{du}{x^2}$$

$$= \frac{1}{3}\int \frac{u^{1/2}\ du}{x^2}$$

Let $u = x^3 - 1$.

$$\frac{du}{dx} = 3x^2$$

$$du = 3x^2\ dx$$

$$\frac{du}{3x^2} = dx$$

Observe that after substituting we are left with a factor of x^2 in the denominator and no matching factor in the numerator with which to cancel it. Also, $\frac{1}{x^2}$ cannot be taken outside the integral because x^2 is *not* a constant. Thus, we are left with an integration problem containing both variables. To finish the problem requires methods beyond the scope of this unit. A method of approach will be discussed in Unit 26.

You should be able by now to integrate a variety of problems—many which contain radicals, exponentials, logarithms, or quotients—using a technique called integration by substitution.

Remember that the technique can be summarized by the following six steps:

1. Let u equal the quantity *inside* the parentheses and/or the most complicated algebraic expression.

2. Take the derivative.

3. "Pretend" to solve for dx.

4. Substitute and simplify.

5. Integrate with respect to u.

6. Change back to the original variable.

Before beginning the next unit, you should do the following exercises. When you have completed them, check your answers against those at the back of the book.

EXERCISES

Find:

1. $\displaystyle\int (x-2)^5 \, dx$

2. $\displaystyle\int e^{3-5x} \, dx$

3. $\displaystyle\int (x^4 + 2x)^3 \, (4x^3 + 2) \, dx$

4. $\displaystyle\int \frac{x}{(5+x^2)^4} \, dx$

5. $\displaystyle\int (2x^5 + 1)^2 \, 10x^4 \, dx$

6. $\displaystyle\int \frac{2x^3}{x^4 + 1} \, dx$

7. $\displaystyle\int (x^3 - 5)^7 x^2 \, dx$

8. $\displaystyle\int (2x-2) \, e^{x^2 - 2x + 1} \, dx$

9. $\displaystyle\int x e^{x^2 + 1} \, dx$

10. $\displaystyle\int 3x^2 \ln (x^3 - 2) \, dx$

11. $\displaystyle\int \frac{(\ln x)^2}{x} \, dx$

12. $\displaystyle\int \frac{1}{x+1} \, dx$

13. $\displaystyle\int \frac{6x^2}{x^3 + 2} \, dx$

14. $\displaystyle\int 36x^2 \sqrt{6x^3 + 1} \, dx$

15. $\displaystyle\int x \sqrt{4x^2 - 2} \, dx$

16. $\displaystyle\int 2x\,(5x^3 - 2x)^7\,dx$

17. $\displaystyle\int \frac{e^x}{7 + e^x}\,dx$

18. $\displaystyle\int \frac{4x^2}{\sqrt{x^3 - 7}}\,dx$

If additional practice is needed:

Barnett, Ziegler, and Byleen, pages 386–388, problems 1–81
Hoffmann and Bradley, pages 394–398, problems 1–45
Waner and Costenoble, pages 333–334, problems 1–66

UNIT 26

Integration Using Formulas

This unit continues our discussion of integration. When you have finished the unit you will be able to integrate a variety of functions using integration formulas.

At the conclusion of Unit 25, Example 9 illustrated a problem that we were unable to integrate with the methods and rules available to us at that point. If your future plans include the need to do extensive integration, I suggest you obtain a good set of integration formulas. At last count, there were over 700 rules for integration. These can be found in almost any mathematical handbook. Do understand, however, that even with a complete table of integration formulas, not all functions can be integrated.

A condensed version of the more frequently used formulas is provided below for your convenience. The seven rules from Unit 24 appear at the beginning. In the list, x is the variable and all other letters are constants. The constant of integration has been omitted. The numbers at the left are for reference purposes in the examples to follow.

SELECTED INTEGRATION FORMULAS

1. $\displaystyle\int x^n \, dx = \frac{x^{n+1}}{n+1}$ provided $n \neq -1$

2. $\displaystyle\int dx = x$

3. $\displaystyle\int [f(x) + g(x)] \, dx = \int f(x) \, dx + \int g(x) \, dx$

4. $\displaystyle\int c \, f(x) \, dx = c \int f(x) \, dx$

5. $\displaystyle\int e^x \, dx = e^x$

6. $\displaystyle\int \frac{dx}{x} = \ln|x|$

7. $\displaystyle\int \ln x \, dx = x \ln x - x$

8. $\displaystyle\int \frac{dx}{x \ln x} = \ln|\ln x|$

9. $\displaystyle\int x \, e^x \, dx = (x - 1)e^x$

10. $\displaystyle\int \frac{dx}{a^2 - x^2} = \frac{1}{2a} \ln \left|\frac{a + x}{a - x}\right|$

11. $\displaystyle\int \frac{dx}{x^2 - a^2} = \frac{1}{2a} \ln \left|\frac{x - a}{x + a}\right|$

12. $\displaystyle\int \frac{dx}{\sqrt{x^2 - a^2}} = \ln\left|x + \sqrt{x^2 - a^2}\right|$

13. $\displaystyle\int \frac{dx}{\sqrt{x^2 + a^2}} = \ln\left(x + \sqrt{x^2 + a^2}\right)$

14. $\displaystyle\int \frac{dx}{x^2 \sqrt{x^2 \pm a^2}} = \mp\frac{\sqrt{x^2 \pm a^2}}{a^2 x}$

Note: If the plus sign appears under the radical to be integrated, which is the top sign, the answer uses the top sign on the right side of the formula, which is the minus in front of the radical and the plus sign under the radical. Example 3 will illustrate the use of this formula.

15. $\displaystyle\int \frac{dx}{(x^2 \pm a^2)^{3/2}} = \frac{\pm x}{a^2 \sqrt{x^2 \pm a^2}}$

16. $\displaystyle\int x^n \ln x \, dx = x^{n+1}\left(\frac{\ln x}{n + 1} - \frac{1}{(n + 1)^2}\right) \qquad n \neq -1$

17. $\displaystyle\int \frac{x^n \, dx}{\sqrt{a + bx}} = \frac{2x^n \sqrt{a + bx}}{b(2n + 1)} - \frac{2an}{b(2n + 1)} \int \frac{x^{n-1} \, dx}{\sqrt{a + bx}}$

18. $\displaystyle\int \frac{dx}{x^n \sqrt{a + bx}} = -\frac{\sqrt{a + bx}}{a(n - 1)x^{n-1}} - \frac{b(2n - 3)}{2a(n - 1)} \int \frac{dx}{x^{n-1}\sqrt{a + bx}} \qquad n \neq -1$

19. $\int \sqrt{(x^2 \pm a^2)^n}\, dx = \dfrac{x\sqrt{(x^2 \pm a^2)^n}}{n+1} \pm \dfrac{na^2}{n+1}\int \sqrt{(x^2 \pm a^2)^{n-2}}\, dx \quad n \neq -1$

20. $\int x^n e^x\, dx = x^n e^x - n\int x^{n-1} e^x\, dx$

21. $\int \dfrac{e^x\, dx}{x^n} = \dfrac{e^x}{(n-1)x^{n-1}} + \dfrac{1}{n-1}\int \dfrac{e^x\, dx}{x^{n-1}} \quad n \neq 1$

22. $\int x^n (\ln x)^m\, dx = \dfrac{x^{n+1}}{n+1}(\ln x)^m - \dfrac{m}{n+1}\int x^n (\ln x)^{m-1}\, dx \quad n \neq -1$

23. $\int e^{ax}\, dx = \dfrac{e^{ax}}{a}$

24. $\int x^n e^{ax}\, dx = \dfrac{x^n e^{ax}}{a} - \dfrac{n}{a}\int x^{n-1} e^{ax}\, dx$

25. $\int \left((ax+b)^n\, dx = \dfrac{(ax+b)^{n+1}}{(n+1)\,a}\right)$

26. $\int x(ax+b)^{m/2}\, dx = \dfrac{2(ax+b)^{(m+4)/2}}{a^2(m+4)} - \dfrac{2b(ax+b)^{(m+2)/2}}{a^2(m+2)}$

27. $\int \dfrac{dx}{x(ax+b)} = \dfrac{1}{b}\ln\left|\dfrac{x}{ax+b}\right|$

As I stated earlier, if you are required to do a great deal of integration, extensive lists of formulas are available in mathematical handbooks. The above list of 27 are the ones I believe that you are most apt to encounter. In most sets of tables, the formulas are grouped into similar types. Grouping was not quite possible with this list because of its limited size.

USING INTEGRATION FORMULAS

To integrate using formulas involves finding the rule that matches the given function to be integrated, determining the values of the various constants, and finally substituting those values into the right-hand side of the formula. Often this is a one-step procedure, as some of the following examples will illustrate.

EXAMPLE 1

Find: $\displaystyle\int \dfrac{dx}{x(2x+3)}$.

Solution: The integral fits the pattern of Rule 27.

$$\int \frac{dx}{x(ax+b)} = \frac{1}{b} \ln \left| \frac{x}{ax+b} \right|$$

$$\int \frac{dx}{x(2x+3)} =$$

With $a = 2$

and $b = 3$,

substitute into the right-hand side of the formula.

$$= \frac{1}{3} \ln \left| \frac{x}{2x+3} \right| + C$$

EXAMPLE 2

Find: $\displaystyle \int \frac{dx}{\sqrt{x^2 + 25}}$.

Solution: The integral fits the pattern of Rule 13.

$$\int \frac{dx}{\sqrt{x^2 + a^2}} = \ln \left(x + \sqrt{x^2 + a^2} \right)$$

$$\int \frac{dx}{\sqrt{x^2 + 25}} =$$

With $a^2 = 25$,

substitute into the right-hand side of the formula.

$$= \ln \left(x + \sqrt{x^2 + 25} \right) + C$$

EXAMPLE 3

Find: $\displaystyle \int \frac{dx}{x^2 \sqrt{x^2 - 9}}$.

Solution: The integral fits the pattern of rule 14 using the bottom signs.

$$\int \frac{dx}{x^2 \sqrt{x^2 \pm a^2}} = \mp \frac{\sqrt{x^2 \pm a^2}}{a^2 x}$$

$$\int \frac{dx}{x^2 \sqrt{x^2 - 9}} =$$

With $a^2 = 9$,

substitute into the right-hand side of the formula.

$$= \frac{\sqrt{x^2 - 9}}{9x} + C$$

You should find Problems 1 and 2 relatively short. Let me help you with one, and then do the second completely on your own.

Problem 1

Find: $\int \frac{dx}{4 - x^2}$.

Solution: The integral fits the pattern of rule 10.

$$\int \frac{dx}{a^2 - x^2} = \frac{1}{2a} \ln \left| \frac{a + x}{a - x} \right|$$

$$\int \frac{dx}{4 - x^2}$$

With $a = 2$, substitute.

Answer: $\frac{1}{4} \ln \left| \frac{2 + x}{2 - x} \right| + C$

This time you're on your own.

Problem 2

Find: $\int \frac{dx}{x(5x + 7)}$.

Solution:

Answer: $\dfrac{1}{7} \ln\left|\dfrac{x}{5x+7}\right| + C$

As you probably noticed, some of the formulas contain integral signs on the right-hand side as well. When using one of those formulas, you must continue to integrate the right-hand side until all the integral signs are removed.

EXAMPLE 4

Find: $\displaystyle\int xe^x \, dx.$

Solution: The integral fits the pattern of rule 20.

$$\int x^n e^x \, dx \quad = x^n e^x - n \int x^{n-1} e^x \, dx$$

$$\int xe^x \, dx \quad =$$

With $n = 1$,
substitute. $= xe^x - \displaystyle\int x^0 e^x \, dx$

$$= xe^x - \int e^x \, dx$$

Use rule 5. $= xe^x - e^x + C$

We will do another example, which requires additional integration steps.

EXAMPLE 5

Find: $\displaystyle\int (\ln x)^2 \, dx.$

Solution: Although it may not be obvious at first glance, the integral fits the pattern of rule 22 because the function can be rewritten using x^0, which equals 1.

$$\int x_1^n (\ln x)^m dx \qquad = \frac{x^{n+1}(\ln x)^m}{n+1} - \frac{m}{n+1}\int x^n (\ln x)^{m-1} dx$$

$$\int x^0 (\ln x)^2 dx \qquad =$$

With $n = 0$
and $m = 2$,
substitute.

$$= \frac{x(\ln x)^2}{1} - \frac{2}{1}\int x^0 (\ln x)^{2-1} dx$$

$$= x(\ln x)^2 - 2\int (\ln x)\, dx$$

Use rule 7.

$$= x(\ln x)^2 - 2(x \ln x - x) + C$$

$$= x(\ln x)^2 - 2x \ln x + 2x + C$$

A word of warning: Now that you have choices regarding which formula to use, the required number of steps and the final form of your answer can vary greatly. Consider the next problem.

Problem 3

Find: $\displaystyle\int \frac{x\, dx}{\sqrt{1+x}}$ using the specified rules.

Solution: Step 1:   Use rule 17 with $n = 1$, $a = 1$, and $b = 1$.

Step 2:   Use rule 25 with $n = -\dfrac{1}{2}$, $a = 1$, and $b = 1$.

Answer: $\displaystyle\frac{2x\sqrt{1+x} - 4\sqrt{1+x}}{3} + C$

For a change of pace I had you do the hard problem; I will take an easy example.

EXAMPLE 6

Find: $\displaystyle\int \frac{x\, dx}{\sqrt{1+x}}$.

Solution: By way of showing you the choices, I am going to solve the same problem that you did, but I will use rule 26 after rewriting the function.

$$\int x(ax+b)^{m/2}\,dx = \frac{2(ax+b)^{(m+4)/2}}{a^2(m+4)} - \frac{2b(ax+b)^{(m+2)/2}}{a^2(m+2)}$$

$$\int x(x+1)^{-1/2}\,dx =$$

With $a = 1$,

and $b = 1$,

and $m = -1$,

substitute.
$$= \frac{2(x+1)^{(-1+4)/2}}{(-1+4)} - \frac{2(x+1)^{(-1+2)/2}}{(-1+2)} + C$$

$$= \frac{2(x+1)^{3/2}}{3} - 2(x+1)^{1/2} + C$$

Observe two things. One, notice how much shorter the second method is than the procedure you used to work the problem. Second, notice the apparently different answers. Actually the two answers are equivalent, but their forms are different. The same situation may very well happen with the exercises at the end of the unit. Remember that you can always check by differentiating.

You should now be able to integrate a variety of functions using a table of integrals. A list of some of the more commonly used formulas was provided at the beginning of this unit. Remember that, even with extensive formulas available, not all functions can be integrated.

Before beginning the next unit, use the formulas provided in this unit to do the following exercises. Check your answers with those in the back of the book. If you used a different formula than I did, verify that your answer is correct by differentiating.

EXERCISES

Find each of the following:

1. $\displaystyle\int \frac{dx}{\sqrt{x^2-7}}$

2. $\displaystyle\int e^{2x}\,dx$

3. $\displaystyle\int xe^x\,dx$

4. $\displaystyle\int \frac{dx}{x^2-1}$

5. $\displaystyle\int \frac{dx}{\sqrt{x^2 + 81}}$

6. $\displaystyle\int (2x - 3)^3 \, dx$

7. $\displaystyle\int \frac{dx}{x \ln x}$

8. $\displaystyle\int \frac{dx}{\sqrt{x^2 - 100}}$

9. $\displaystyle\int \frac{dx}{x(10x + 3)}$

10. $\displaystyle\int x^5 \ln x \, dx$

11. $\displaystyle\int \frac{14 \, dx}{x(2 + 3x)}$

12. $\displaystyle\int x^2 e^x \, dx$

13. $\displaystyle\int \frac{dx}{x^2 \sqrt{x^2 - 9}}$

14. $\displaystyle\int x(x - 6)^{1/2} \, dx$

15. $\displaystyle\int \frac{dx}{2x \ln x}$

16. $\displaystyle\int x^3 (\ln x)^2 \, dx$

17. $\displaystyle\int x^2 e^{8x} \, dx$

18. $\displaystyle\int x^2 e^{3x} \, dx$

If additional practice is needed:

Barnett, Ziegler, and Byleen, pages 482–483, problems 1–56
Harshbarger and Reynolds, pages 970–971, problems 1–42

UNIT 27

Integration—Definite Integrals

In this unit you will be introduced to the concept of a definite integral. Upon completion of this unit you will be able to evaluate a definite integral and to determine the area under a curve or between two curves.

We start with the Fundamental Theorem of Calculus.

Fundamental Theorem of Calculus

If $f(x)$ is a continuous function from $x = a$ to $x = b$, then

$$\int_a^b f(x)\ dx = F(x)\Big|_a^b = F(b) - F(a), \quad \text{where } F'(x) = f(x)$$

$\int_a^b f(x)\ dx$ is read "the definite integral of $f(x)$ from $x = a$ to $x = b$," with a and b called the **limits of integration**.

In other words, the definite integral is the numerical value of the antiderivative evaluated at the upper limit minus the antiderivative evaluated at the lower limit.

PROCEDURE FOR DETERMINING THE DEFINITE INTEGRAL

To calculate a definite integral, which as stated above results in a numerical value, requires a straight-forward four-step procedure.

1. Integrate the function.

2. Evaluate the antiderivative at the upper limit.

3. Evaluate the antiderivative at the lower limit.

4. Subtract the value found in step 3 from the value found in step 2.

Examples 1–4 will illustrate the procedure.

EXAMPLE 1

Find: $\displaystyle\int_{1}^{2} 2x\ dx$.

Solution: $\displaystyle\int_{1}^{2} f(x)\ dx = F(x)\Big|_{1}^{2} = F(2) - F(1)$

$\displaystyle\int_{1}^{2} 2x\ dx = \frac{2x^2}{2} + C\Big|_{1}^{2} = [(2)^2 + C] - [(1)^2 + C]$

$= [4 + C] - [1 + C]$

$= 4 + C - 1 - C$

$= 3$

Observe that the constant of integration cancels out, as will always happen. Therefore it is acceptable to omit C entirely from the calculations when determining a definite integral.

EXAMPLE 2

Find: $\displaystyle\int_{-1}^{2} x^4\ dx$.

Solution: $\displaystyle\int_{-1}^{2} f(x)\ dx = F(x)\Big|_{-1}^{2} = F(2) - F(-1)$

$\displaystyle\int_{-1}^{2} x^4\ dx = \frac{x^5}{5}\Big|_{-1}^{2} = \frac{[(2)^5]}{5} - \frac{[(-1)^5]}{5}$

$= \frac{32}{5} - \frac{(-1)}{5}$

$= \frac{32 + 1}{5}$

$= \frac{33}{5}$

The operation we are performing is referred to as integrating between limits.

EXAMPLE 3

Find: $\displaystyle\int_0^1 (1-x)^2 \, dx$.

Solution: $\displaystyle\int_0^1 f(x) \, dx = F(x) \Big|_0^1 = F(1) - F(0)$

$$\int_0^1 (1-x^2) \, dx = x - \frac{x^3}{3} \Big|_0^1 = \left(1 - \frac{1}{3}\right) - [0 - 0]$$

$$= \frac{2}{3}$$

EXAMPLE 4

Find: $\displaystyle\int_1^5 (x+2) \, dx$.

Solution: $\displaystyle\int_1^5 f(x) \, dx = F(x) \Big|_1^5 = F(5) - F(1)$

$$\int_1^5 (x+2) \, dx = \frac{x^2}{2} + 2x \Big|_1^5 = \left(\frac{25}{2} + 10\right) - \left(\frac{1}{2} + 2\right)$$

$$= \frac{25}{2} + 10 - \frac{1}{2} - 2$$

$$= 12 + 8$$

$$= 20$$

When there are fractions present as in Example 4, I tend to remove the parentheses before collecting terms. Typically it makes the problem easier by allowing you to combine the fractions with the same denominator first. If you are working with decimals, it doesn't make any difference.

Are you ready to try a few? Work *all* of the following problems. Don't skip any because I intend to use the problems as illustrations of basic properties of a definite integral after you have finished.

Problem 1

Find: $\displaystyle\int_1^2 x \, dx$.

Solution:

Answer: 1.5

Problem 2

Find: $\displaystyle\int_{-1}^{1} x^3 \, dx$.

Solution:

Answer: 0

Problem 3

Find: $\displaystyle\int_{1}^{3} (x^3 + x) \, dx$.

Solution:

Answer: 24

Problem 4

Find: $\displaystyle\int_{3}^{1} (x^3 + x) \, dx$.

Solution:

Problem 5

Find: $\displaystyle\int_{1}^{1}(-2x^3 + 5x^2 + 3x + 1)\,dx.$

Solution:

Answer: 0

I will do one more example before going on to something new. Why am I doing Example 5 rather than you? The reason is that the limits are in terms of a constant, a, rather than being numerical values. The technique, however, is the same.

EXAMPLE 5

Find: $\displaystyle\int_{a}^{2a}(a + x)\,dx.$

Solution: $\displaystyle\int_{a}^{2a} f(x)\,dx = F(x)\Big|_{a}^{2a} = F(2a) - F(a)$

$$\int_{a}^{2a}(a + x)\,dx = ax + \frac{x^2}{2}\Big|_{a}^{2a} = \left[a(2a) + \frac{(2a)^2}{2}\right] - \left[a^2 + \frac{a^2}{2}\right]$$

$$= 2a^2 + \frac{4a^2}{2} - a^2 - \frac{a^2}{2}$$

$$= a^2 + \frac{3a^2}{2}$$

$$= \frac{5a^2}{2}$$

PROPERTIES OF A DEFINITE INTEGRAL

A definite integral has the following basic properties.

1. $\displaystyle\int_a^b f(x)\, dx = -\int_b^a f(x)\, dx$

2. $\displaystyle\int_a^a f(x)\, dx = 0$

3. $\displaystyle\int_a^b f(x)\, dx = \int_a^c f(x)\, dx + \int_c^b f(x)\, dx \quad \text{where } a \le c \le b$

The basic properties can, if recognized and used correctly, save time during the calculations for a definite integral.

The first property states that if the limits are interchanged the absolute value of the definite integral remains the same, but the sign changes. In Problem 3, you found that the definite integral of $(x^3 + x)$ integrated from $x = 1$ to $x = 3$ was 24. In Problem 4 the same function integrated from $x = 3$ to $x = 1$ was -24, or the same absolute value, but with the opposite sign.

The second property should be obvious. If the limits of integration are equal, the difference between the antiderivatives evaluated at the upper limit and the lower limit is zero. Problem 5 was an illustration of this property.

The third property indicates that the value of a definite integral integrated from its lower limit to its upper limit is the same as if an intermediate value is used between the two limits, with each definite integral calculated separately and then their values combined.

The next problem will be used to verify the third property.

Problem 6

Verify: $\displaystyle\int_{-1}^5 2x\, dx = \int_{-1}^1 2x\, dx + \int_1^5 2x\, dx.$

Solution:

Answer: $24 = 0 + 24$

APPLICATIONS

If this were a book for mathematics majors or pre-engineering students, which it is not, we would just be introducing the real strength of integral calculus—the solution of application problems. However, for the majority of readers of this book, application problems requiring integration are somewhat limited. Two examples should be sufficient.

EXAMPLE 6

Professor Haymaker is the director of the art gallery at a private college. Tomorrow morning is the scheduled opening of a student show, followed by a luncheon for the artists. The professor is trying to decide whether or not an attendant is needed at the gallery while he is attending the luncheon. He estimates that t hours after the doors open at 9 A.M. visitors will be entering at the rate of $-4t^3 + 54t^2$ people per hour.

How many visitors does Professor Haymaker estimate will visit the gallery during the luncheon between noon and 2 P.M.?

Solution: The problem states that visitors will be entering at the **rate** of $-4t^3 + 54t^2$, which indicates that the function is a derivative, say $F'(t)$.

We need to determine the antiderivative, $F(t)$, which would represent the total number of visitors entering since 9 A.M., and to estimate the number that will visit between noon and 2 P.M. by evaluating $F(5) - F(3)$. ($x = 3$ corresponds to noon, that is, 3 hours from 9 A.M., and $x = 5$ corresponds to 2 P.M.) $F(5) - F(3)$ is nothing more than the definite integral evaluated from $x = 3$ to $x = 5$.

What we need to find, then, is

$$\int_3^5 (-4t^3 + 54t^2)\,dt = \left. \frac{-4t^4}{4} + \frac{54t^3}{3} \right|_3^5$$

$$= \left. -t^4 + 18t^3 \right|_3^5$$

$$= [-(5)^4 + 18(5)^3] - [-(3)^4 + 18(3)^3]$$

$$= (-625 + 2{,}250) - (-81 + 486)$$

$$= -625 + 2{,}250 + 81 - 486$$

$$= 1{,}220$$

Answer: Professor Haymaker estimates that 1,220 visitors will enter the gallery between noon and 2 P.M. on opening day.

EXAMPLE 7

A manufacturing firm knows that its marginal cost is given by the function

$$MC = C'(x) = 12x + 8$$

where x = the number of units manufactured

and MC = the marginal cost per unit, in dollars.

Determine the total change in cost if production rises from 20 to 25 units.

Solution: The total change in cost is determined by evaluating $C(25) - C(20)$, but that is nothing more than the definite integral evaluated from $x = 20$ to $x = 25$.

$$\text{The change in cost} = \int_{20}^{25} C'(x)\, dx$$

$$= \int_{20}^{25} (12x + 8)\, dx$$

$$= \frac{12x^2}{2} + 8x \,\Big|_{20}^{25}$$

$$= 6x^2 + 8x \,\Big|_{20}^{25}$$

$$= [6(25)^2 + 8(25)] - [6(20)^2 + 8(20)]$$

$$= (3{,}750 + 200) - (2{,}400 + 160)$$

$$= 1{,}390$$

Answer:　　If production rises from 20 to 25 units, the total change in cost will be $1,390.

AREA UNDER A CURVE

We can now state a theorem that relates the area under a curve to the definite integral.

The definite integral can be interpreted as the area under a continuous and nonnegative curve $y = f(x)$, bounded by the x–axis on the bottom, and by the vertical lines $x = a$ on the left and $x = b$ on the right.

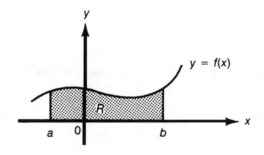

THEOREM

The area of $R = \int_{a}^{b} f(x)\, dx$

$$= F(x) \,\Big|_{a}^{b}$$

$$= F(b) - F(a)$$

provided that $f(x)$ is continuous and nonnegative over the interval $a \le x \le b$.

Observe that the theorem requires that $f(x)$ be nonnegative over the interval. Therefore graphs will be required to verify that the function is, indeed, nonnegative.

EXAMPLE 8

Find the area between the x-axis and the curve $f(x) = 5x$ from $x = 0$ to $x = 4$.

Solution: A sketch of $f(x)$ and the desired area appear below. The shaded area is computed by the definite integral.

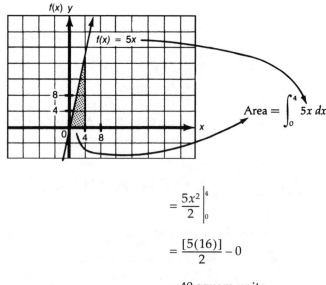

$$= \frac{5x^2}{2} \Big|_0^4$$

$$= \frac{[5(16)]}{2} - 0$$

$$= 40 \text{ square units}$$

Note that the answers to problems of this type will be in square units because we are determining areas.

What would have been the answer to Example 8 if you had used the formula from geometry for finding the area of a triangle?

EXAMPLE 9

Find the area bounded by the curve $f(x) = 3x^2$, the x-axis, and the vertical line $x = 2$.

Solution: A sketch of $f(x)$ and the indicated area is provided. Observe that $f(x)$ is nonnegative over the desired interval $0 \leq x \leq 2$ as required. The shaded area can be determined by the definite integral.

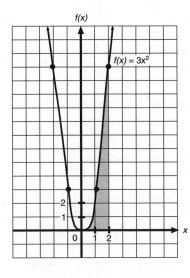

$$\text{Area} = \int_0^2 3x^2\, dx$$

$$= \frac{3x^3}{3}\bigg|_0^2$$

$$= x^3\bigg|_0^2$$

$$= (2)^3 - 0$$

$$= 8 \text{ square units}$$

Does the answer to Example 9 appear reasonable? As best you can, count the square units in the shaded area and compare answers.

Using the graph provided in Example 9, compute the area specified in Problems 7 and 8.

Problem 7

Find the area between the x-axis and the curve $f(x) = 3x^2$ from $x = 1$ to $x = 3$.

Solution:

Answer: 26 square units

Problem 8

Find the area bounded by the curve $f(x) = 3x^2$, the x-axis, $x = -1$, and $x = 1$.

Solution:

Answer: 2 square units

It's time for a review example on graphing a cubic that we can then use for Examples 11 and 12.

EXAMPLE 10

Sketch: $y = f(x) = x^3 + x^2 - 6x$.

Solution: 1.   This is a cubic with $a = 1$ and $d = 0$.

2.   The curve opens up:

3.   The y-intercept is 0.

4.   The x-intercepts are 0, –3, and 2 because

$$0 = x^3 + x^2 - 6x$$

$$= x(x^2 + x - 6)$$

$$= x(x + 3)(x - 2)$$

$$x = 0 \qquad x = -3 \qquad x = 2$$

Answer:

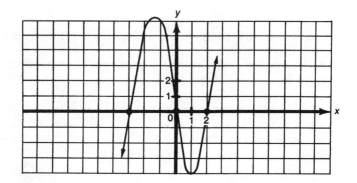

As stated earlier, the theorem for area requires that the function be nonnegative over the interval from $x = a$ to $x = b$. If the function is negative over the interval, the definite integral results in a negative value that is the opposite of the area. In such a case, the area is found by taking the absolute value of the result.

The next example will illustrate what I mean.

EXAMPLE 11

Find the area between the x-axis and the curve $y = f(x) = x^3 + x^2 - 6x$ from $x = 1$ to $x = 2$.

Solution: A sketch of $f(x)$ and the indicated area is repeated below. Note that the curve is negative over the interval $1 \le x \le 2$, and thus the shaded area appears below the x-axis.

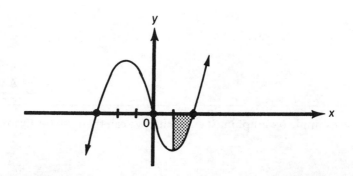

To determine the area, we begin by computing the definite integral:

$$\int_1^2 (x^3 + x^2 - 6x)\, dx = \frac{x^4}{4} + \frac{x^3}{3} - \frac{6x^2}{2} \Big|_1^2$$

$$= \left(\frac{16}{4} + \frac{8}{3} - \frac{24}{2}\right) - \left(\frac{1}{4} + \frac{1}{3} - \frac{6}{2}\right)$$

$$= 4 + \frac{8}{3} - 12 - \frac{1}{4} - \frac{1}{3} + 3$$

$$= -5 + \frac{7}{3} - \frac{1}{4}$$

$$= \frac{-60 + 28 - 3}{12}$$

$$= -2\frac{11}{12}$$

The definite integral results in a negative number because the area is located below the x-axis. The area of the shaded area is found by taking the absolute value of the definite integral.

Answer: Area = $\left| -2\frac{11}{12} \right|$ = $2\frac{11}{12}$ square units.

You might conclude from the above example that, anytime the definite integral results in a negative number, you merely take the absolute value of it for the area and never have to bother with a graph. Unfortunately that shortcut won't work all the time, as the next example will show.

Example 12 illustrates the situation where some of the desired area is above the x-axis and some is below the x-axis.

EXAMPLE 12

Find the area between the x-axis and the curve $f(x) = x^3 + x^2 - 6x$ from $x = -3$ to $x = 3$.

Solution: A sketch of $f(x)$ and the indicated area shows that some of the area appears above the x-axis and some of the area appears below the x-axis. In such a situation, each part must be determined separately and the answer added for the total area.

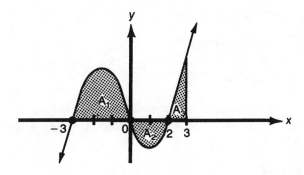

$$\text{Area of } A_1 = \int_{-3}^{0} (x^3 + x^2 - 6x) \, dx$$

$$= \frac{x^4}{4} + \frac{x^3}{3} - 3x^2 \Big|_{-3}^{0}$$

$$= (0 + 0 + 0) - \left(\frac{81}{4} - \frac{27}{3} - 27 \right)$$

$$= 0 - 20.25 + 9 + 27$$

$$= 15.75 \text{ square units}$$

$$\text{Area of } A_2 = \left| \int_{0}^{2} (x^3 + x^2 - 6x) \, dx \right|$$

$$= \left| \frac{x^4}{4} + \frac{x^3}{3} - 3x^2 \Big|_{0}^{2} \right|$$

$$= \left| \left(\frac{16}{4} + \frac{8}{3} - 12 \right) - (0 + 0 - 0) \right|$$

$$= | 4 + 2.67 - 12 |$$

$$= | -5.33 |$$

$$= 5.33 \text{ square units}$$

$$\text{Area of } A_3 = \int_{2}^{3} (x^3 + x^2 - 6x) \, dx$$

$$= \frac{x^4}{4} + \frac{x^3}{3} - 3x^2 \Big|_{2}^{3}$$

$$= \left(\frac{81}{4} + \frac{27}{3} - 27 \right) - \left(\frac{16}{4} + \frac{8}{3} - 12 \right)$$

$$= 20.25 + 9 - 27 - 4 - 2.67 + 12$$

$$= 7.58 \text{ square units}$$

Answer: The total area is $A = A_1 + A_2 + A_3$

$$= 15.75 + 5.33 + 7.58$$

$$= 28.66 \text{ square units.}$$

In effect, when we have areas that appear both above and below the x-axis, the total area can be found by adding the areas above the x-axis and subtracting the areas that appear below the x-axis rather than using absolute values.

Your turn to do an easy problem.

Problem 9

Find: $\displaystyle\int_{-3}^{3} (x^3 + x^2 - 6x)\, dx.$

Solution:

Answer: 18

Why does your answer to Problem 9 differ from the answer to Example 12? What does your answer represent? Do you now see the necessity for sketching the graph whenever an area is being computed?

AREA BETWEEN TWO CURVES

Suppose we need to find the area between two curves, as shown in the diagram. What can we do?

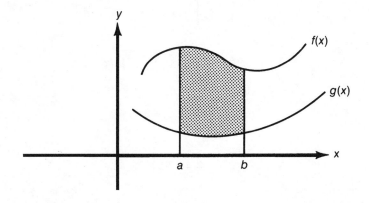

Using the theorem for area, we can find the area bounded by $f(x)$ and the x-axis. Then we find the area under $g(x)$ and above the x-axis. We subtract them to determine the shaded area.

That is basically what the next theorem does, except that the functions are subtracted before calculating the area.

THEOREM

If $f(x)$ and $g(x)$ are continuous with $f(x) \geq g(x)$ on the interval $a \leq x \leq b$, then the area between the two curves is equal to

$$\int_{a}^{b} [\, f(x) - g(x)]\, dx.$$

Notice that the theorem requires that $f(x)$ be above $g(x)$ on the interval. There is no requirement, however, that either function be nonnegative.

EXAMPLE 13

Find the area between $y = f(x) = x^2 + 5$ and $y = g(x) = -x^2$ from $x = 1$ to $x = 2$.

Solution: A sketch of both functions and the desired area appears below.

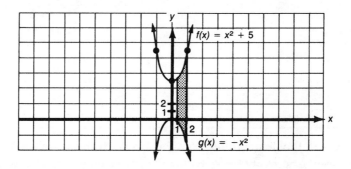

The shaded area is computed by the definite integral of the difference of the functions, $f(x) - g(x)$, with $f(x)$ being the function on the top.

$$\text{Area} = \int_1^2 [f(x) - g(x)] \, dx$$

$$= \int_1^2 [(x^2 + 5) - (-x^2)] \, dx$$

$$= \int_1^2 (x^2 + 5 + x^2) \, dx$$

$$= \int_1^2 (2x^2 + 5) \, dx$$

$$= \frac{2x^3}{3} + 5x \Big|_1^2$$

$$= \left[\frac{2(8)}{3} + 10\right] - \left(\frac{2}{3} + 5\right)$$

$$= \frac{16}{3} + 10 - \frac{2}{3} - 5$$

$$= \frac{14}{3} + 5$$

$$= 4.67 + 5$$

$$= 9.67 \text{ square units}$$

The final example is the hardest, of course.

EXAMPLE 14

Find the area bounded by the curves $y = f(x) = x + 1$ and $y = g(x) = x^2 - 2x + 1$.

Solution: $f(x)$ is a linear function with slope of 1 and y-intercept of 1.

$g(x)$ is a quadratic with the parabola opening up, y-intercept of 1, and x-intercept of 1.

A sketch of both functions appears below.

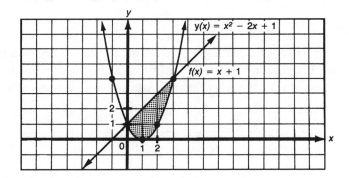

We must determine the x-coordinates of the points of intersection.

To find the points of intersections, set

$$f(x) = g(x)$$

$$x + 1 = x^2 - 2x + 1$$

$$0 = x^2 - 3x$$

$$0 = x(x - 3)$$

$$x = 0 \qquad x = 3$$

Using the x-coordinates just calculated, we find that the area between the two curves is from $x = 0$ to $x = 3$. This area can be determined by the definite integral of the difference of the two functions integrated from 0 to 3.

$$\text{Area} = \int_0^3 [f(x) - g(x)]\, dx$$

$$= \int_0^3 [(x + 1) - (x^2 - 2x + 1)]\, dx$$

$$= \int_0^3 (x + 1 - x^2 + 2x - 1)\, dx$$

$$= \int_0^3 (-x^2 + 3x)\, dx$$

$$= \frac{-x^3}{3} + \frac{3x^2}{2} \Big|_0^3$$

$$= \left(\frac{-(27)}{3} + \frac{27}{2} \right) - 0$$

$$= -9 + 13.5$$

$$= 4.5 \text{ square units}$$

CALCULUS AND THE CALCULATOR (Optional)

Evaluating a Definite Integral

Much like evaluating derivatives, evaluating a definite integral can be done quickly and easily with the graphing calculator. Integration is not even required, as the next example will show. We will rework various examples from this unit so that you can compare the answers with those already obtained.

EXAMPLE 15 (Example 6 Revisited)

Evaluate $\displaystyle\int_{3}^{5} (-4x^3 + 54x^2)\, dx$.

Solution: Graph $f(x) = -4x^3 + 54x^2$ with a viewing window of $[0, 10]$ by $[0, 2{,}200]$.

To evaluate the definite integral, press (2nd) (CALC) (7).

At the <u>Lower Limit?</u> prompt, press (3) (ENTER).

At the <u>Upper Limit?</u> prompt, press (5) (ENTER).

The answer appears at the bottom: $\displaystyle\int f(x)\, dx = 1220$

Verify for yourself that this is, indeed, the correct answer by returning to Example 6, where the same problem was worked manually.

You should also have noticed that on your screen the portion under the curve from $x = 3$ to $x = 5$ was shaded. In other words, another interpretation of this problem is that the area under the curve bounded by the x-axis and the limits of integration is 1,220 square units.

Can we use this feature of the calculator to find areas? The answer is yes, *provided* the conditions of the theorem are met, specifically that the function is continuous and nonnegative over the limits of integration as they were in Example 15.

EXAMPLE 9 Revisited

Find the area bounded by the curve $y = f(x) = 3x^2$, the x-axis, and the vertical line $x = 2$.

Solution: Graph $y = f(x) = 3x^2$ on the standard viewing window.

Observe that $f(x)$ is nonnegative over the desired interval $0 \le x \le 2$ as required. Thus the area can be determined by the definite integral.

To evaluate the definite integral, press (**2nd**) (**CALC**) (**7**).

At the <u>Lower Limit?</u> prompt, press (**0**) (**ENTER**).

At the <u>Upper Limit?</u> prompt, press (**2**) (**ENTER**).

The answer appears at the bottom: $\displaystyle\int f(x)\, dx = 8$

Answer: The desired area is 8 square units.

A comparison with Example 9 will show that the two answers are identical.

The next is a problem for you to work with a discussion to follow.

Problem 10

Find: $\displaystyle\int_{-3}^{3} (x^3 + x^2 - 6x)\, dx.$

Solution:

Answer: 18

Now compare your answer with that for Example 12. Your graph with its shaded area should be identical to the sketch shown with Example 12; however, your answer is 18 and the answer for Example 12 is 28.66. Why are they different? Which one is correct?

Both answers are correct because different questions were asked. Eighteen is the value of the definite integral evaluated from −3 to 3 as requested in Problem 7. It does not represent the area even though it is shown shaded on the calculator screen. Why not? Because the function did not meet the condition of the theorem that it be nonnegative over the limits of integration. To find the area would require setting up three separate definite integrals for evaluation exactly as was done in Example 12.

Thus, when determining areas, it is absolutely necessary to view the graph before setting up the definite integral.

You should now be able to evaluate a definite integral. In addition, you should be able to compute the area bounded by a curve and the x-axis, or the area between two curves. Again, remember that, when determining areas, it is necessary to make a sketch before setting up the definite integral.

EXERCISES

Evaluate the following definite integrals:

1. $\displaystyle\int_1^3 (8x + 2)\, dx$

2. $\displaystyle\int_1^2 x^2\, dx$

3. $\displaystyle\int_0^1 e^x\, dx$

4. $\displaystyle\int_2^{10} (-x + 10)\, dx$

5. $\displaystyle\int_2^5 \frac{dx}{x}$

6. $\displaystyle\int_0^1 (x - x^2)\, dx$

7. $\displaystyle\int_1^2 2x(x^2 + 1)\, dx$

8. $\displaystyle\int_1^e \ln x\, dx$ (This one may surprise you.)

9. $\displaystyle\int_{-2}^1 (x + 3)^3\, dx$

10. $\displaystyle\int_1^5 (x^2 - 6x + 5)\, dx$

11. $\displaystyle\int_{-1}^3 2xe^{x^2-1}\, dx$

12. Find the area bounded by the curve $f(x) = x^3 + 8$, the x-axis, and the vertical lines at $x = -1$ and $x = 4$.

13. Find the area bounded by the curve $f(x) = x^3 + 8$, the x-axis, and the vertical lines at $x = -2$ and $x = 2$.

14. Find the area between the x-axis and the curve $f(x) = x^2 - 4x$.

15. Find the area bounded by the curve $f(x) = x^2 - 4x$, the x-axis, and the vertical lines at $x = -1$ and $x = 2$.

16. Find the area between the curve $f(x) = x^3 - 4x$ and the x-axis from $x = -1$ to $x = 3$.

17. Find the area between the curve $f(x) = x^3 - 4x$ and the x-axis from $x = -2$ to $x = 2$.

18. Find the area bounded by the curve $f(x) = 4$ and $g(x) = x^2$.

19. Find the area between $f(x) = x + 1$ and $g(x) = x^2 - 2x + 1$ from $x = -2$ to $x = 0$. Hint: Use the graph from Example 14.

20. Maintenance costs for a home tend to increase as the house gets older. From past records, Penny, a real estate agent, determined that the rate of increase in maintenance costs, in dollars per year, for a small house is given by

$$M'(x) = 9x^2 + 50$$

where $x =$ the age of the house, in years,

and $M(x) =$ the total accumulated cost of maintenance for x years.

Write a definite integral that will give the total maintenance costs for the period from 5 to 10 years after the house was built, and evaluate it.

C21. Find the area bounded by the curve $f(x) = \dfrac{1}{x^2}$ and the x-axis from $x = 0.5$ to $x = 4$.

C22. Find the area bounded by the curve $f(x) = 2e^x + \dfrac{6}{x} - 7x^2$ and the x-axis from $x = 0.5$ to $x = 1$.

If additional practice is needed:

   Barnett, Ziegler, and Byleen, pages 429–431, problems 1–44 and 77–80; pages 454–455, problems 1–60
   Harshbarger and Reynolds, pages 940–942, problems 1–53; page 952, problems 1–23
   Hoffmann and Bradley, pages 440–441, problems 1–35
   Waner and Costenoble, page 351, problems 1–10; page 362, problems 17–24 and 59–62; page 390, problems 1–24

UNIT 28

Functions of Several Variables

The purpose of this last unit is to provide you with a working knowledge of differentiating functions of several variables. When you have finished this unit, you will be able to differentiate such functions and, for some, to locate and classify their critical points.

DERIVATIVES REVISITED

Because it has been a number of units since we used derivatives, the formulas are repeated below. However, the $\frac{dy}{dx}$ notation has been used rather than $f'(x)$. In the formulas c and n are real numbers. The variables u and v denote functions of x. In other words, rather than writing out $u(x)$ and $v(x)$ the notation is shortened to u and v.

Here are the rules for differentiation:

$$\frac{d(c)}{dx} = 0$$

$$\frac{d(x^n)}{dx} = nx^{n-1}$$

$$\frac{d(cf(x))}{dx} = c\,\frac{d(f(x))}{dx}$$

$$\frac{d(u^n)}{dx} = nu^{n-1}\,\frac{du}{dx}$$

$$\frac{d(uv)}{dx} = u\,\frac{dv}{dx} + v\,\frac{du}{dx}$$

$$\frac{d\left(\dfrac{u}{v}\right)}{dx} = \frac{v\,\dfrac{du}{dx} - u\,\dfrac{dv}{dx}}{v^2}$$

$$\frac{d(e^u)}{dx} = e^u\,\frac{du}{dx}$$

$$\frac{d(\ln u)}{dx} = \frac{1}{u}\cdot\frac{du}{dx}$$

PARTIAL DERIVATIVES

When f is a function of x and y, defined by $z = f(x, y)$, it is useful to speak of the instantaneous rate of change in f with respect to x when y is held constant, and vice versa.

Definition:　The instantaneous rate of change of f with respect to x (y is held constant) is called the **partial derivative of f with respect to x,** or simply the **first partial.**

Notation:　$\dfrac{\partial f}{\partial x} = \dfrac{\partial z}{\partial x} = f_x(x, y) = f_x$, all denote the partial derivative of f with respect to x.

The partial of f with respect to x is found by treating x as the only variable, treating the remaining independent variables as constants, and applying the differentiation rules. The process is known as partial differentiation.

Since f and z are interchangeable, to speak of the partial of f with respect to x is the same as to refer to the partial of z with respect to x.

Definition:　The instantaneous rate of change of f with respect to y (x is held constant) is called the **partial derivative of f with respect to y,** or simply the **first partial.**

Notation:　$\dfrac{\partial f}{\partial y} = \dfrac{\partial z}{\partial y} = f_y(x, y) = f_y$, all denote the partial derivative of f with respect to y.

The partial of f with respect to y is found by treating y as the only variable, treating the remaining independent variables as constants, and applying the differentiation rules.

The notation expands, as you would expect, if there are additional independent variables. All of the rules for differentiation we have used remain valid; and, you will be happy to learn, there are no new rules. The only difficulty is keeping in mind which variables are being held constant during the differentiation.

EXAMPLE 1

If $z = f(x, y) = x^2 - y^4$, find $\dfrac{\partial f}{\partial x}$ and $\dfrac{\partial f}{\partial y}$.

Solution: To find $\dfrac{\partial f}{\partial x}$, we treat y as a constant and differentiate with respect to x.

$$z = f(x, y) = x^2 - y^4$$

$$\frac{\partial f}{\partial x} = f_x(x, y) = 2x$$

To find $\dfrac{\partial f}{\partial y}$, we treat x as a constant and differentiate with respect to y.

$$z = f(x, y) = x^2 - y^4$$

$$\frac{\partial f}{\partial y} = f_y(x, y) = -4y^3$$

EXAMPLE 2

If $z = f(x, y) = 5x + 2x^2y + y^2$, find f_x and f_y.

Solution: To find f_x, we treat y as a constant and differentiate with respect to x. To show you what is intended and why, I will put an extra step in before actually differentiating.

$$z = f(x, y) = 5x + 2x^2y + y^2$$

$$f_x(x, y) = 5\frac{\partial(x)}{\partial x} + 2y\frac{\partial(x^2)}{\partial x} + \frac{\partial(y^2)}{\partial x}$$

$$= 5(1) + 2y(2x) + 0$$

$$= 5 + 4xy$$

Observe that in the middle term both 2 and y were brought out in front because both are being considered as constants, and only x^2 was differentiated.

To find f_y, we treat x as a constant and differentiate with respect to y.

$$z = f(x, y) = 5x + 2x^2y + y^2$$

$$f_y(x, y) = \frac{\partial(5x)}{\partial y} + 2x^2\frac{\partial(y)}{\partial y} + \frac{\partial(y^2)}{\partial y}$$

$$= 0 + 2x^2(1) + 2y$$

$$= 2x^2 + 2y$$

Notice the similarity of the notation. With functions of a single independent variable, $\dfrac{dy}{dx}$ is used. With functions of several variables, $\dfrac{\partial f}{\partial x}$ is used.

I'll do one more example, and then you can try two problems. If you prefer, cover up the answer to Example 3 and try it yourself.

EXAMPLE 3

If $z = f(w, x, y) = ye^x + xe^{2y} - 7w^5$, find f_w, f_x, and f_y.

Solution: To find f_w, we treat x and y as constants and differentiate with respect to w.

$$z = f(w, x, y) = ye^x + xe^{2y} - 7w^5$$

$$\frac{\partial z}{\partial w} = f_w(w, x, y) = \frac{\partial(ye^x)}{\partial w} + \frac{\partial(xe^{2y})}{\partial w} - 7\frac{\partial(w^5)}{\partial w}$$

$$= 0 + 0 - 7(5w^4)$$

$$= -35w^4$$

To find f_x, we treat w and y as constants and differentiate with respect to x.

$$z = f(w, x, y) = ye^x + xe^{2y} - 7x^5$$

$$\frac{\partial z}{\partial x} = f_x(w, x, y) = y\frac{\partial(e^x)}{\partial x} + e^{2y}\frac{\partial(x)}{\partial x} - \frac{\partial(7w^5)}{\partial x}$$

$$= y(e^x) + e^{2y}(1) + 0$$

$$= ye^x + e^{2y}$$

To find f_y, we treat w and x as constants and differentiate with respect to y.

$$z = f(w, x, y) = ye^x + xe^{2y} - 7w^5$$

$$\frac{\partial z}{\partial y} = f_y(w, x, y) = e^x\frac{\partial(y)}{\partial y} + x\frac{\partial(e^{2y})}{\partial y} - \frac{\partial(7w^5)}{\partial y}$$

$$= e^x(1) + x(2e^{2y}) + 0$$

$$= e^x + 2xe^{2y}$$

Ready to try a problem? I will make it easier.

Problem 1

If $z = f(w, x, y) = x^2 - 4x + 3y + 20w^3 - 2w^2x^3y^5$, find f_w, f_x, and f_y.

Solution:

Answers: $f_w = 60w^2 - 4wx^3y^5$;

$f_x = 2x - 4 - 6w^2x^2y^5$;

$f_y = 3 - 10w^2x^3y^4$

Try one more.

Problem 2

If $z = f(x, y) = e^{x^2 - 3y}$, find $\dfrac{\partial z}{\partial x}$ and $\dfrac{\partial z}{\partial y}$.

Solution:

Answers: $\dfrac{\partial z}{\partial x} = 2xe^{x^2 - 3y}$;

$\dfrac{\partial z}{\partial y} = -3e^{x^2 - 3y}$

APPLICATIONS

A few application problems may help your understanding of the meaning and the importance of partial derivatives.

Recall that in economics the term *marginal analysis* refers to the practice of using a derivative to estimate the change in the value of a function resulting from a unit increase in the independent variable. Partial derivatives are used in the same way, that is, to estimate the change in the value of a function resulting from a one-unit increase in one of the variables.

EXAMPLE 4

W. Spaniel does the production planning for two fiber-manufacturing facilities. At the South Carolina plant he estimates that the number of pounds of acetate fibers that can be produced each week is given by the function

$$z = f(x, y) = 12{,}000x + 5{,}000y + 10x^2y - 10x^3$$

where z = the number of pounds of acetate fiber per week,

x = the number of skilled workers at the plant,

and y = the number of unskilled workers at the plant.

Currently the work force at the South Carolina plant consists of 40 skilled and 60 unskilled workers.

a.   Find and interpret $f(40, 60)$.

b.   Express the rate of change of output with respect to the number of skilled workers.

c.   Find $f_x(40, 60)$.

d.   Interpret the answer to part c.

e.   Compute the actual change in output that will result if one additional skilled worker is hired.

Solution: a. $z = f(x, y) = 12,000x + 5,000y + 10x^2y - 10x^3$

$z = f(40, 60) = 12,000(40) + 5,000(60) + (10)(40)^2(60) - 10(40)^3$

$= 480,000 + 300,000 + 960,000 - 640,000$

$= 1,100,000$

With a work force of 40 skilled and 60 unskilled workers, it is estimated that 1,100,000 pounds of acetate fiber can be produced per week.

b.   The rate of change of output, z, with respect to the number of skilled workers, x, is the partial derivative f_x.

$$z = f(x, y) = 12,000x + 5,000y + 10x^2y - 10x^3$$

$$\frac{\partial z}{\partial x} = f_x(x, y) = 12,000\,\frac{\partial(x)}{\partial x} + \frac{\partial(5,000y)}{\partial x} + 10y\,\frac{\partial(x^2)}{\partial x} - 10\,\frac{\partial(x^3)}{\partial x}$$

$$= 12,000(1) + 0 + 10y(2x) - 10(3x^2)$$

$$= 12,000 + 20xy - 30x^2$$

For any values of x and y, this partial derivative is an approximation of the *additional* output that will result from adding one *additional* skilled worker.

c. $f_x(40, 60)$ denotes that f_x is to be evaluated at $x = 40$ and $y = 60$.

$$f_x(x, y) = 12,000 + 20xy - 30x^2$$

$$f_x(40, 60) = 12,000 + 20(40)(60) - 30(40)^2$$

$$= 12,000 + 48,000 - 48,000$$

$$= 12,000$$

d.   Currently the work force is 40 skilled and 60 unskilled workers. If one additional skilled worker is hired, it is estimated that output will increase by approximately 12,000 pounds of acetate per week.

e.   The actual change is the difference between the output with 41 skilled and 60 unskilled workers and the output with 40 skilled and 60 unskilled workers.

$$z = f(x, y) = 12,000x + 5,000y + 10x^2y - 10x^3$$

$$z = f(41, 60) = 12,000(41) + 5,000(60) + (10)(41)^2(60) - 10(41)^3$$

$$= 492,000 + 300,000 + 1,008,600 - 689,210$$

$$= 1,111,390$$

$$z = f(40, 60) = 1,100,000$$

Actual change $= f(41, 60) - f(40, 60)$

$$= 1,111,390 - 1,100,000$$

$$= 11,390$$

In geometric terms, the difference between these two quantities is the difference between using the tangent line to estimate the next point one unit away and using the actual function to determine the point.

The wording of Example 4 should have sounded familiar. The only difference from the earlier units is that we have expanded our functions to include those with more than one independent variable. The concepts remain the same.

Problem 3

Continue with W. Spaniels' employment analysis.

a.   Find the rate of change of output with respect to the number of unskilled workers.

b.   Find $f_y(40, 60)$.

c.   Interpret the answer to part b.

d.   What would you advise W. Spaniel to do with regard to hiring an additional person?

Solution:

Answers: a. $f_y(x, y) = 5{,}000 + 10x^2$.   b. $f_y(40, 60) = 21{,}000$.   c. If the number of unskilled workers is increased from 60 to 61 while the number of skilled workers remains at 40, it is estimated that output will increase by 21,000 pounds per week.   d. If one more person is hired, the worker should be unskilled rather than skilled.

Here's another short problem for you to do.

Problem 4

At the Virginia fibers plant, W. Spaniel has determined that output is a function of skilled, unskilled, and part-time workers. For this plant, output is estimated by the function

$$z = f(w, x, y) = 1{,}200x + 500y + x^2y - x^3 - y^2 + 2w^3$$

where z = the number of pounds of acetate fiber produced per week,

w = the number of part-time workers,

x = the number of skilled workers,

and y = the number of unskilled workers.

Currently the work force at the Virginia plant consists of 10 part-time, 31 skilled, and 53 unskilled workers. Use marginal analysis to estimate the change in output if one additional part-time person is hired.

Solution:

Answer: $f_w(10, 31, 53) = 6(10)^2 = 600$

Aren't you glad I asked about a part-time worker rather than either of the other two types?

SECOND PARTIAL DERIVATIVES

Definition: Partial derivatives themselves can be differentiated. The resulting functions are called **second-order partial derivatives**, or simply **second partials**.

If $z = f(x, y)$, then there are four second partials.

1.   The partial of f_x with respect to x,

$$\text{denoted by } \frac{\partial^2 z}{\partial x^2} = \frac{\partial^2 f}{\partial x^2} = f_{xx}(x, y) = f_{xx}$$

2.   The partial of f_y with respect to y,

$$\text{denoted by } \frac{\partial^2 z}{\partial y^2} = \frac{\partial^2 f}{\partial y^2} = f_{yy}(x, y) = f_{yy}$$

3.   The partial of f_y with respect to x,

$$\text{denoted by } \frac{\partial^2 z}{\partial x\, \partial y} = \frac{\partial^2 f}{\partial x\, \partial y} = f_{yx}(x, y) = f_{yx}$$

4.   The partial of f_x with respect to y,

$$\text{denoted by } \frac{\partial^2 z}{\partial y\, \partial x} = \frac{\partial^2 f}{\partial y\, \partial x} = f_{xy}(x, y) = f_{xy}$$

EXAMPLE 5

If $z = f(x, y) = 3xy - x^2 + 5y^3$, find all first and second partials.

Solution: The two first-order partial derivatives:

$$z = f(x, y) = 3xy - x^2 + 5y^3$$

$$\frac{\partial z}{\partial x} = f_x(x, y) = 3y - 2x$$

$$z = f(x, y) = 3xy - x^2 + 5y^3$$

$$\frac{\partial z}{\partial y} = f_y(x, y) = 3x + 15y^2$$

The four second-order partial derivatives:

1. To find $f_{xx}(x, y)$, we treat y as a constant and differentiate f_x with respect to x.

$$\frac{\partial z}{\partial x} = f_x(x, y) = 3y - 2x$$

$$\frac{\partial^2 z}{\partial x^2} = f_{xx}(x, y) = \frac{\partial(3y)}{\partial x} - 2\frac{\partial(x)}{\partial x}$$

$$= 0 - 2(1)$$

$$= -2$$

2. To find $f_{yy}(x, y)$, we treat x as a constant and differentiate f_y with respect to y.

$$\frac{\partial z}{\partial y} = f_y(x, y) = 3x + 15y^2$$

$$\frac{\partial^2 z}{\partial y^2} = f_{yy}(x, y) = \frac{\partial(3x)}{\partial y} + 15\frac{\partial(y^2)}{\partial y}$$

$$= 0 + 15(2y)$$

$$= 30y$$

3. To find $f_{xy}(x, y)$, we treat x as a constant and differentiate f_x with respect to y.

$$\frac{\partial z}{\partial x} = f_x(x, y) = 3y - 2x$$

$$\frac{\partial^2 z}{\partial y\, \partial x} = f_{xy}(x, y) = 3\frac{\partial(y)}{\partial y} - \frac{\partial(2x)}{\partial y}$$

$$= 3(1) - 0$$

$$= 3$$

4. To find $f_{yx}(x, y)$, we treat y as a constant and differentiate f_y with respect to x.

$$\frac{\partial z}{\partial y} = f_y(x, y) = 3x + 15y^2$$

$$\frac{\partial^2 z}{\partial x\, \partial y} = f_{yx}(x, y) = 3\frac{\partial(x)}{\partial x} + \frac{\partial(15y^2)}{\partial x}$$

$$= 3(1) + 0$$

$$= 3$$

The last two partials, f_{xy} and f_{yx}, are sometimes referred to as **mixed partials** or **cross partials**. The reason should be obvious—the first partial derivative is taken with respect to one variable, and the second partial is taken with respect to the other variable. However, if you are like me, you may tend to get the notation for the mixed partials confused. The comforting part, though, is that, for all of the functions you will ever see, the mixed partials will be equal, as they are in Example 5.

Here is a practice problem for you. You may want to cover the answer before doing the work.

Problem 5

If $z = f(x, y) = 2x^3y^4$, find all first- and second-order partials.

Solution:

Answers: $f_x = 6x^2y^4$; $f_y = 8x^3y^3$; $f_{xx} = 12xy^4$;
$f_{yy} = 24x^3y^2$; $f_{xy} = f_{yx} = 24x^2y^3$

RELATIVE MAXIMA AND MINIMA FOR BIVARIATE FUNCTIONS

We will conclude this unit with a brief discussion of the criteria for maximum and minimum values for bivariate functions. As always, we will allow geometric examples to guide us.

Recall from Unit 1 that a bivariate function has two independent variables and a multivariate function has more than two independent variables. If $z = f(x, y)$ is a function of two variables, x and y, it is natural to regard the function as a surface in three-dimensional space. We can use this surface as an aid in studying the properties of the functions, meaning that we can continue to draw pictures to help us.

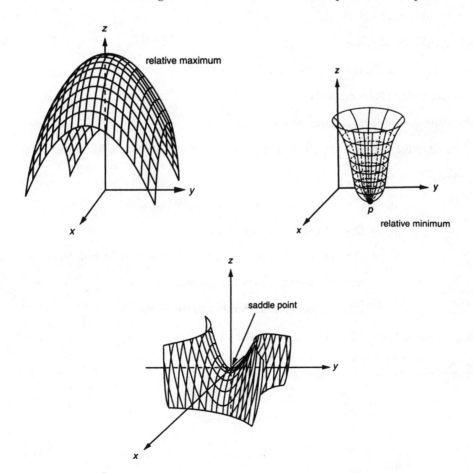

Relative maxima and relative minima of bivariate functions are defined in a manner similar to that used for a single-variable function. The graphs illustrate each. Notice that the point marked "relative maximum" is higher than any of its neighboring points, and that the point labeled relative minimum is lower than any of its neighboring points. The only difference in terminology is the use of the term **saddle point**, rather than "stationary inflection point," to denote a leveling-off point that is neither higher nor lower than all of its neighboring points.

You will be happy to know we are only doing a brief introduction to this material, and you will *not* be asked to make any three-dimensional drawings.

For a function $f(x, y)$ to have a relative maximum or minimum, it is necessary that

$$f_x = 0 \quad \text{and}$$

$$f_y = 0 \quad \text{be satisfied simultaneously.}$$

These two conditions are used to determine the critical points of the function.

Definition: A point (a, b) for which both $f_x(a, b) = 0$ and $f_y(a, b) = 0$ is said to be a **critical point** of f.

One procedure used to classify critical points of a bivariate function is the second-derivative test.

SECOND-DERIVATIVE TEST FOR CLASSIFYING CRITICAL POINTS OF A BIVARIATE FUNCTION

Let $D(x, y) = f_{xx}(x, y) f_{yy}(x, y) - [f_{xy}(x, y)]^2$.

Evaluate D at the critical point.

1.   If D is negative, the critical point is a saddle point.

2.   If D is zero, the test is inconclusive.

3.   If D is positive, evaluate f_{xx} at the critical point.

   If f_{xx} is negative, the critical point is a maximum; otherwise it is a minimum.

   In symbols:

$$\text{If } D < 0, \text{ the critical point is a saddle point.}$$

$$\text{If } D = 0, \text{ the test is inconclusive.}$$

$$\text{If } D > 0, \text{ the critical point is either a maximum or minimum}$$

$$\text{if } f_{xx} < 0, \text{ the critical point is a maximum}$$

$$\text{if } f_{xx} > 0, \text{ the critical point is a minimum}$$

Two examples should be sufficient to illustrate the procedure.

EXAMPLE 6

Find and classify all critical points of $f(x, y) = x^2 + xy + y^2 - 3x + 2$.

Solution: The first step is to find all the critical points.

Critical points, by definition, occur where $f_x = 0$ and $f_y = 0$ are satisfied simultaneously. This means that we take the first partials, set them equal to zero, and solve the equations simultaneously. Notice the similarity to our earlier work.

$$f(x, y) = x^2 + xy + y^2 - 3x + 2$$

$$f_x(x, y) = 2x + y - 3$$

$$f_y(x, y) = x + 2y$$

$$0 = 2x + y - 3$$

$$0 = x + 2y$$

$$3 = 2x + y$$

$$0 = x + 2y$$

$$-6 = -4x - 2y$$

$$\underline{0 = \quad x + 2y}$$

$$-6 = -3x$$

$$x = 2, \text{ and since } 0 = x + 2y$$

by substitution, $0 = 2 + 2y$

$$-2y = 2$$

$$y = -1$$

If $x = 2$ and $y = -1$

with $f(x, y) = x^2 + xy + y^2 - 3x + 2$,

then $f(2, -1) = (2)^2 + (2)(-1) + (-1)^2 - 3(2) + 2 = -1$

and the function has a critical point at $(2, -1, -1)$.

To classify the critical point, use the second-derivative test for bivariate functions with

$$f_{xx}(x, y) = 2$$

$$f_{yy}(x, y) = 2$$

$$f_{xy}(x, y) = 1$$

Let $D(x, y) = f_{xx}(x, y)f_{yy}(x, y) - [f_{xy}(x, y)]^2$

$$= (2)(2) - (1)^2$$

$$= 4 - 1$$

$$= 3$$

At point $(2, -1, -1)$ with

$$D(x, y) = 3,$$

$$D(2, -1) = 3$$

Since the value of D is positive, f has either a relative maximum or a relative minimum at the critical point. To determine which one it is, determine the value of f_{xx} at the critical point.

$$f_{xx}(2, -1) = 2$$

Since the value of f_{xx} is positive, the critical point is a relative minimum.

Answer: The critical point $(2, -1, -1)$ is a relative minimum. In other words, -1 is the minimum value for z, and it occurs when $x = 2$ and $y = -1$.

Example 6 was a rather unique situation because all of the second-order partials were constants. That is rarely the case, and we will consider a final example where the numerical value of D must be calculated.

I realize that these are long, long problems, but Example 7 is the last one!

EXAMPLE 7

Find all relative maxima or minima of $f(x, y) = x^2 - 2x + y^3 - 3y^2 - 9y + 6$.

Solution: The first step is to find all the critical points.

Critical points, by definition, occur where $f_x = 0$ and $f_y = 0$ are satisfied simultaneously. This means that we take the first partials, set them equal to zero, and solve the equations simultaneously.

$$f(x, y) = x^2 - 2x + y^3 - 3y^2 - 9y + 6$$

$$f_x(x, y) = 2x - 2$$

$$f_y(x, y) = 3y^2 - 6y - 9$$

$$0 = 2x - 2$$

$$0 = 3y^2 - 6y - 9$$

This time we have two separate equations to be solved. The first one is in terms of x and the second one is in terms of y.

$$0 = 2x - 2 \quad 0 = 3y^2 - 6y - 9$$

$$2 = 2x \qquad 0 = y^2 - 2y - 3$$

$$x = 1 \qquad 0 = (y - 3)(y + 1)$$

$$y - 3 = 0 \qquad y + 1 = 0$$

$$y = 3 \qquad\quad y = -1$$

If $x = 1$ and $y = 3$

and $f(x, y) = x^2 - 2x + y^3 - 3y^2 - 9y + 6$,

then $f(1, 3) = 1 - 2 + 27 - 27 - 27 + 6$

and (1, 3, –22) is a critical point.

If $x = 1$ and $y = -1$

and $f(x, y) = x^2 - 2x + y^3 - 3y^2 - 9y + 6$,

then $f(1, -1) = 1 - 2 - 1 - 3 + 9 + 6 = 10$

and (1, –1, 10) is another critical point.

To classify the critical points, use the second-derivative test for bivariate functions with

$$f_{xx}(x, y) = 2$$

$$f_{yy}(x, y) = 6y - 6$$

$$f_{xy}(x, y) = 0$$

Let $D(x, y) = f_{xx}(x, y)f_{yy}(x, y) - [f_{xy}(x, y)]^2$

$$= 2(6y - 6) - (0)^2$$

$$= 12y - 12$$

At point (1, 3, –22) with

$D(x, y) = 12y - 12$,

$D(1, 3) = 12(3) - 12 = 36 - 12 = 24$

Since the value of D is positive, f has either a relative maximum or a relative minimum at the critical point. To determine which one it is, determine the value of f_{xx} at the critical point.

$$f_{xx}(1, 3) = 2$$

Since the value of f_{xx} is positive, the critical point is a relative minimum.

The critical point (1, 3, –22) is a relative minimum. In other words, –22 is the minimum value for z and it occurs when $x = 1$ and $y = 3$.

At point (1, –1, 10) with

$D(x, y) = 12y - 12$,

$D(1, -1) = 12(-1) - 12 = -12 - 12 = -24$

Since the value of D is negative, the critical point is a saddle point.

The critical point (1, –1, 10) is neither a high point nor a low point, but a saddle point.

Congratulations; you have completed the book.

You should now be able to find all partial derivatives for multivariate functions and use them to estimate the change in the value of the function for a unit change in one of the independent variables. In addition, for some bivariate functions, you should be able to find and classify all critical points.

I can no longer say, "before continuing to the next unit," but do the following exercises anyway. There are only ten.

EXERCISES

1. If $z = f(x, y) = 3x^2 + 4y^5 + 1$, find:

    a. f_x

    b. f_y

    c. $f(1, -1)$

    d. $f_x(1, -1)$

2. If $f(x, y) = (3x - 7y)^5$, find:

    a. f_x

    b. f_y

3. If $z = f(x, y) = 4x^6 - 8x^3 - 7x + 6xy + 8y + x^3y^5$, find:

    a. f_x

    b. f_y

4. If $f(w, x, y) = 2xy + 3x^2yw + w^2 + 10$, find:

    a. f_w

    b. f_x

    c. f_y

5. If $z = f(x, y) = x^3 + y^2 - 8xy + 2x^3y^2 + 11$, find:

    a. $\dfrac{\partial z}{\partial x}$

    b. $\dfrac{\partial z}{\partial y}$

    c. $\dfrac{\partial^2 z}{\partial x^2}$

    d. $\dfrac{\partial^2 z}{\partial y^2}$

    e. $\dfrac{\partial^2 z}{\partial x\, \partial y}$

    f. $\dfrac{\partial^2 z}{\partial y\, \partial x}$

6. If $f(x, y) = 3x^4y^3 + 2xy$, find all first- and second-order partials.

7. Find and classify all critical points for

$$f(x, y) = 3x^2 + 4y^2 + 2xy - 30x - 32y + 50.$$

8. Find and classify all critical points for

$$f(x, y) = 3 - x^2 - y^2.$$

9. Toby Poplin, an economics professor, spends his summers farming. Currently he farms 5 acres and spends a meager $10 on advertising in the weekly newspaper. During the winter Professor Poplin determined by using past records that his annual profit is given by

$$z = f(x, y) \quad = 10x^3 + 20y^2 - 10xy$$

where x = the weekly advertising budget, in dollars,

and y = the number of acres used for farming.

a. Find $f(10, 5)$ and interpret.

Find each of the following:

b. f_x

c. f_y

d. $f_x(10, 5)$

e. $f_y(10, 5)$

f. f_{xx}

g. $f_{xx}(10, 5)$

Use marginal analysis to answer the following questions:

h. Alice, the professor's wife, suggests that he increase the weekly advertising budget by $1 a week. How would this affect the profit?

i. Toby's youngest son suggests that, rather than spend more money on advertising, the family farm an additional acre. How would this affect the profit?

j. Toby's oldest son, who knows nothing about marginal analysis, thinks the best solution is to adapt both Alice's and the youngest's son's suggestions and thereby increase profit by $3,050. Is he right?

10. The annual profit of a small inn in New England is given by

$$f(x, y) = 4x^2 + 100y^2 + 2x + 5y + 5,000$$

where x = the weekly advertising budget, in dollars,

and y = the number of rooms available for rent.

Currently the inn has 10 rooms available and an advertising budget of $100 per week.

a. If an additional room is constructed in an unfinished area, how will annual profit be affected?

b. If an additional dollar is added to the weekly advertising budget, how will annual profit be affected?

If additional practice is needed:

Barnett, Ziegler, and Byleen, pages 508–509, problems 1–68; pages 528–529, problems 1–30

Harshbarger and Reynolds, pages 1013–1014, problems 1–48; page 1032, problems 1–16; page 1041, problems 1–15

Hoffmann and Bradley, pages 528–529, problems 1–31; pages 546–547, problems 1–28; page 562, problems 1–15

Waner and Costenoble, page 460, problems 1–44; pages 470–472, problems 1–40; pages 482–484, problems 1–39

Answers to Exercises

UNIT 1

1. $f(0) = 3$

2. $f(10) = (10)^2 - 2(10) + 3$

    $= 100 - 20 + 3$

    $= 83$

3. $f(-2) = 11$

4. $f(a) = a^2 - 2a + 3$

5. $f(x + 1) = x^2 + 2$

6. $f(-x) = (-x)^2 - 2(-x) + 3$

    $= x^2 + 2x + 3$

7. The domain is the set of all reals.

8. $g(x) = \dfrac{1}{x - 3}$

    $g(5) = \dfrac{1}{5 - 3}$

    $= \dfrac{1}{2}$

9. $g(c) = \dfrac{1}{c - 3}$

10. $g(a + 3) = \dfrac{1}{(a + 3) - 3}$

    $= \dfrac{1}{a}$

11. The domain is the set of all real numbers except 3.

12. $q = h(p) = \sqrt{13 - p}$

    The independent variable is p.

13. The domain is the set of all p such that $p \leq 13$ or, using set notation: $D = \{p : p \leq 13\}$.

14. $h(12) = \sqrt{13 - 12}$

    $= \sqrt{1}$

    $= 1$

15. $h(13) = 0$

16. $h(-3) = \sqrt{13 - (-3)}$

    $= \sqrt{16}$

    $= 4$

17. $h(2) = \sqrt{11}$

    There is no real advantage to changing this to a decimal approximation at this time.

18. $\alpha(x, y, z) = x - 3z + xy$

    $\alpha(5, 3, -2) = 5 - 3(-2) + (5)(3)$

    $= 5 + 6 + 15$

    $= 26$

19. $H(7) = 6$

20. $H(0) = 0 + 1 = 1$

21. $H(-4) = -3$

22. $H(3) = 3 + 1 = 4$

23. $D = \{5, 8\}$

24. $R = \{5, 6, 7\}$

25. No.

26. $\{1, 4\}$

27. m

28. $z(-2) = 3(-2)^2 - 10$

    $= 3(4) - 10$

    $= 12 - 10$

    $= 2$

29. $z(x - 1) = 3x^2 - 6x - 7$

30. The domain is the set of all reals.

C31. 1217; 20.15; −7.625

C32. Keystrokes for function: (Y=) (CLEAR) (−) (•)
(6) (X,T,θ,n) (∧) (5) (−) (4) (X,T,θ,n) (∧) (3)
(+) (2) (X,T,θ,n) (ENTER)

$f(5) = -2{,}365$; $f(1.02) = -2.867$; $f(0) = 0$; $f(-2.3) = 82.686$; $f(-4.1) = 962.62$

UNIT 2

1. $(f + g)(5) = f(5) + g(5) = 9$

2. $(H - f)(5) = H(5) - f(5)$

    $= (5^2 + 5) - (5 + 1)$

    $= 30 - 6$

    $= 24$

3. $(GF)(2) = G(2) \, F(2) = -2$

4. $g(F(G(2))) = g(F(1))$

    $= g(-2)$

    $= (-2) - 2$

    $= -4$

    $G(2) = 3(2) - 5$

    $= 6 - 5$

    $= 1$

    $F(1) = (1)^2 - 3(1)$

    $= 1 - 3$

    $= -2$

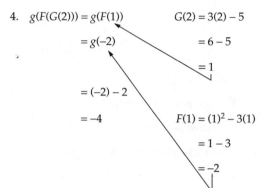

5. $F(w + 1) = w^2 - w - 2$

6. $f(h(x)) = f(2x + 3)$

    $= (2x + 3) + 1$

    $= 2x + 3 + 1$

    $= 2x + 4$

7. $2f(x) + g(x) - h(x) = x - 3$

8. $h(f(x)) = h(x + 1)$

    $= 2(x + 1) + 3$

    $= 2x + 2 + 3$

    $= 2x + 5$

9. $H(2x) - h(x) = 4x^2 - 3$

10. $G(F(x)) = G(x^2 - 3x)$

    $= 3(x^2 - 3x) - 5$

    $= 3x^2 - 9x - 5$

11. $N(t) = -t^2 + 400t + 50{,}000$

    a. The current population of the community is 50,000 people.

    b. The population of the community 10 years from now is estimated to be 53,900 people.

12. $C(x) = 6{,}452 + 72x \quad x \geq 0$

    a. $C(10) = 6{,}452 + 72(10)$

        $= 6{,}452 + 720$

        $= \$7{,}172$

b. $C(100) = 6{,}452 + 72(100)$

    $= 6{,}452 + 7{,}200$

    $= 13{,}652$

    The total cost of producing 100 rugs will be $13,562.

c. $C(0) = 6{,}452 + 72(0)$

    $= 6{,}452$

13. Total cost = fixed cost +
    (variable cost/unit)(number of units)

    a. $C(x) = 13{,}200 + 43x$

    b. The total cost of producing no braided rugs is $13,200. Or, the company has fixed costs of $13,200.

        The total cost of producing 10 braided rugs is $13,630.

        The total cost of producing 100 braided rugs is $17,500.

14. $R(x) = 153x$

    a. $R(50) = 153(50)$

        $= 7{,}650$

    b. $R(100) = 153(100)$

        $= 15{,}300$

        The total revenue gained from selling 100 crocheted rugs is $15,300.

        $R(0) = 153(0) = 0$

        The total revenue gained from selling 0 crocheted rug is $0.

15. Total revenue = (selling price per unit) ×
    (number of units sold)

    a. $R(x) = 78x$

    b. The total revenue gained from selling 1,000 braided rugs is $78,000.

    c. Restricted domain: $x \geq 0$

16. a. $P(x) = 81x - 6{,}452 \quad x \geq 0$

        $P(200) = 81(200) - 6{,}452$

        $= 16{,}200 - 6{,}452$

        $= 9{,}748$

b. $P(50) = 81(50) - 6{,}452$

$\qquad = 4{,}050 - 6{,}452$

$\qquad = -2{,}402$

The company would lose \$2,402 on the production and sale of only 50 crocheted rugs.

17. $C(x_1, x_2, x_3) = 1{,}750 + 7.5x_1 + 6x_2 + 10.6x_3$

a.   The total cost of producing 10 toy soldiers, 25 teddy bears, and 100 dolls is \$3,035.

b.   Fixed costs are \$1,750.

The variable cost per unit for toy soldiers is \$7.50.

The variable cost per unit for teddy bears is \$6.00.

The variable cost per unit for dolls is \$10.60.

18.   a. $y = f(5)$

$\qquad = 4(5)$

$\qquad = 20$ If the gardener works 5 hours he will earn \$20 for the day.

b. $x = g(32)$

$\qquad = (32 - 30)/2$

$\qquad = 1$ If the temperature is 32°F at 8 A.M., H. Jones will work 1 hour.

c. $y = f(g(32))$

$\qquad = f(1)$

$\qquad = 4(1)$

$\qquad = 4$ If the temperature is 32°F at 8 A.M., H. Jones will earn \$4 for the day.

d. $y = h(t) = f(g(t))$

$\qquad = f\left(\dfrac{t - 30}{2}\right)$

$\qquad = \dfrac{4(t - 30)}{2}$

$\qquad = 2t - 60$

The salary earned per day by H. Jones is a function of the temperature, in F°, at 8 A.M.

e. $y = h(50)$

$\qquad = 2(50) - 60$

$\qquad = 100 - 60$

$\qquad = 40$

19.   a.   The independent variable for f is x.

b.   Restricted domain for f: $0 \le x \le 24$.

c.   The independent variable for g is t.

d.   Restricted domain for g: $30 \le t \le 78$.

e.   Restricted range for f: $0 \le y \le 96$.

C20. Keystrokes for function: (**Y=**) (**CLEAR**) (**0**) (**·**) (**0**) (**0**) (**0**) (**2**) (**X,T,θ,n**) (**^**) (**3**) (**+**) (**1**) (**0**) (**X,T,θ,n**) (**ENTER**)

$P(51) = 536.53$

If Carlos washes 51 cars per week, his weekly profit will be \$536.53.

UNIT 3

1. $y = f(x) = 2x + 5$

The slope is 2 or $m = \dfrac{\Delta y}{\Delta x} = \dfrac{2}{1}$

The y-intercept is 5.

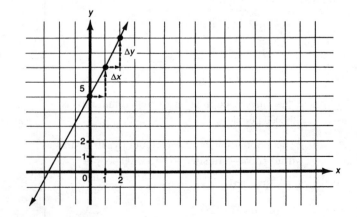

2. $y = f(x) = \dfrac{7}{3}x - 1$

The slope is $\dfrac{7}{3}$ or $m = \dfrac{\Delta y}{\Delta x} = \dfrac{7}{3}$

The y-intercept is -1.

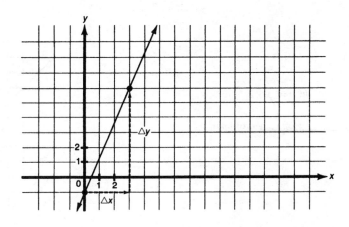

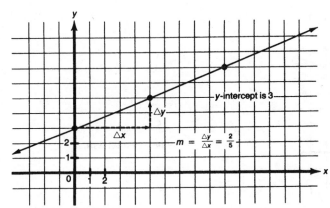

3. $y = \dfrac{2}{5}x + 3$

slope   y-intercept

4. $f(x) = x - 4$

slope   y-intercept

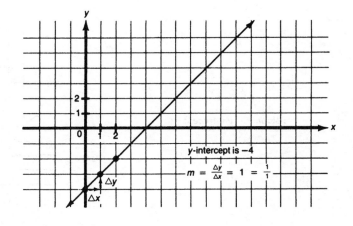

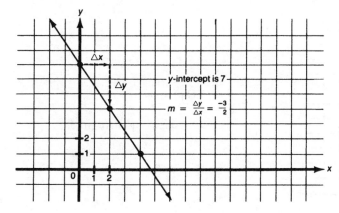

5. $g(x) = -\dfrac{3}{2}x + 7$

slope   y-intercept

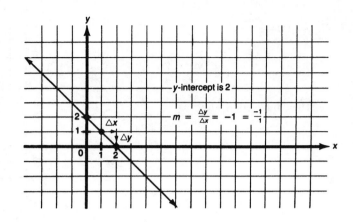

6. $h(x) = -x + 2$

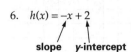

slope   y-intercept

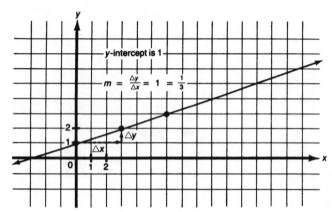

7. $y = f(x) = \dfrac{x}{3} + 1$

$y = f(x) = \dfrac{1}{3}x + 1$

slope   y-intercept

8. $y = f(x) = 0.25x + 1.50$

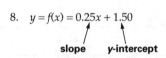

slope   y-intercept

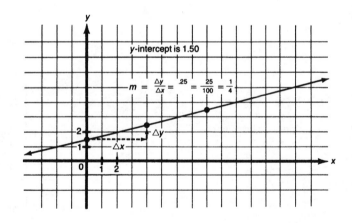
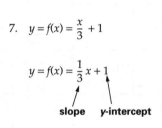

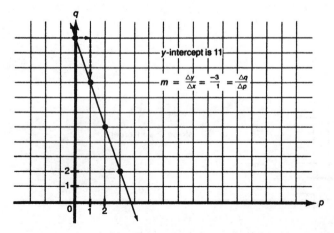

9. $q = D(p) = -3p + 11$

slope   y-intercept

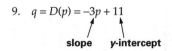

10. a. $x - 3y = -12$

$-3y = -x - 12$

$y = \dfrac{1}{3}x + 4$ slope-intercept form

slope   y-intercept

b.   The slope is $\dfrac{1}{3}$, meaning that, if x increases by 3, the y value will increase by 1.

c.   The y-intercept is 4.

The x-intercept is the value of x when $y = 0$.

$x - 3y = -12$

$x - 3(0) = -12$

$x = -12$

The x-intercept is -12 or point $(-12, 0)$.

d.

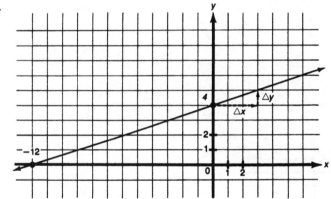

C11. $y = f(x) = 2.53x + 1.75$

slope   y-intercept

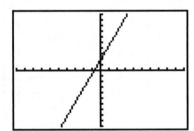

C12. $y = f(x) = -3.08x - 7.15$

a.   press (**Y=**) (**(−)**) (**3**) (**·**) (**0**) (**8**) (**X,T,θ,n**)
(**(−)**) (**7**) (**·**) (**1**) (**5**) (**ENTER**) (**GRAPH**).

b.   The slope is -3.08.

c.   The y-intercept is -7.15

To find the x-intercept, press (**2nd**) (**CALC**) (**2**).

At the <u>Left Bound?</u> prompt, move the cursor to the left of
the x-intercept; press (**ENTER**).

At the <u>Right Bound?</u> prompt, move the cursor to the right of the x-intercept; press (**ENTER**).

At the <u>Guess?</u> prompt, move the cursor between the boundaries; press (**ENTER**).

The x-intercept is -2.32.

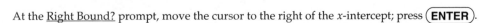

d. To find y, press (**2nd**)(**1**)(**2**)(**•**)(**4**)(**5**)(**ENTER**).

The y value is -14.696.

13. a. $y = f(x) = -\dfrac{2}{3}x + 7$

b. $-\dfrac{2}{3}$

c. $10\dfrac{1}{2}$

d. 7

e. -12

14. a. $5p - 4q = 20$

$-4q = -5p + 20$

$q = \dfrac{5}{4}p - 5$

$\uparrow$

slope

b. $m = \dfrac{\Delta y}{\Delta x} = \dfrac{5}{4} = \dfrac{\Delta q}{\Delta p}$

The slope is $\dfrac{5}{4}$, meaning that, if p increases by 4, the q value will increase by 5.

15.

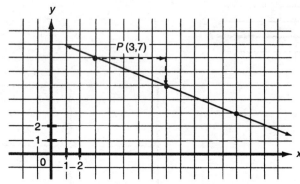

16.

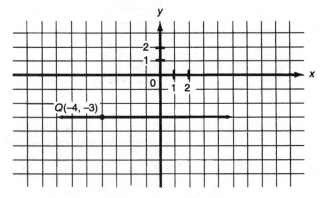

17.

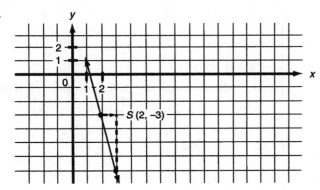

UNIT 4

1. $m = \dfrac{-1}{2}$

2. $y - y_1 = m(x - x_1)$

    $y - (-3) = 5(x - 4)$

      $y + 3 = 5x - 20$

         $y = 5x - 23$

3. $y = \dfrac{-4}{5}x - 1$

4. $y - y_1 = m(x - x_1)$

    $y - 7 = \dfrac{-1}{6}(x - 6)$

    $y - 7 = \dfrac{-1}{6}x + 1$

       $y = \dfrac{-1}{6}x + 8$

5. $y = \dfrac{-1}{5}x + 7\dfrac{2}{5}$

6. $(7, 5)$ $m = \dfrac{\Delta y}{\Delta x} = \dfrac{0}{10} = 0$

    $\dfrac{(-3, 5)}{10, 0}$

    $\Delta x \uparrow \quad \uparrow \Delta y$

    To write the function: $y - y_1 = m(x - x_1)$

                       $y - 5 = 0(x - 7)$

                       $y - 5 = 0$

                           $y = 5$

7. $y = -8$

8. $y - y_1 = m(x - x_1)$

    $y - 100 = 0.15(x - 2.35)$

    $y - 100 = 0.15x - 0.3525$

         $y = 0.15x + 99.6475$

9. $y = -2$

10. $(-5, -2)$ $m = \dfrac{\Delta y}{\Delta x} = \dfrac{6}{-2} = -3$

    $\dfrac{(-3, -8)}{-2, \ \ 6}$

        $y - y_1 = m(x - x_1)$

     $y - (-2) = -3(x - (-5))$

        $y + 2 = -3(x + 5)$

        $y + 2 = -3x - 15$

          $y = -3x - 17$

11. $y = -5x$

12. $(3{,}600, \ 1{,}000)$ $m = \dfrac{\Delta y}{\Delta x} = \dfrac{800}{3{,}200} = \dfrac{1}{4}$

    $\dfrac{(400, \ \ 200)}{3{,}200, \ \ 800}$

        $y - y_1 = m(x - x_1)$

     $y - 200 = \dfrac{1}{4}(x - 400)$

     $y - 200 = \dfrac{1}{4}x - 100$

        $y = \dfrac{1}{4}x + 100$

13. A linear function does not exist, but the equation of the line is $x = -9$.

14. (3.95, 435) $m = \dfrac{\Delta y}{\Delta x} = \dfrac{135}{1.20} = 112.5$

$\dfrac{(2.75, 300)}{1.20, 135}$

$y - y_1 = m(x - x_1)$

$y - 300 = 112.5(x - 2.75)$

$y - 300 = 112.5x - 309.375$

$y = 112.5x - 9.375$

15. $x = 15$ is the equation, but a linear function does not exist.

16. Use the two points to calculate the slope.

$(-2, \ 5)$

$\dfrac{(-2, 13)}{0, -8}$ $m = \dfrac{\Delta y}{\Delta x} = \dfrac{-8}{0} = \text{undefined}$

Since the slope of the line connecting the points is undefined, the line must be vertical. The equation of the vertical line through point $(-2, 5)$ is $x = -2$.

17. The slope is undefined.

18. If the line is to be parallel to $y = 5x + 2$, it must have a slope of 5.

$y - y_1 = m(x - x_1)$

$y - 4 = 5(x - 6)$

$y - 4 = 5x - 30$

$y = 5x - 26$

19. $y = -3x$

20. If the line is to be perpendicular to $y = 3x - 1$, its slope must be the negative reciprocal of 3, or $m = -\dfrac{1}{3}$.

$y - y_1 = m(x - x_1)$

$y - 2 = -\dfrac{1}{3}(x - (-7))$

$y - 2 = -\dfrac{1}{3}(x + 7)$

$y - 2 = -\dfrac{1}{3}x - \dfrac{7}{3}$

$y = -\dfrac{1}{3}x - \dfrac{7}{3} + 2$

$= -\dfrac{1}{3}x - \dfrac{1}{3}$

21. $y = 4$

22. $x = 4$

23. $y = -4$

UNIT 5

1. a. $L(t) = 1.5t - 60$

$L(86) = 1.5(86) - 60$

$= 129 - 60$

$= 69$

If the temperature is 86°F, the street vendor should expect to sell 69 lemonades at lunch.

b. $L(t) = 1.5t - 60$

$L(40) = 1.5(40) - 60$

$= 60 - 60$

$= 0$

If the temperature is 40°F or lower, the street vendor does not expect to sell any lemonade.

c. $m = 1.5$

For each increase of 1° in the temperature, the number of lemonades sold is expected to increase by 1.5.

However, the following is a better answer, since $m = 1.5 = \dfrac{3}{2}$.

For each increase of 2° in the temperature, the number of lemonades sold is expected to increase by 3.

2. a. Break even occurs where $R(x) = C(x)$

$$10x = 4.5x + 38.5$$

$$5.5x = 38.5$$

$$x = 7$$

Marguerite needs to work 7 hours per week to break even.

b. $P(x) = R(x) - C(x)$

$$= 10x - (4.5x + 38.5)$$

$$= 10x - 4.5x - 38.5$$

$$= 5.5x - 38.5$$

$$22 = 5.5x - 38.5$$

$$5.5x = 60.5$$

$$x = 11$$

Marguerite needs to work 11 hours next week in order to earn enough profit to pay for a concert ticket costing $22.

3.

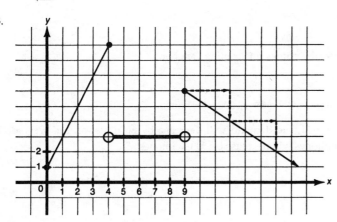

$f(x) = 2x + 1$ if $0 \le x \le 4$

| lower limit | if $x = 0$, $f(0) = 2(0) + 1 = 1$ | $(0, 1)$ |
| upper limit | if $x = 4$, $f(4) = 2(4) + 1 = 9$ | $(4, 9)$ |

$f(x) = 3$	if $4 < x < 9$	
lower limit	if $x = 4$, $f(4) = 3$	$(4, 3)$ open
upper limit	if $x = 9$, $f(9) = 3$	$(9, 3)$ open

| $f(x) = \dfrac{-2}{3} x + 12$ | if $9 \le x$ | |
| lower limit | if $x = 9$, $f(9) = \dfrac{-2}{3} (9) + 12 = 6$ | $(9, 6)$ |

There is no upper limit. This section of the graph is a half line. It can be graphed by using the slope, which is $\dfrac{-2}{3}$ to plot additional points.

4. a. *f(t)* denotes that *t* is the independent variable.

    Compare: $y = mx + b$

    $$n = 322t + 1,850$$

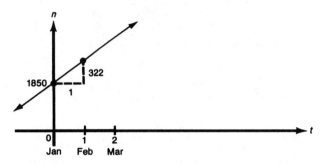

    b. The *y*-intercept is 1,850. In January of this year, the number of reported crimes was estimated to be 1,850.

    c. The slope is 322. For each additional month starting from January of this year, the number of reported crimes is estimated to increase by 322.

    d. $n = f(t) = 322t + 1,850$ and October corresponds to $t = 9$.

        $n = f(9) = 322(9) + 1,850$

        $\qquad = 4,748$

        It is estimated that there will be 4,748 reported crimes this coming October.

5. a. $q = D(100) = 12,000$

        If the price is $100 per telephone, the demand will be for 12,000 telephones.

    b. The slope is −30. If the price increases by $1, the demand will decrease by 30 units; or, for every dollar increase in price, the demand for the telephones will decrease by 30 telephones.

    c. 30

6. a. $p + 2q - 5,000 = 0$

        $2q \qquad = -p + 5,000$

        $q = -\dfrac{1}{2}p + 2,500$

        $q = D(p) = -\dfrac{1}{2}p + 2,500$

    b. $D(p) = -\dfrac{1}{2}p + 2,500$

        $D(2,000) = -\dfrac{1}{2}(2,000) + 2,500$

        $\qquad = -1,000 + 2,500$

        $\qquad = 1,500$

        If the price is $2,000, the demand will be for 1,500 units.

    c. The *y*-intercept is 2,500

        If free, the company could give away 2,500.

d. $D(5,000) = -\frac{1}{2}(5,000) + 2,500 = 0$

If the price is \$5,000, the demand is 0. And further, at a price of \$5,000 or more, there will be no demand.

e. Slope $= \frac{-1}{2} = -0.5$

For every \$1 increase in price, demand will drop by $\frac{1}{2}$ unit. Or, for every \$2 increase in price, demand will drop by 1 unit.

7. a. If the milk is priced at \$0.70 per quart, the weekly demand for milk is expected to be 2,100 quarts.

   b. If the milk were free, the ABC Dairy Company could give away 4,200 quarts of milk.

   c. If the milk is priced at \$1.40 a quart or more, there will be no demand.

   d. If the ABC Dairy Company wants to sell 1,000 quarts per week, it should price the milk at \$1.07 per quart.

8. a. If $p = 5, q = 46$ (5, 46)

   If $p = 9, q = 86$ (9, 86) $m = \frac{\Delta y}{\Delta x} = \frac{-40}{-4} = 10$

   $y - y_1 = m(x - x_1)$

   $y - 46 = 10(x - 5)$

   $\quad\quad = 10x - 50$

   $\quad y = 10x - 4$

   $q = S(p) = 10p - 4$

   b. $q = S(15) = 10(15) - 4$

   $\quad\quad = 150 - 4$

   $\quad\quad = 146$

   If city maps are priced at \$15, suppliers are willing to produce 146 maps.

   c. The x-intercept is where $y = 0$. In this instance we are looking for p when $q = 0$.

   $0 = 10p - 4$

   $10p = 4$

   $p = 0.40$

   If the price of maps is \$0.40 or less, suppliers are unwilling to produce any maps.

   d. Restricted domain: $p \geq 0.40$

9. a. \$3,400

   b. $C(200) = 5,070$

   If the Spaniel Company manufactures 200 cordless toy telephones, its total cost will be \$5,070.

   c. Restricted domain: $0 \leq x \leq 300$

10. a. $R(x) = 17.95x$

    b. $R(200) = 17.95(200) = 3,590$

    If the company sells 200 cordless toy telephones, its total revenue will be \$3,590.

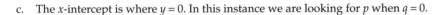

11. a. $P(x) = 9.6x - 3,400$

    b. If the Spaniel Company produces and sells 200 cordless toy telephones, it will have a loss of $1,480.

12. a. If $p = 5$, then $q = 65$ $(5, 65)$
    If $p = 10$, then $q = 40$ $(10, 40)$
    $$\frac{-5 \quad 25}{} \qquad m = \frac{25}{-5} = -5$$

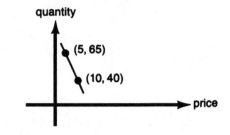

$$y - y_1 = m(x - x_1)$$

$$y - 40 = -5(x - 10)$$

$$= -5x + 50$$

$$y = -5x + 90$$

$$q = D(p) = -5p + 90$$

    b. y-intercept is 90

    If the maps are free, the city could give away 90.

    x-intercept: $0 = -5p + 90$

    $$5p = 90$$

    $$p = 18$$

    At a price of $18 per map, the demand is 0.

    c. Restricted domain: $0 \leq p \leq 18$

    d. Equilibrium occurs where $S(p) = D(p)$

    $$10p - 4 = -5p + 90$$

    $$15p = 94$$

    $$p = 6.27$$

    and $q = 10p - 4 = 10(6.27) - 4 = 58.7$

    At a price of $6.27 each, the supply and demand will be equal for 59 city maps.

13. No

C14. Press $\boxed{\textbf{WINDOW}}$ and reset the viewing window as suggested.

Key in *both* functions and view the graph.

$\backslash Y_1$ $\boxed{\textbf{Y=}}\,\boxed{0}\,\boxed{\cdot}\,\boxed{5}\,\boxed{\textbf{X,T,θ,n}}\,\boxed{+}\,\boxed{5}\,\boxed{0}\,\boxed{0}\,\boxed{\textbf{ENTER}}$

$\backslash Y_2$ $\boxed{\textbf{Y=}}\,\boxed{0}\,\boxed{\cdot}\,\boxed{7}\,\boxed{5}\,\boxed{\textbf{X,T,θ,n}}\,\boxed{\textbf{ENTER}}$

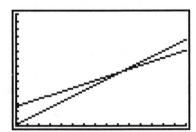

To find the point of intersection, press $\boxed{\textbf{2nd}}\,\boxed{\textbf{CALC}}\,\boxed{5}$

At the <u>First curve?</u> prompt, move the cursor along the first line to where the intersection appears to be and press $\boxed{\textbf{ENTER}}$.

At the <u>Second curve?</u> prompt, move the cursor along the second line to where the intersection appears to be and press $\boxed{\textbf{ENTER}}$.

At the <u>Guess?</u> prompt, move the cursor to between the boundaries and press (**ENTER**).

Intersection X = 2000   Y = 1500 appears at the bottom of screen.

Two thousand items should be produced and sold daily to break even, at which level daily costs and revenues will be equal at $1,500.

UNIT 6

1.

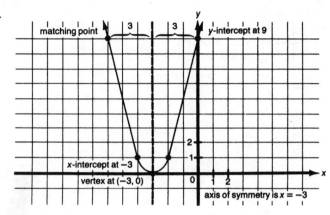

2. $y = f(x) = x^2 + 2x + 1$

1.   This is a quadratic with $a = 1$, $b = 2$, $c = 1$.

2.   The parabolic curve opens up.

3.   The y-intercept is 1.

4.   The x-intercept is -1 because

$$0 = x^2 + 2x + 1$$

$$0 = (x + 1)^2$$

$$x + 1 = 0$$

$$x = -1$$

5.   The vertex is located at $(-1, 0)$ because the

axis of symmetry is $x = -1$

and $f(-1) = (-1)^2 + 2(-1) + 1$

$$= 1 - 2 + 1$$

$$= 0$$

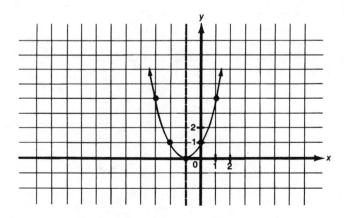

3.

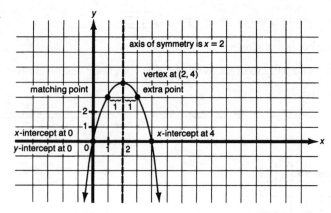

4. $y = f(x) = -x^2 + 2x + 3$

1.   This is a quadratic with $a = -1$, $b = 2$, $c = 3$.

2.   The parabolic curve opens down.

3.   The y-intercept is 3.

4.   The x-intercepts are -1 and 3 because

$$0 = -x^2 + 2x + 3$$

$$0 = x^2 - 2x - 3$$

$$0 = (x - 3)(x - 1)$$

$$x = 3, x = -1$$

5.   The vertex is located at (1, 4) because

$$\frac{-b}{2a} = \frac{-2}{2(-1)} = \frac{-2}{-2} = 1$$

and $f(1) = (-1) + 2(1) + 3$

$$= -1 + 2 + 3$$

$$= 4$$

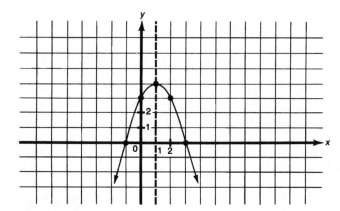

5.

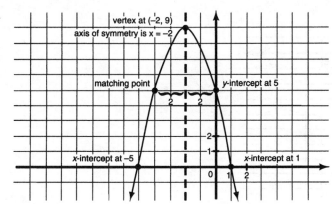

6. $y = f(x) = -2x^2 + 4x$

   1. This is a quadratic with $a = -2$, $b = 4$, $c = 0$.

   2. The parabolic curve opens down.

   3. The y-intercept is 0.

   4. The x-intercepts are 0 and 2 because

      $0 = -2x^2 + 4x$

      $0 = 2x^2 - 4x$

      $0 = 2x(x - 2)$

      $x = 0 \qquad x = 2$

   5. The vertex is located at (1, 2) because

      the axis of symmetry is $x = 1$ and

      $f(1) = -2(1)^2 + 4(1)$

      $\qquad = -2 + 4$

      $\qquad = 2$

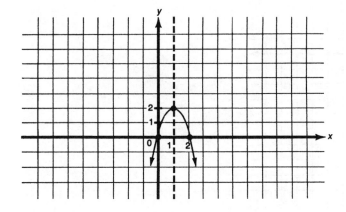

7.

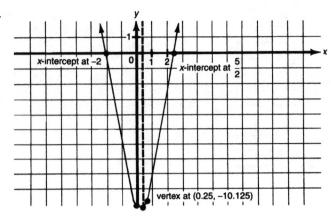

8. $y = f(x) = -x^2 - 7$

     1.   This is a quadratic with $a = -1$, $b = 0$, $c = -7$.

     2.   The parabolic curve opens down.

     3.   The y-intercept is -7.

     4.   There are no x-intercepts because

         $0 = -x^2 - 7$

         $x^2 = -7$ has no solution.

     5.   The vertex is located at $(0, -7)$ because

$$\frac{-b}{2a} = \frac{0}{2(-1)} = 0$$

         and $f(0) = (0)^2 - 7 = -7$

     6.   Locate one point in either side of vertex

         let $x = 1$, then $f(1) = -1 - 7 = -8$

         let $x = -1$, then $f(-1) = -(-1)^2 - 7 = -8$

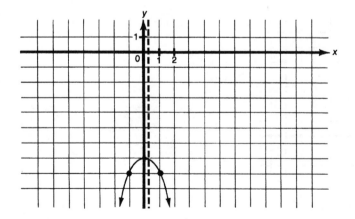

9. Omit step 4 because $0 = x^2 - 4x + 2$ is not easily factored and problem does not ask that the x-intercepts be found.

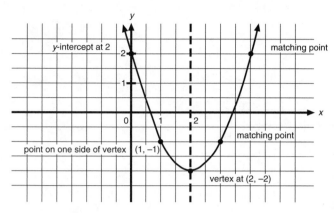

10. $y = f(x) = 3x^2 - 9x$

1. This is a quadratic with $a = 3, b = -9, c = 0$.

2. The parabolic curve opens up.

3. The y-intercept is 0.

4. The x-intercepts are 0 and 3 because

$$0 = 3x^2 - 9x$$

$$0 = 3x(x - 3)$$

$$3x = 0 \qquad x - 3 = 0$$

$$x = 0 \qquad x = 3$$

5. The vertex is located at (1.5, −6.75) because

the axis of symmetry is $x = 1.5$

and $f(1.5) = 3(1.5)^2 - 9(1.5)$

$$= 3(2.25) - 13.5$$

$$= 6.75 - 13.5$$

$$= -6.75$$

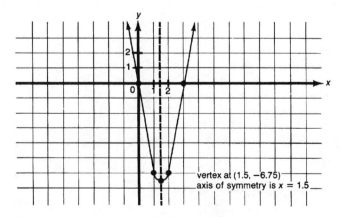

C11. The two x-intercepts are 0.59 and 3.41. Answers may vary slightly because of differences in where the boundaries were set.

UNIT 7

1. $y = f(x) = 0.01x^2 - 8x$

    a.

    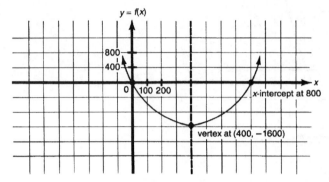

    b. At a price of \$800 or less, the Albemarle Manufacturing Company would be unwilling to supply any of its stress-reducing comfort recliners.

    c. Restricted domain: $800 \le x$

2. a. $q = S(p) = p^2 - 100$

    1. This is a quadratic with $a = 1$, $b = 0$, $c = -100$.

    2. The parabolic curve opens up.

    3. The y-intercept is -100.

    4. The x-intercepts are 10 and -10 because

        $0 = p^2 - 100$

        $p^2 = 100$

        $p = \pm 10$

    5. The vertex is located at $(0, -100)$

    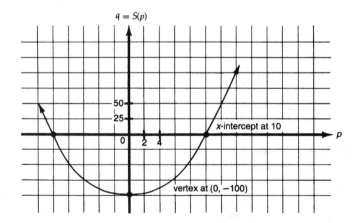

    b. $S(18.50) = (18.50)^2 - 100 = 342.25 - 100 = 242.25$

    At a price of \$18.50, farmers are willing to supply 242.25 pounds of seed.

c. If $S(p) = 800$, then

$$800 = p^2 - 100$$

$$p^2 = 900$$

$$p = \pm 30$$

The price per pound would have to be $30 before farmers would be willing to supply 800 pounds.

3. a.

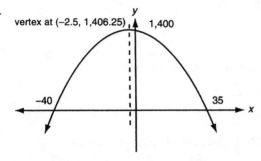

vertex at (−2.5, 1,406.25)

b. If the coffee is free, you will give away 1,400 pounds.

c. $D(15.75) = 1,073.19$

If the coffee is priced at $15.75 per pound, you will sell 1,073 pounds.

d. If the coffee is priced at $35 or more per pound, you won't sell any.

e. Restricted range: $0 \leq y \leq 1,400$

f. If you want to sell all 1,334 pounds of coffee, the highest price you should charge is $6 per pound.

4. Equilibrium occurs where supply and demand are equal. It does not change the definition just because the functions are quadratics.

Equilibrium is where $S(p) = D(p)$

$$2p^2 + 5p - 250 = 2p^2 - 297p + 10,960$$

$$297p + 5p = 10,960 + 250$$

$$302p = 11,210$$

$$p \approx 37.12$$

To find q when $p = 37.12$

$$S(p) = 2p^2 + 5p - 250$$

$$S(37.12) = 2(37.12)^2 + 5(37.12) - 250$$

$$= 2(1,377.89) + 185.60 - 250$$

$$= 2,691.38$$

$$= 2,691$$

At a price of $37.12 per unit, the supply and demand for the item will be equal at approximately 2,691 units.

5. a.

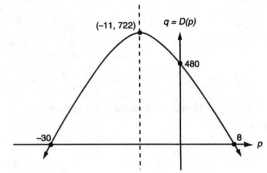

b. If the Creatures are priced at $5 each, Cathy would expect to sell 210 of them.

c. If the Creatures are free, she could give away 480 of them.

d. Restricted domain: $0 \le p \le 8$

e. At a price of $8 or more, customers would be unwilling to buy any of Cathy's Creatures.

C6. Create a table of values for the function with the following keystrokes:

$$(Y=)(8)(1)(3)(5)(+)(4)(\cdot)(7)(5)(X,T,\theta,n)(+)(0)(\cdot)(0)(2)(X,T,\theta,n)(\wedge)(2)(ENTER).$$

a. To calculate $C(312)$, press $(2nd)(TABLE)$ and key in 312.

$C(312) = 11{,}564$

b. If the Dooley Company manufactures 312 trivets, its total cost will be $11,564.

7. a. 10,000

b. 12,000

8. a. $H(t) = -16t^2 + 128t + 320$

1. This is a quadratic with $a = -16$, $b = 128$, $c = 320$.

2. The parabolic curve opens down.

3. The y-intercept is 320.

4. The x-intercepts are -2 and 10 because

$0 = -16t^2 + 128t + 320$

$0 = 16t^2 - 128t - 320$

$0 = t^2 - 8t - 20$

$0 = (t - 10)(t + 2)$

$t - 10 = 0 \qquad t + 2 = 0$

$\qquad t = 10 \qquad\qquad t = -2$

5. The vertex is located at (4, 576).

$$\frac{-b}{2a} = \frac{-128}{2(-16)} = \frac{-128}{-32} = 4$$

and

$$H(4) = -16(4)^2 + 128(4) + 320$$

$$= -16(16) + 512 + 320$$

$$= -256 + 512 + 320$$

$$= 576$$

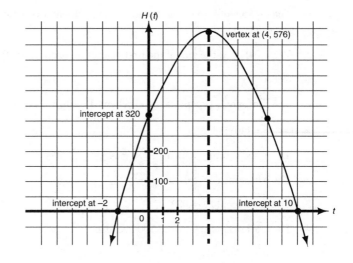

b. (2, 512)

Two seconds after the ball is thrown upward, it is 512 feet from the ground.

c. $H(t) = -16t^2 + 128t + 320$

$H(9) = -16(9)^2 + 128(9) + 320$

$= -16(81) + 1,152 + 320$

$= -1,296 + 1,152 + 320$

$= 176$

Nine seconds after the ball is thrown, it will be only 176 feet from the ground.

d. The ball will hit the ground in 10 seconds.

e. The ball will rise 576 feet.

f. The building was 320 feet high.

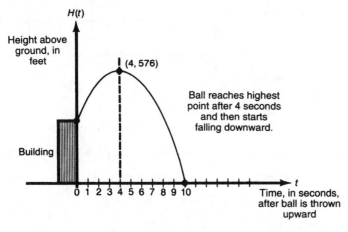

C9. a. Using the standard viewing window:

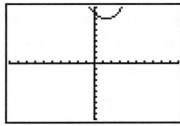

b. The restricted domain, as stated in the problem, does not include $t = 0$. Therefore, the vertical intercept does not exist.

c. $V(1, 8.04)$ At the end of the first year of incorporation, the Smith Company's annual profit was $8,040.

d. $9,640

e. The company's prospects are excellent. Profits are expected to increase every year.

10. $P(x) = -x^2 + 140x - 4,000$

1. This is a quadratic with $a = -1$, $b = 140$, $c = -4,000$.

2. The parabolic curve opens down.

3. The y-intercept is $-4,000$.

4. The x-intercepts are 40 and 100 because

$$0 = -x^2 + 140x - 4,000$$

$$0 = x^2 - 140x + 4,000$$

$$0 = (x - 40)(x - 100)$$

$$x - 40 = 0 \qquad x - 100 = 0$$

$$x = 40 \qquad\qquad x = 100$$

5. The vertex is located at $(70, 900)$ because the axis of symmetry is $x = 70$

and $P(70) = -(70)^2 + 140(70) - 4,000$

$$= -4,900 + 9,800 - 4,000$$

$$= 900$$

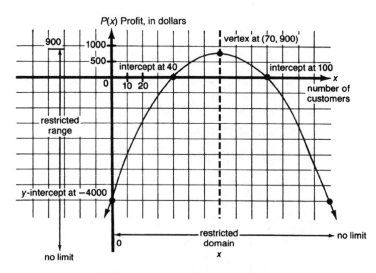

$(0, -4,000)$ If Jim has no customers, he will have losses of $4,000 per day.

$(40, 0)$ If there are 40 customers per day, he will break even; that is, his profit will be 0, but no losses.

$(100, 0)$ If there are 100 customers per day, he will also break even; that is, profit will be 0, but no losses.

$(70, 900)$ If there are 70 customers, he will make a profit of $900. This is the best he can achieve in one day.

If there are between 39 and 99 customers per day, inclusive, Jim will make a profit.

If there are fewer than 40 customers per day, Jim will lose money.

If there are more than 100 customers per day, for whatever reason, Jim will also lose money.

Restricted domain: $x \geq 0$ or, if you prefer, $0 \leq x$

Restricted range: $P(x) \leq 900$

This time x, the number of people, must be nonnegative; but $P(x)$, which is the profit, could be negative, denoting losses.

The relevant portion of the graph is in quadrants I and IV.

UNIT 8

1. $y = f(x) = x^3 - 4x^2 - 5x$

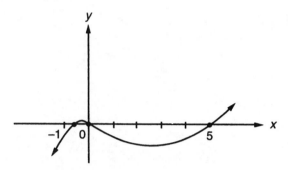

2. $y = f(x) = x^3 - 9x^2 + 18x$

    1. This is a cubic with $a = 1$, $d = 0$.

    2. The curve opens up: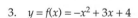

    3. The y-intercept is 0.

    4. The x-intercepts are 0, 3, and 6 because

        $0 = x^3 - 9x^2 + 18x$

        $0 = x(x^2 - 9x + 18)$

        $0 = x(x - 3)(x - 6)$

        $x = 0 \qquad x - 3 = 0 \qquad x - 6 = 0$

        $\qquad\qquad\quad x = 3 \qquad\quad x = 6$

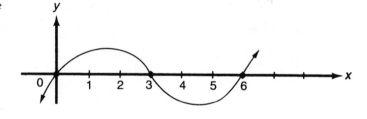

3. $y = f(x) = -x^2 + 3x + 4$

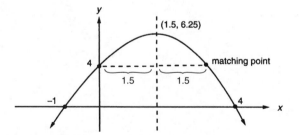

4. $y = f(x) = 2x^3 + 6x^2 - 20x$

    1. This is a cubic with $a = 2$, $d = 0$.

    2. The curve opens up: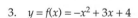

    3. The y-intercept is 0.

4. The x-intercepts are 0, –5, and 2 because

    $0 = 2x^3 + 6x^2 - 20x$

    $0 = 2x(x^2 + 3x - 10)$

    $0 = 2x(x + 5)(x - 2)$

    $2x = 0 \quad x + 5 = 0 \quad x - 2 = 0$

    $x = 0 \qquad x = -5 \qquad x = 2$

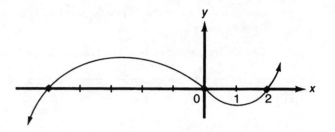

5. $y = f(x) = -x + 3$

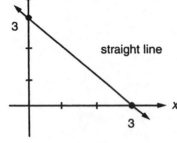

straight line

6. $y = f(x) = 12x^3 - 24x^2 + 12x$

    1. This is a cubic with $a = 12$, $d = 0$.

    2. The curve opens up:

    3. The y-intercept is 0.

    4. The x-intercepts are 0 and 1 because

        $0 = 12x^3 - 24x^2 + 12x$

        $0 = 12x(x^2 - 2x + 1)$

        $0 = 12x(x - 1)^2$

        $12x = 0 \qquad x - 1 = 0$

        $x = 0 \qquad\qquad x = 1$

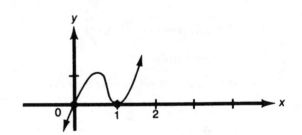

7. $y = f(x) = (x - 2)(x - 3)$

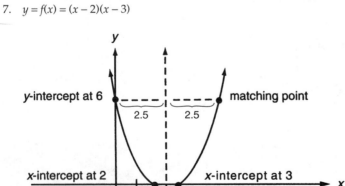

y-intercept at 6   matching point

2.5   2.5

x-intercept at 2 x-intercept at 3

vertex at (2.5, –0.25)

8. $y = f(x) = (x - 2)(x + 3)(x - 1)$

   $$= x^3 + \cdots + 6$$

   1. This is a cubic with $a = 1$, $d = 6$.

   2. The curve opens up: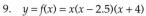

   3. The y-intercept is 6.

   4. The x-intercepts are 2, –3, and 1 because

   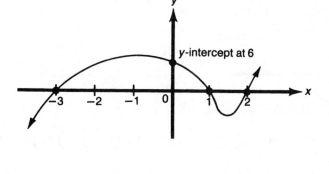

   $$0 = (x - 2)(x + 3)(x - 1)$$

   $x - 2 = 0 \qquad x + 3 = 0 \qquad x - 1 = 0$

   $x = 2 \qquad\qquad x = -3 \qquad\quad x = 1$

9. $y = f(x) = x(x - 2.5)(x + 4)$

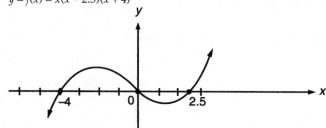

10. $y = f(x) = -x^3 + 3x^2$

    1. This is a cubic with $a = -1$, $d = 0$.

    2. The curve opens down: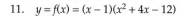

    3. The y-intercept is 0.

    4. The x–intercepts are 0 and 3 because

    $$0 = -x^3 + 3x^2$$

    $$0 = -x^2(x - 3)$$

    $-x^2 = 0 \qquad x - 3 = 0$

    $x = 0 \qquad\quad x = 3$

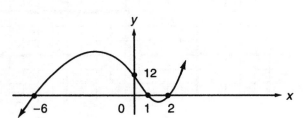

11. $y = f(x) = (x - 1)(x^2 + 4x - 12)$

12. $y = f(x) = x^3 - 4x$

    1. This is a cubic with $a = 1$, $d = 0$.

    2. The curve opens up: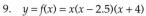

    3. The y-intercept is 0.

4. The x-intercepts are 0, 2, and –2 because

$$0 = x^3 - 4x$$

$$0 = x(x^2 - 4)$$

$$0 = x(x + 2)(x - 2)$$

$x = 0$ $x + 2 = 0$ $x - 2 = 0$

                $x = -2$ $x = 2$

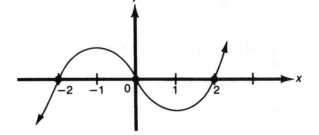

C13. The x-intercepts are 7.73 and 45.27.

C14. $\backslash Y_1$ (Y=) (2) (X,T,θ,n) (∧) (3) (+) (7) (X,T,θ,n) (∧) (2) (−) (1) (5) (X,T,θ,n) (+) (2) (0) (ENTER)

Set the viewing window: [–10, 10] by [–100, 100] and graph. Use (TRACE).

Move the cursor near the x-intercept; press (ZOOM) (2) (ENTER).

Press (2nd) (CALC) (2) and set the left bound, right bound, and guess prompts.

The x-intercept is –5.279.

15.

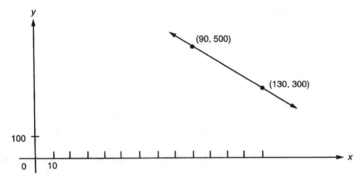

16. $f(x) = 3$ if $5 < x$

linear—half line—slope 0—horizontal

lower limit	if $x = 5$, $f(5) = 3$	(5, 3) open
$f(x) = x^2$	if $x \leq 2$	

quadratic $\smile$, vertex at (0, 0)

upper limit   if $x = 2$, $f(2) = (2)^2 = 4$ (2, 4)

$f(x) = x + 1$ if $3 \leq x \leq 5$

linear—line segment

lower limit   if $x = 3$, $f(3) = 3 + 1 = 4$ (3, 4)

upper limit   if $x = 5$, $f(5) = 5 + 1 = 6$ (5, 6)

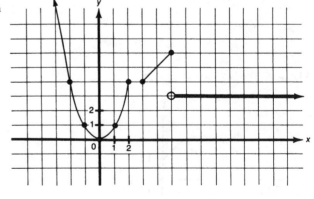

UNIT 9

1. $f'(x) = 45x^{14}$

2. $f(x) = -7x^2$

   $f'(x) = -7(2x)$

   $\qquad = -14x$

3. $f'(x) = 147x^6$

4. $f(x) = x$

   $f'(x) = 1$

5. $f'(x) = -99x^8$

6. $f(x) = 10$

   $f'(x) = 0$

7. $f'(x) = 0$

8. $f(x) = 4x + 3$

   $f'(x) = 4(1) + 0$

   $\qquad = 4$

9. $f'(x) = -7$

10. $f(x) = -3x + 17$

    $f'(x) = -3(1) + 0$

    $\qquad = -3$

11. $f'(x) = \dfrac{2}{5}$

12. $f(x) = 3x^2 + 2x - 6$

    $f'(x) = 3(2x) + 2(1) + 0$

    $\qquad = 6x + 2$

13. $f'(x) = -24x^2 - 14x + 11$

14. $f(x) = 5x^3 - 4x^2 + 3x + 20$

    $f'(x) = 5(3x^2) - 4(2x) + 3(1) + 0$

    $\qquad = 15x^2 - 8x + 3$

15. $f'(x) = -42x - 21x^2 + 40x^3$

16. $f(x) = 34x^2 + x^5$

    $f'(x) = 34(2x) + 5x^4$

    $\qquad = 68x + 5x^4$

17. $f'(x) = x^3 - x - 1$

18. $f(x) = \dfrac{1}{3}x^3 + \dfrac{1}{2}x^2 - 2x + 13$

    $f'(x) = \dfrac{1}{3}(3x^2) + \dfrac{1}{2}(2x) - 2(1) + 0$

    $\qquad = x^2 + x - 2$

19. $f'(x) = 4x^{11} - 6x + \sqrt{5}$

20. $f(x) = ax^2 + bx + c$

    $f'(x) = a(2x) + b(1) + 0$

    $\qquad = 2ax + b$

21. $f'(x) = 5ax^4 + x^2 - 2$

22. $f(x) = mx + b$

    $f'(x) = m(1) + 0$

    $\qquad = m$

23. $f(x) = ax^3 + bx^2 + cx + d$

    $f'(x) = a(3x^2) + b(2x) + c(1) + 0$

    $\qquad = 3ax^2 + 2bx + c$

24. $f(x) = x^2 - 3x + 1$

    $f'(x) = 2x - 3(1) + 0$

    $\qquad = 2x - 3$

    $f'(2) = 2(2) - 3$

    $\qquad = 4 - 3$

    $\qquad = 1$

25. $f'(x) = -3x^2 - 1$

    $f'(2) = -13$

26. $f(x) = ax^3 + x$

    $f'(x) = a(3x^2) + 1$

    $\qquad = 3ax^2 + 1$

    $f'(2) = 3a(4) + 1$

    $\qquad = 12a + 1$

27. $f'(x) = 5$

 $f'(2) = 5$

28. $f(x) = x^3 - 2x + 1$

 $f'(x) = 3x^2 - 2(1) + 0$

 $\quad = 3x^2 - 2$

 $f(3) = (3)^3 - 2(3) + 1$

 $\quad = 27 - 6 + 1$

 $\quad = 22$

 $f'(3) = 3(3)^2 - 2$

 $\quad = 25$

29. $f'(x) = -10x$

 $f(3) = -43$

 $f'(3) = -30$

30. $f(x) = x^2 - 6x + 12$

 $f'(x) = 2x - 6(1) + 0$

 $\quad = 2x - 6$

 $f(3) = (3)^2 - 6(3) + 12$

 $\quad = 3$

 $f'(3) = 2(3) - 6$

 $\quad = 0$

UNIT 10

1. $\dfrac{df}{dx} = -10x + 2$

2. $y = 11 + 2x - x^3$

 $\dfrac{dy}{dx} = 2 - 3x^2$

3. $D_x(7x + 2 - 3x^5) = 7 - 15x^4$

4. $D_x(x^7 - 7x) = 7x^6 - 7$

5. $\dfrac{dy}{dx} = 4x^3 - 24x^2 - 3$

6. $f(x) = 3 - 7x^2 + 21x + 2x^5$

 $f'(x) = -14x + 21 + 10x^4$

 $f'(1) = -14(1) + 21 + 10(1)^4$

 $\quad = -14 + 21 + 10$

 $\quad = 17$

7. $f' = 11$

 The slope of the line is 11.

8. $\dfrac{d(1 - 3x^2)}{dx} = -6x$

9. $\dfrac{d(x^2 - 5)}{dx} = 2x$

10. $y = \dfrac{x^6}{3} - 2x$

 $\quad = \dfrac{1}{3}x^6 - 2x$ (rewrite first)

 $\dfrac{dy}{dx} = \dfrac{1}{3}(6x^5) - 2$

 $\quad = 2x^5 - 2$

11. $f'(x) = 3 - x$

12. $f(x) = \dfrac{x^4}{4} - \dfrac{x^3}{3} + 5x^2$

 $\quad = \dfrac{1}{4}x^4 - \dfrac{1}{3}x^3 + 5x^2$ (rewritten)

 $f'(x) = \dfrac{1}{4}(4x^3) - \dfrac{1}{3}(3x^2) + 5(2x)$

 $\quad = x^3 - x^2 + 10x$

13. $f'(x) = -x$

14. $f(x) = 4x^2 - 2x + 9$

 a. $f'(x) = 4(2x) - 2(1)$

   $\quad = 8x - 2$

 b. $f'(0) = 8(0) - 2$

   $\quad = -2$

 c. $f'(2) = 8(2) - 2$

   $\quad = 16 - 2$

   $\quad = 14$

 d. Find x such that $f'(x) = 0$ means solve

   $0 = 8x - 2$

   $8x = 2$

   $x = \dfrac{2}{8} = \dfrac{1}{4}$

15.   a. $f'(x) = 2x + 48$

    b. $f'(0) = 48$

    c. $f'(1) = 50$

    d. $x = -24$

16. $y = f(x) = -x^2 + 140x - 10$

    a. $f'(x) = -2x + 140$

    b. $f(0) = -0 + 140(0) - 10$

          $= -10$

    c. $f'(0) = -2(0) + 140$

          $= 140$

    d. $f'(x) = 0$

      $-2x + 140 = 0$

        $-2x = -140$

          $x = 70$

17.   a. $f'(x) = 8x - 2$

    b. $f(1) = 11$

    c. $y = 6x + 5$

18. $y = f(x) = x^9 - 5x^8 + x + 12$

    $f'(x) = 9x^8 - 40x^7 + 1$

  $f'(-1) = 9(-1)^8 - 40(-1)^7 + 1$

      $= 9(1) - 40(-1) + 1$

      $= 9 + 40 + 1$

      $= 50$

$$y - y_1 = m(x - x_1)$$

$$y - 5 = 50(x - (-1))$$

$$y - 5 = 50(x + 1)$$

$$y - 5 = 50x + 50$$

$$y = 50x + 55$$

An equation of the line is $y = 50x + 55$.

C19. $f(-1.25) = -26.5$, $f'(-1.25) = 245.38$

The slope of the tangent line to the curve at point $(-1.25, -26.5)$ is 245.38.

20. $f(x) = x^4 - 3x^3 + 2x^2 - 6$

    a. $f'(x) = 4x^3 - 9x^2 + 4x$

    b. Need coordinates of the point at $x = 2$

      $f(2) = (2)^4 - 3(2)^3 + 2(2)^2 - 6$

        $= 16 - 24 + 8 - 6$

        $= -6$, which is point $(2, -6)$

Need the slope at $x = 2$

    $f'(2) = 4(2)^3 - 9(2)^2 + 4(2)$

        $= 4(8) - 9(4) + 8$

        $= 32 - 36 + 8$

        $= 4$

To write equation: $y - y_1 = m(x - x_1)$

        $y - (-6) = 4(x - 2)$

          $y + 6 = 4x - 8$

            $y = 4x - 14$

An equation of the line is $y = 4x - 14$.

21.

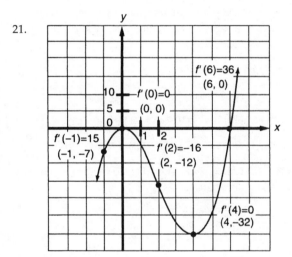

C22. $f(x) = 2x^4 - 5x^3 + x^2 - 4x + 1$

    $f'(x) = 8x^3 - 15x^2 + 2x - 4$

$\backslash Y_1$ (Y=) (2) (X,T,θ,n) (∧) (4) (−) (5) (X,T,θ,n) (∧) (3) (+) (X,T,θ,n) (∧) (2) (−) (4) (X,T,θ,n) (+) (1) (ENTER)

$\backslash Y_2$ (8) (X,T,θ,n) (∧) (3) (−) (1) (5) (X,T,θ,n) (∧) (2) (+) (2) (X,T,θ,n) (−) (4) (ENTER)

To create a table of values, press (2nd) (TABLE) and key in the values for x.

X	Y₁	Y₂
0	1	-4
2	-11	4
4	193	276
-3	319	-361
X=		

$f(0) = 1$,
$f'(0) = -4$

$f(2) = -11$,
$f'(2) = 4$

$f(4) = 193$,
$f'(4) = 276$

$f(-3) = 319$,
$f'(-3) = -361$

C23. $dy/dx = 6.22$

C24. Key in the function, $f(x)$, and view the graph.

To find $f'(0)$, press (2nd) (CALC) (6) followed by (0) (ENTER).

To find $f'(-2)$, press (2nd) (CALC) (6) followed by (−) (2) (ENTER).

Answers appear at the bottom of each screen:

$\frac{dy}{dx} = 17$, $\frac{dy}{dx} = -1.259$

Thus, $f'(0), = 17, f'(-2) = -1.259$

UNIT 11

1. a. 50,000

    b. $P'(x) = 400x$

    c. It is predicted that the city's population will increase by approximately 4,000 people in the year 2013.

2. $N(t) = -t^2 - 5t + 750$

    a. $N'(t) = -2t - 5$

    b. $N'(10) = -2(10) - 5 = -20 - 5 = -25$

    c. $N(0) = -(0)^2 - 5(0) + 750 = 750$

C3. a. $C(0) = 1,500$. The ABC Company's fixed costs are $1,500.

    b. $MC = C'(x) = 0.516x$

    c. $C'(24) = \$12.384$

    d. The actual cost of producing the 25th unit is $12.70.

4. $C(x) = 10 - 2.5x^2 + x^3 \qquad x \geq 2$

    a. $MC = C'(x) = 0 - 2.5(2x) + 3x^2$

    $= -5x + 3x^2$

    b. $MC = C'(10) = -5(10) + 3(10)^2$

    $= -50 + 3(100)$

    $= -50 + 300$

    $= 250$

    c. The marginal cost of the 11th unit is $250.

5. a. $P(x) = -2x^2 + 210x - 5000$

    b. $MP = P'(x) = -4x + 210$

    c. $MP = P'(44) = \$34$

    d. The actual profit from producing and selling the 45th clock is $32.

6. $\qquad C(q) = 25q^2 + q + 100$

    a. $MC = C'(q) = 25(2q) + 1$

    $= 50q + 1$

    To estimate the cost of 101st unit, use 100.

    $MC = C'(100) = 50(100) + 1$

    $= 5,000 + 1$

    $= 5,001$

    The approximate cost of manufacturing the next unit (101st) is $5,001.

    b. $C(q) = 25q^2 + q + 100$

    $C(101) = 25(101)^2 + (101) + 100$

    $= 255,025 + 201$

    $= \$255,226$

    $C(100) = 25(100)^2 + 100 + 100$

    $= 250,000 + 200$

    $= 250,200$

    Cost of 101st unit $= C(101) - C(100)$

    $= 255,226 - 250,200$

    $= \$5,026$

C7. a. $P'(x) = -5x + 50$

    b. Before the start of the advertising campaign, the T.C. Mellott Company earned $30,000 profit per day.

    c. On day 5 of the advertising campaign the company's profit will be $30,187.50.

    d. The approximate increase in profit for day 6 of the campaign will be $25.

8. $N = \frac{1}{3}t^3 - 6t^2 + t + 1,757$

a. $N' = \frac{1}{3}(3t^2) - 6(2t) + 1$

$= t^2 - 12t + 1$

b. $N'(6) = (6)^2 - 12(6) + 1$

$= 36 - 72 + 1$

$= -35$

Sales were decreasing at that point in time.

c. $N'(12) = (12)^2 - 12(12) + 1$

$= 144 - 144 + 1$

$= 1$

Sales were increasing at that point in time.

9. a. $MR = R'(x) = -4x + 80$

b. $R'(5) = 60$

c. The additional revenue obtained by selling the 11th unit will be approximately $60.

10. $Q(x) = 0.1x^3 + 3x^2 + 1$

a. $Q'(x) = 0.3x^2 + 6x$

$Q'(10) = 0.3(10)^2 + 6(10)$

$= 0.3(100) + 60$

$= 30 + 60$

$= 90$

It is estimated that, if the work force was increased by one employee, bringing the total to 11, daily output would increase by 90 units.

b. Output if there are 11 workers is

$Q(11) = 0.1(11)^3 + 3(11)^2 + 1$

$= 0.1(1,331) + 3(121) + 1$

$= 133.1 + 363 + 1$

$= 497.1$ units

Output if there are 10 workers is

$Q(10) = 0.1(10)^3 + 3(10)^2 + 1$

$= 0.1(1,000) + 3(100) + 1$

$= 100 + 300 + 1$

$= 401$ units

Actual change in output $= Q(11) - Q(10)$

$= 497.1 - 401$

$= 96.1$ units

11. a. The revenue obtained from selling 20 units is $200.

b. The marginal revenue of the 21st unit is −$10. Or, at 20 units sold, selling one more unit will *decrease* revenue by approximately $10.

c. $R(21) = \$189$

12. $P(x) = -x^2 + 140x - 4{,}000$

a. Marginal profit equals the derivative of the profit function; therefore

$MP = P'(x) = -2x + 140$

b. $MP = P'(50) = -2(50) + 140$

$= -100 + 140$

$= 40$

c. Profit is increasing.

The marginal profit of the 51st customer is $40. Or, at 50 customers per day, one more customer would increase profits by approximately $40.

13. a. There were 4,914 students enrolled in 2003.

b. The number of students enrolled in 2004 decreased by approximately 61.

c. The fact that enrollment is decreasing is not a good sign.

14. $y = H(t) = -16t^2 + 64t + 80$

a. Velocity $= H'(t) = -32t + 64$

$H'(1) = -32 + 64$

$= 32$

After 1 second, the ball's velocity is 32 feet per second.

b. Distance $= H(t) = -16t^2 + 64t + 80$

$H(4) = -16(4)^2 + 64(4) + 80$

$= -256 + 256 + 80$

$= 80$

The ball will be 80 feet above the ground after 4 seconds.

c. Velocity $= H'(t) = -32t + 64$

$H'(4) = -32(4) + 64$

$= -128 + 64$

$= -64$

After 4 seconds, the ball's velocity will be −64 feet per second. Or, after 4 seconds, the ball will be falling at the rate of 64 feet per second.

d. Initial velocity occurs when $t = 0$.

Velocity $= H'(t) = -32t + 64$

$$H'(0) = 0 + 64$$

$$= 64$$

The ball was tossed upward with an initial velocity of 64 feet per second.

e. $y = H(t) = -16t^2 + 64t + 80$

The function is a quadratic with $a = -16$.

The parabolic curve opens down

The y-intercept is 80.

The x-intercepts are −1 and 5 because

$$0 = -16t^2 + 64t + 80$$

$$0 = -16(t^2 - 4t - 5)$$

$$0 = -16(t - 5)(t + 1)$$

$$t - 5 = 0 \qquad t + 1 = 0$$

$$t = 5 \qquad t = -1$$

The vertex is located at (2, 144) because the axis of symmetry is $x = 2$.

$$H(2) = -16(2)^2 + 64(2) + 80$$

$$= -64 + 128 + 80$$

$$= 144$$

$$y = H(t) = -16t^2 + 64t + 80$$

From the graph, $t = 2$.

The ball will reach its maximum height after 2 seconds.

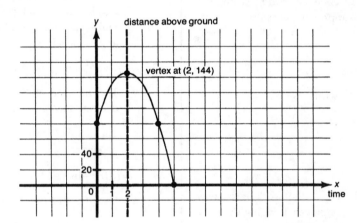

f. From the graph, $y = 144$.

The ball will rise to a maximum height of 144 feet.

g. From the graph, $t = 5$.

The ball will hit the ground in 5 seconds.

h. Velocity $= H'(t) = -32t + 64$

$$H'(5) = -32(5) + 64$$

$$= -160 + 64$$

$$= -96$$

The ball will hit the ground with a velocity of −96 feet per second.

UNIT 12

1. $f'(x) = 9x^2 + 6x - 15$

2. $f(x) = (2x + 1)(x^2 - 3)$

$f'(x) = (2x + 1)\dfrac{d}{dx}(x^2 - 3) + (x^2 - 3)\dfrac{d}{dx}(2x + 1)$

$$= (2x + 1)(2x) + (x^2 - 3)(2)$$

$$= 4x^2 + 2x + 2x^2 - 6$$

$$= 6x^2 + 2x - 6$$

3. $f'(x) = 8x^3 - 9x^2 + 16x - 5$

4. $y = (x^2 + 1)(2 - x^3)$

$$\frac{dy}{dx} = (x^2 + 1)\frac{d}{dx}(2 - x^3) + (2 - x^3)\frac{d}{dx}(x^2 + 1)$$

$$= (x^2 + 1)(-3x^2) + (2 - x^3)(2x)$$

$$= -3x^4 - 3x^2 + 4x - 2x^4$$

$$= -5x^4 - 3x^2 + 4x$$

5. $y' = 3(7x^6 + 10x^4 - 3x^2 + 6)$

6. $f(u) = (u^3 + 5)^7$

$$f'(u) = 7(u^3 + 5)^6 \frac{d}{dx}(u^3 + 5)$$

$$= 7(u^3 + 5)^6(3u^2)$$

$$= 21u^2(u^3 + 5)^6$$

7. $y' = 5(3x^7 - x + 13)^4\,(21x^6 - 1)$

8. $f(x) = \dfrac{-1}{x^2}$

$$f'(x) = \frac{x^2 \frac{d}{dx}(-1) - (-1)\frac{d}{dx}(x^2)}{x^4}$$

$$= \frac{x^2 \cdot 0 + 1(2x)}{x^4}$$

$$= \frac{2x}{x^4}$$

$$= \frac{2}{x^3}$$

9. $f' = \dfrac{-100}{x^2}$

10. $D_x\!\left(\dfrac{3}{x+4}\right) = \dfrac{(x+4)\frac{d}{dx}(3) - 3\frac{d}{dx}(x+4)}{(x+4)^2}$

$$= \frac{(x+4)\cdot 0 - 3\,(1)}{(x+4)^2}$$

$$= \frac{-3}{(x+4)^2}$$

11. $y' = \dfrac{-3x^2 + 12x + 1}{(3x^2 + 1)^2}$

12. $y = \dfrac{x^2 + 2}{x - 3}$

$$y' = \frac{(x-3)\frac{d}{dx}(x^2 + 2) - (x^2 + 2)\frac{d}{dx}(x-3)}{(x-3)^2}$$

$$= \frac{(x-3)\,(2x) - (x^2 + 2)\,(1)}{(x-3)^2}$$

$$= \frac{2x^2 - 6x - x^2 - 2}{(x-3)^2}$$

$$= \frac{x^2 - 6x - 2}{(x-3)^2}$$

13. $\dfrac{dy}{dx} = 2(11x^3 - 7x + \pi)(33x^2 - 7)$

14. $N = \dfrac{100t}{t - 9}$

$$\frac{dN}{dt} = \frac{(t-9)\frac{d}{dt}(100t) - 100t\frac{d}{dt}(t-9)}{(t-9)^2}$$

$$= \frac{(t-9)\,(100) - 100t\,(1)}{(t-9)^2}$$

$$= \frac{100t - 900 - 100t}{(t-9)^2}$$

$$= \frac{-900}{(t-9)^2}$$

15. $f'(x) = \dfrac{-2(x+3)}{x^4}$

16. $f(x) = (6x^2 + 12x + 1)^5$

     a. $f(-2) = (6(-2)^2 + 12(-2) + 1)^5$

$$= (24 - 24 + 1)^5$$

$$= (1)^5$$

$$= 1$$

     Point $(-2, 1)$ is on the graph of the function. Or, when $x = -2$, $f(-2) = 1$.

     b. $f'(x) = 5(6x^2 + 12x + 1)^4 \frac{d}{dx}(6x^2 + 12x + 1)$

$$= 5(6x^2 + 12x + 1)^4(12x + 12)$$

$$f'(-2) = 5[6(-2)^2 + 12(-2) + 1]^4(12(-2) + 12)$$

$$= 5(1)^4(-12)$$

$$= -60$$

     The slope of the tangent line at point $(-2, 1)$ is -60.

17.   a.   Point $(-2, 0)$ is on the graph of the function. Or, when $x = -2$, $f(-2) = 0$.

     b.   The slope of the tangent line at point $(-2, 0)$ is $\dfrac{1}{2}$.

18. $R(x) = (x+1)(2x^2 + x + 3)$ $x \geq 0$

    a. $MR = R'(x)$

    $$= (x+1)\frac{d}{dx}(2x^2 + x + 3) + (2x^2 + x + 3)\frac{d}{dx}(x+1)$$

    $$= (x+1)(4x+1) + (2x^2 + x + 3)(1)$$

    $$= 4x^2 + x + 4x + 1 + 2x^2 + x + 3$$

    $$= 6x^2 + 6x + 4$$

    b. $MR = 6x^2 + 6x + 4$

    $R'(10) = 6(10)^2 + 6(10) + 4$

    $$= 6(100) + 60 + 4$$

    $$= 664$$

    Revenue is increasing because $R'(10)$ is positive.

19. a. $A(x) = \dfrac{10x + 5{,}000}{x}$

    b. $A'(x) = \dfrac{-5{,}000}{x^2}$

    c. The total cost of producing 100 units is $6,000.

    d. At a production level of 100 units, the average cost per unit is $60.

    e. At a production level of 100 units, total cost is increasing by $10 per unit.

    f. At a production level of 100 units, the average cost per unit is decreasing by $0.50 per unit.

20. $D(p) = 6 + p - p^2$ $0 \leq p \leq 3$

    a. $D'(p) = 1 - 2p$

    b. $D(1.5) = 6 + (1.5) - (1.5)^2$

    $$= 6 + 1.5 - 2.25$$

    $$= 5.25$$

    At a price of $1.50 per pound, 5.25 pounds of jelly beans will be sold per day.

    $D'(1.5) = 1 - 2(1.5)$

    $$= 1 - 3$$

    $$= -2$$

    If the price of a pound of jelly beans is increased by $1 to $2.50, the demand for jelly beans will drop by approximately 2 pounds per day.

21. a. $A'(t) = \dfrac{300 - 3t^2}{(100 + t^2)^2}$

    b. The concentration of alcohol will increase by about 0.027%.

22. $P(t) = 15 - \dfrac{6}{t+1}$ million

    a. $P'(t) = 0 - \dfrac{(t+1)\frac{d}{dt}(6) - 6\frac{d}{dt}(t+1)}{(t+1)^2}$

    $$= -\frac{(t+1)(0) - 6(1)}{(t+1)^2}$$

    $$= +\frac{6}{(t+1)^2}$$

    b. $P'(1) = \dfrac{6}{(1+1)^2}$

    $$= \frac{6}{4}$$

    $$= 1.5$$

    One year from now the population will be growing at the rate of 1.5 million per year.

    c. $P(2) = 15 - \dfrac{6}{2+1}$

    $$= 15 - \frac{6}{3}$$

    $$= 13 \text{ million people 2 years from now}$$

    $P(1) = 15 - \dfrac{6}{1+1}$

    $$= 15 - 3$$

    $$= 12 \text{ million}$$

    $P(2) - P(1) = 13 - 12 = 1$

    The population will actually increase by only 1 million during the second year.

    d. $P'(t) = \dfrac{6}{(t+1)^2}$

    $P'(9) = \dfrac{6}{(9+1)^2}$

    $$= \frac{6}{(10)^2}$$

    $$= \frac{6}{100}$$

    $$= 0.06$$

Nine years from now the population will be growing at the rate of 0.06 million, or 60,000, per year.

e. In the long run, as t gets very large, the rate of population growth will become very small.

C23. $f'(-1) = -1,162$, $f'(0) = -75$, $f'(1) = 425.69$, $f'(2) = 75,332.92$, $f'(3) = 5,003,779.2$

C24. a. $T = \dfrac{600}{x^2 + 4x + 10}$

$T' = \dfrac{-600(2x + 4)}{(x^2 + 4x + 10)^2}$

$= \dfrac{-1200(x + 2)}{(x^2 + 4x + 10)^2}$

The keystrokes are shown here so that you can check your entries in the event that your answers were wrong.

$\backslash Y_1$ (Y=) (6) (0) (0) (÷) (() (X,T,θ,n) (^) (2) (+) (4) (X,T,θ,n) (+) (1) (0) ()) (ENTER)

$\backslash Y_2$ (Y=) (() ((−)) (1) (2) (0) (0) (() (X,T,θ,n) (+) (2) ()) ()) (÷) (() (X,T,θ,n) (^) (2) (+) (4) (X,T,θ,n) (+) (1) (0) ()) (^) (2) (ENTER)

b. Create a table of x-values of 0, 1, 10, 24, 48, and 72.

$T(0) = 60$, $T'(0) = -24$

The temperature of the food when it was placed in the freezer was 60°F, and it will cool by about 24°F the first hour.

$T(1) = 40$, $T'(1) = -16$

The temperature of the food 1 hour after it was placed in the freezer was 40°F, and it will cool by about another 16°F during the second hour.

$T(10) = 4$, $T'(10) = -0.64$

The temperature of the food 10 hours after it was placed in the freezer was 4°F, and it will cool by about another 0.64°F during the eleventh hour.

c. $T(24) = 0.88$, $T'(24) = -0.067$

$T(48) = 0.24$, $T'(48) = -0.010$

$T(72) = 0.11$, $T'(72) = -0.003$

d. After 72 hours the temperature of the food is very close to 0°F, and the rate of change of the temperature is also very close to 0°F. It appears that the temperature of the food will cool to about 0°F.

C25. $y = 6x + 5$

C26. Key in the function and view the graph.

To write the equation of the tangent line at $x = 2$,

press (2nd) (DRAW) (5) followed by (2) (ENTER) for the specified value of $x = 2$.

$y = 4x - 14$

UNIT 13

1. $f'(x) = \dfrac{-7}{x^2}$

2. $f(x) = 3x^{-4}$

$f'(x) = 3(-4x^{-4-1})$

$= -12x^{-5}$

$= \dfrac{-12}{x^5}$

3. $f'(x) = \dfrac{10}{x^3}$

4. $f(x) = \dfrac{11}{x^3}$

$= 11x^{-3}$

$f'(x) = 11(-3x^{-3-1})$

$= -33x^{-4}$

$= \dfrac{-33}{x^4}$

5. $f'(x) = 6x$

6. $y = 12x^{-3} + 3x$

$y' = 12(-3x^{-3-1}) + 3(1)$

$= -36x^{-4} + 3$

$= \dfrac{-36}{x^4} + 3$

7. $y' = \dfrac{1}{2\sqrt{x}}$

8. $f(x) = 2x^{3/2}$

$$f'(x) = 2\left(\frac{3}{2}x^{(3/2)-1}\right)$$

$$= \cancel{2}\cdot\frac{3}{\cancel{2}}x^{1/2}$$

$$= 3\sqrt{x}$$

9. $y' = \dfrac{2}{5\sqrt[5]{x^3}}$

10. $f(x) = x^{-1/3}$

$$f'(x) = -\frac{1}{3}x^{(-1/3)-1}$$

$$= \frac{1}{3}x^{-4/3}$$

$$= -\frac{1}{3}\cdot\frac{1}{x^{4/3}}$$

$$= \frac{-1}{3\sqrt[3]{x^4}}$$

$$= \frac{-1}{3x\sqrt[3]{x}}$$

11. $f'(x) = \dfrac{-4}{3\sqrt[3]{x}}$

12. $f(x) = \dfrac{5}{x^2} - \dfrac{1}{x}$

$$= 5x^{-2} - x^{-1}$$

$$f'(x) = 5(-2x^{-2-1}) - (-1x^{-1-1})$$

$$= -10x^{-3} + x^{-2}$$

$$= \frac{-10}{x^3} + \frac{1}{x^2}$$

13. $y' = \dfrac{1}{\sqrt{x}}$

14. $y = \sqrt{2x}$

$$= (2x)^{1/2}$$

$$\frac{dy}{dx} = \frac{1}{2}(2x)^{(1/2)-1}\frac{d}{dx}(2x) \quad \text{Chain Rule}$$

$$= \frac{1}{2}(2x)^{-1/2}\cdot 2$$

$$= \frac{1}{\cancel{2}}\cdot\frac{1}{(2x)^{1/2}}\cdot\cancel{2}$$

$$= \frac{1}{\sqrt{2x}}$$

15. $f'(x) = \dfrac{2}{3\sqrt[3]{x^2}}$

16. $y = \dfrac{2}{\sqrt{x}}$

$$= 2x^{-1/2}$$

$$\frac{dy}{dx} = 2\left(-\frac{1}{2}x^{(-1/2)-1}\right)$$

$$= 2\cdot-\frac{1}{2}x^{-3/2}$$

$$= \cancel{2}\cdot\frac{-1}{\cancel{2}}\cdot\frac{1}{x^{3/2}}$$

$$= \frac{-1}{\sqrt{x^3}} \quad \text{or} \quad \frac{-1}{x\sqrt{x}}$$

17. $f'(x) = \dfrac{4}{3\sqrt[3]{x^5}} \quad \text{or} \quad \dfrac{4}{3x\sqrt[3]{x^2}}$

18. $g(x) = \dfrac{1}{3x^2}$

$$= \frac{1}{3}x^{-2}$$

$$g'(x) = \frac{1}{3}(-2x^{-2-1})$$

$$= \frac{1}{3}\cdot-2x^{-3}$$

$$= \frac{1}{3}\cdot-2\cdot\frac{1}{x^3}$$

$$= \frac{-2}{3x^3}$$

19. $f'(x) = \dfrac{-1}{2x^2}$

20. $y = \sqrt{1 - 3x^2}$

    $= (1 - 3x^2)^{1/2}$

    $\dfrac{dy}{dx} = \dfrac{1}{2}(1 - 3x^2)^{(1/2)-1} \dfrac{d}{dx}(1 - 3x^2)$ Chain Rule

    $= \dfrac{1}{2}(1 - 3x^2)^{-1/2}(-6x)$

    $= \dfrac{1}{2} \cdot \dfrac{1}{(1 - 3x^2)^{1/2}} \cdot -6x$

    $= \dfrac{-3x}{\sqrt{1 - 3x^2}}$

21. $D'(x) = \dfrac{2}{\sqrt{4x - 3}}$

22. $D(p) = \sqrt[3]{p^2 + 1}$

    $= (p^2 + 1)^{1/3}$

    $D'(p) = \dfrac{1}{3}(p^2 + 1)^{(1/3)-1} \dfrac{d}{dp}(p^2 + 1)$ Chain Rule

    $= \dfrac{1}{3}(p^2 + 1)^{-2/3}(2p)$

    $= \dfrac{1}{3} \cdot \dfrac{1}{(p^2 + 1)^{2/3}} \cdot 2p$

    $= \dfrac{2p}{3\sqrt[3]{(p^2 + 1)^2}}$

23. $f'(x) = \dfrac{5x^4}{2\sqrt{x^5 - 8}}$

24. $g(u) = 2u - 4u^{-1}$

    $g'(u) = 2 - 4(-1u^{-1-1})$ $g'(2) = 2 + \dfrac{4}{(2)^2}$

    $= 2 + 4u^{-2}$ $= 2 + \dfrac{4}{4}$

    $= 2 + 4\left(\dfrac{1}{u^2}\right)$ $= 2 + 1$

    $= 2 + \dfrac{4}{u^2}$ $= 3$

25. $f'(13) = \dfrac{1}{5}$

C26. $f(x) = \sqrt[3]{3x^2 - 1}$

     $= (3x^2 - 1)^{1/3}$

     $f'(x) = \dfrac{1}{3}(3x^2 - 1)^{-2/3}(6x)$

     $= \dfrac{2x}{(3x^2 - 1)^{2/3}}$

     $= \dfrac{2x}{\sqrt[3]{(3x^2 - 1)^2}}$

Key in the derivative and create a table of values for $x = 0, 1, 2,$ and 3.

$\backslash Y_1$ (Y=) (( ) (2) (X,T,θ,n) ( )) (+) (( ) (3)
(X,T,θ,n) (^) (2) (−) (1) ( )) (^) (( ) (2)
(÷) (3) ( )) (ENTER)

$f'(0) = 0, f'(1) = 1.26, f'(2) = 0.81,$ and $f'(3) = 0.68$

Key in the function, view the graph, and use (2nd) (CALC) (5) to verify the answers.

C27. $f'(x) = \dfrac{2x + 5}{2\sqrt{x^2 + 5x - 2}}$

$\backslash Y_1$ (Y=) (2nd) (√) (X,T,θ,n) (^) (2) (+) (5)
(X,T,θ,n) (−) (2) ( )) (ENTER)

$\backslash Y_2$ (Y=) (( ) (2) (X,T,θ,n) (+) (5) ( )) (÷) (( )
(2) (2nd) (√) (X,T,θ,n) (^) (2) (+) (5)
(X,T,θ,n) (−) (2) ( )) ( )) (ENTER)

$f(-10) = 6.93, f'(-10) = -1.08, f(-5) = $ ERROR, $f(5) = 6.93, f'(5) = 1.08, f(10) = 12.17,$ and $f'(10) = 1.03$

Do you understand why there was an error message for $f(-5)$?

Recall that the quantity under the square root symbol must be nonnegative. $f(-5)$ results in the quantity under the radical being -2, which is undefined in the set of real numbers.

UNIT 14

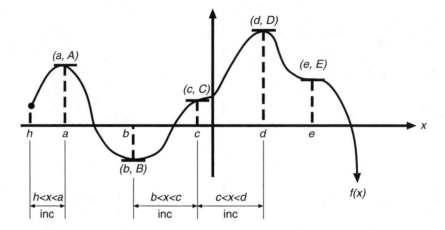

1. There are five critical points.

2. They are located as indicated on the drawing.

3. See drawing.

4. (a, A)—relative maximum

   (b, B)—relative minimum

   (c, C)—stationary inflection point

   (d, D)—relative maximum, also global maximum

   (e, E)—stationary inflection point

5. Yes, the absolute maximum value for y is D and it occurs when $x = d$.

6. There is no absolute minimum.

7. Decreasing when $a < x < b$

   $d < x < e$

   $e < x$

8. Increasing when $h \leq x < a$

   $b < x < c$

   $c < x < d$

C9. $f(x) = x^4 - 12x^3 + 48x^2 - 64x$

   a. Graph:

   b. There are two critical points.

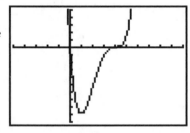

   c. One critical point is located at what appears to be $(1, -27)$, and the second one is located at what appears to be $(4, 0)$.

   d. The point $(1, -27)$ is a relative minimum point, and $(4, 0)$ is a stationary inflection point.

   e. There is no absolute maximum value.

   f. The absolute minimum value appears to be -27 and occurs when $x = 1$.

UNIT 15

1. $(1, -2)$ is an absolute minimum. In other words, the minimum value for y is -2 and it occurs when $x = 1$.

2. $f(x) = -x^2 - 1$

   $f'(x) = -2x$

   $0 = -2x$

   $x^* = 0$

$f(0) = -0 - 1 = -1$ and $(0, -1)$ is a critical point. The function is quadratic with $a = -1$, opening ⌢. The critical point $(0, -1)$ is an absolute maximum.

3. There are no critical points.

4. $f(x) = 6x^2 + x - 1$

$f'(x) = 12x + 1$

$0 = 12x + 1$

$-12x = 1$

$x^* = -\dfrac{-1}{12}$

$f\left(\dfrac{-1}{12}\right) = 6\left(\dfrac{-1}{12}\right)^2 + \left(\dfrac{-1}{12}\right) - 1$

$= \dfrac{6}{144} + \dfrac{-1}{12} - 1$

$= \dfrac{1}{24} + \dfrac{-2}{24} - \dfrac{24}{24}$

$= \dfrac{-25}{24}$; thus $\left(\dfrac{-1}{12}, \dfrac{-25}{24}\right)$ is a critical point.

The function is quadratic with $a = 6$, opening ⌣. The critical point $\left(\dfrac{-1}{12}, \dfrac{-25}{24}\right)$ is an absolute minimum.

5. The critical point $(4, -70)$ is an absolute minimum.

6. $f(x) = x^3 + 3x^2$

$f'(x) = 3x^2 + 6x$

$0 = 3x^2 + 6x$

$= 3x(x + 2)$

$3x = 0 \qquad\qquad x + 2 = 0$

$x^* = 0 \qquad\qquad x^* = -2$

$f(0) = 0 + 0 = 0 \qquad f(-2) = -8 + 12 = 4$

Critical points at $(0, 0)$ and $(-2, 4)$.

The function is a cubic with $a = 1$, opening ⌢⌣ .

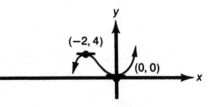

The critical point $(-2, 4)$ is a relative maximum and the critical point $(0, 0)$ is a relative minimum.

7. The critical point $(-10, 1,500)$ is a relative maximum, and the critical point $(0, 1,000)$ is a relative minimum.

8. $f(x) = -x^3 + 7$

$f'(x) = -3x^2$

$0 = -3x^2$

$x^* = 0$

$f(0) = 0 + 7 = 7$; thus $(0, 7)$ is a critical point.

The function is a cubic with one critical point; therefore it must be a stationary inflection point.

9. There are no critical points.

10. $f(x) = x^3 - 3x^2 - 9x + 5$

$f'(x) = 3x^2 - 6x - 9$

$0 = 3x^2 - 6x - 9$

$\quad = x^2 - 2x - 3$

$\quad = (x - 3)(x + 1)$

$x^* = 3 \qquad\qquad\qquad\qquad x^* = -1$

$f(3) = 27 - 3(9) - 9(3) + 5 \qquad f(-1) = -1 - 3(1) - 9(-1) + 5$

$\quad = 27 - 27 - 27 + 5 \qquad\qquad\qquad = -1 - 3 + 9 + 5$

$\quad = -22 \qquad\qquad\qquad\qquad\qquad = 10$

There are critical points at $(3, -22)$ and $(-1, 10)$.

The function is a cubic with $a = 1$, opening 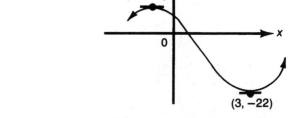 .

The critical point $(-1, 10)$ is a relative maximum and $(3, -22)$ is a relative minimum.

11. The critical point $(0, -25)$ is a stationary inflection point.

12. $f(x) = x^6 + 2$

$f'(x) = 6x^5$

$0 = 6x^5$

$x^* = 0$

$f(0) = 0 + 2 = 2$; thus $(0, 2)$ is a critical point.

To classify, use the original function test.

On left: Let $x = -1$, $f(-1) = (-1)^6 + 2 = 1 + 2 = 3$

At the point: $x = 0$, $f(0) = \qquad\qquad = 2$

On right: Let $x = 1$, $f(1) = 1 + 2 \qquad = 3$

The critical point $(0, 2)$ is a relative minimum.

In fact, it is an absolute minimum.

13. The critical point $(-2, 51)$ is a relative maximum, and the critical point $(3, -74)$ is a relative minimum.

14. $f(x) = 5x^7 - 35x^6 + 63x^5$

$f'(x) = 35x^6 - 210x^5 + 315x^4$

$0 = 35x^6 - 210x^5 + 315x^4$

$0 = x^6 - 6x^5 + 9x^4$

$0 = x^4(x^2 - 6x + 9)$

$0 = x^4(x - 3)^2$

$x^4 = 0 \qquad\qquad (x - 3)^2 = 0$

$x^* = 0 \qquad\qquad x - 3 = 0$

$\qquad\qquad\qquad\qquad x^* = 3$

$f(0) = 0 - 0 + 0 = 0$ $f(3) = 5(2,187) - 35(729) + 63(243)$

$$= 10,935 - 25,515 + 15,309$$

$$= 729$$

There are critical points at $(0, 0)$ and $(3, 729)$.

For point $(0, 0)$:

On left: Let $x = -1$, $f(-1) = 5(-1) - 35(1) + 63(-1) = -103$

At the point: $x = 0$, $f(0) =$ $= 0$

On right: Let $x = 1$, $f(1) = 5 - 35 + 63$ $= 33$

For point $(3, 729)$:

On left: Let $x = 2$, $f(2) = 640 - 2,240 + 2,016$ $= 416$

At the point: $x = 3$, $f(0) =$ $= 729$

On right: Let $x = 4$, $f(4) = 81,920 - 143,360 + 64,512 = 3,072$

Both critical points, $(0, 0)$ and $(3, 729)$, are stationary inflection points.

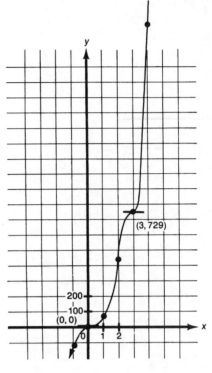

15. The critical point $(0, 5)$ is a relative maximum, and the critical point $(1, 4)$ is a relative minimum.

16. $f(x) = \dfrac{1}{3} x^3 - 3x^2 + 5x + 2$

$f'(x) = \dfrac{1}{3}(3x^2) - 6x + 5$

$\qquad = x^2 - 6x + 5$

$\quad 0 = (x - 1)(x - 5)$

$\quad x^* = 1$ $x^* = 5$

$f(1) = \dfrac{1}{3} - 3 + 5 + 2$ $f(5) = \dfrac{1}{3}(125) - 3(25) + 5(5) + 2$

$\qquad = 4\dfrac{1}{3}$ $= 41\dfrac{2}{3} - 75 + 25 + 2$

$\qquad\qquad\qquad\qquad\qquad\qquad = -6\dfrac{1}{3}$

There are critical points at $\left(1, 4\dfrac{1}{3}\right)$ and $\left(5, -6\dfrac{1}{3}\right)$

The function is a cubic with $a = \dfrac{1}{3}$ and opening

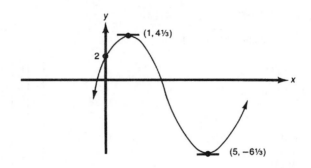

The critical point $\left(1, 4\frac{1}{3}\right)$ is a relative maximum, and the critical point $\left(5, -6\frac{1}{3}\right)$ is a relative minimum.

17. The critical point (0.5, –3.125) is a relative minimum, and the critical point (–4, 88) is a relative maximum.

18. $f(x) = \dfrac{-5}{(x-4)}$

$$f'(x) = \frac{(x-4) \cdot 0 + - (-5)(1)}{(x-4)^2}$$

$$= \frac{0+5}{(x-4)^2}$$

$$= \frac{5}{(x-4)^2}$$

$$0 = \frac{5}{(x-4)^2}$$

$0 = 5$ has no solution.

There are no critical points.

19. The critical point (9, 243) is an absolute maximum.

20.

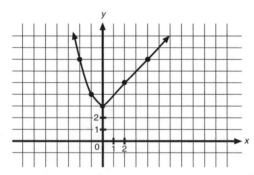

C21. The critical point (–2, 30) is a relative maximum and the critical point (1, 3) is a relative minimum.

C22. a. Graph:

There is absolute minimum point. It appears to be located at about (8.94, –2,108.50). There appears to be an inflection point somewhere near the y-axis, but I was unable to satisfactorily determine its exact location from the graph.

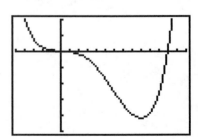

b. Using analytical methods:

$$y = f(x) = x^4 - 12x^3 + x^2 - 2$$

$$y' = f'(x) = 4x^3 - 36x^2 + 2x$$

$$0 = 4x^3 - 36x^2 + 2x$$

$$0 = 2x^3 - 18x^2 + x$$

$$0 = x(2x^2 - 18x + 1)$$

$$x = 0 \qquad 2x^2 - 18x + 1 = 0$$

$$x = \frac{18 \pm \sqrt{(18)^2 - 4(2)(1)}}{2(2)}$$

$$x^* = 0 \qquad x^* = 8.944 \qquad x^* = 0.056$$

$$f(0) = -2 \qquad f(8.944) = -2{,}108.501 \qquad f(0.056) = -1.999$$

In reality there are three, not two, critical points, none of which are inflection points. The critical point (8.944, −2,108.501) is the absolute minimum. The critical point (0.056, −1.999) is a relative maximum, and the critical point (0, −2) is a relative minimum. An enlarged sketch is provided to show you the relative position of the two points and why they are classified as they are.

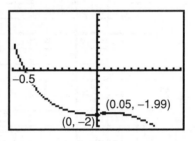

(0.05, −1.99)

(0, −2)

−0.5

UNIT 16

1. The critical point (6, 39) is an absolute maximum.

2. $f(x) = x^4 - 4x^3 + 10$

$$f'(x) = 4x^3 - 12x^2$$

$$= 4x^2(x - 3)$$

Critical points occur where $f'(x) = 0$.

$$0 = 4x^2(x - 3)$$

$$4x^2 = 0 \qquad x - 3 = 0$$

$$x^* = 0 \qquad x^* = 3$$

$$f(0) = 10 \qquad f(3) = 81 - 108 + 10 = -17$$

Critical points at (0, 10) and (3, −17)

Use first-derivative test with $f'(x) = 4x^3 - 12x^2$.

For point (0, 10)		the curve:
On the left:	Let $x = -1, f'(-1) = -4 - 12 = -$	is decreasing
At point (0, 10):		levels off
On the right:	Let $x = 1, f'(1) = 4 - 12 = -$	is decreasing

The critical point is a stationary inflection point.

For point (3, –17):   the curve:

On the left: Let $x = 2$, $f'(2) = 32 - 48 = -$ is decreasing

At point (3, –17):   levels off

On the right: Let $x = 4$, $f'(4) = 256 - 192 = +$ is increasing

The critical point is a relative minimum.

3.

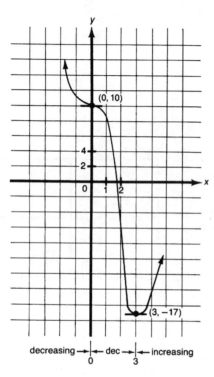

4. a.   The absolute minimum value of $f(x) = -17$ and occurs when $x = 3$. $f(x)$ does not have an absolute maximum.

   b.   The function is increasing over the interval $3 < x$.

   c.   The function is decreasing over the intervals $x < 0$ and $0 < x < 3$.

5. The critical point $(0, 0)$ is a stationary inflection point.

6. $f(x) = \dfrac{1}{4}x^4 + x$

   $f'(x) = x^3 + 1$

     $0 = x^3 + 1$

    $x^3 = -1$

    $x^* = -1$

   $f(-1) = \dfrac{1}{4}(1) - 1 = -\dfrac{3}{4}$

   There is one critical point at $\left(-1, -\dfrac{3}{4}\right)$.

   Use the first-derivative test with $f'(x) = x^3 + 1$.

For point $\left(-1, -\dfrac{3}{4}\right)$ the curve:

On the left: Let $x = -2$, $f'(-2) = -8 + 1 = -$ is decreasing

At point $\left(-1, -\dfrac{3}{4}\right)$:   levels off

On the right: Let $x = 0$, $f'(0) = 0 + 1 = +$ is increasing

The critical point $\left(-1, -\dfrac{3}{4}\right)$ is a relative minimum.

7. The critical point $(0, 2)$ is a relative minimum.

8. $f(x) = x^5 - 5x$

$f'(x) = 5x^4 - 5$

$\quad = 5(x^4 - 1)$

$\quad = 5(x^2 + 1)(x^2 - 1)$

$\quad = 5(x^2 + 1)(x + 1)(x - 1)$

$0 = 5(x^2 + 1)(x + 1)(x - 1)$

$x - 1 = 0$

$x^* = 1$

$f(1) = 1 - 5 = -4$

d $(1, -4)$.

$- 5$.

the curve:

is decreasing

levels off

is increasing

num.

the curve:

$= +$ is increasing

levels off

is decreasing

mum.

9.

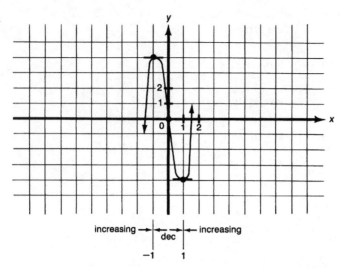

10. a. $f(x)$ has neither an absolute maximum nor an absolute minimum.

   b. The function is increasing over the intervals $x < -1$ and $1 < x$.

   c. The function is decreasing over the interval $-1 < x < 1$.

11. The function has three critical points: $(0, 0)$ is a relative maximum and $(-2, -16)$ and $(2, -16)$ are relative minima.

12. $f(x) = \dfrac{1}{x - 2}$

$$f'(x) = \frac{(x-2)\frac{d}{dx}(1) - (1)\frac{d}{dx}(x-2)}{(x-2)^2}$$

$$= \frac{(x-2)(0) - (1)(1)}{(x-2)^2}$$

$$= \frac{-1}{(x-2)^2}$$

Critical points occur where $f'(x) = 0$.

To find all values of x such that $f'(x) = 0$, set the numerator of $f'(x)$ equal to 0.

Since a variable does not appear in the numerator, there is no value of x that makes $f'(x) = 0$.

The function has no critical points.

13. The critical point $(3, 6)$ is a relative minimum, and the critical point $(-3, -6)$ is a relative maximum.

14. $f(x) = \dfrac{x}{x^2 + 1}$

$$f'(x) = \frac{(x^2 + 1)\frac{d}{dx}(x) - x\frac{d}{dx}(x^2 + 1)}{(x^2 + 1)^2}$$

$$= \frac{(x^2 + 1)(1) - x(2x)}{(x^2 + 1)^2}$$

$$= \frac{x^2 + 1 - 2x^2}{(x^2 + 1)^2}$$

$$= \frac{1 - x^2}{(x^2 + 1)^2}$$

$$= \frac{(1 - x)(1 + x)}{(x^2 + 1)^2}$$

Critical points occur where $f'(x) = 0$.

$$0 = \frac{(1 - x)(1 + x)}{(x^2 + 1)^2} \quad \text{Clear of fractions.}$$

$$0 = (1 - x)(1 + x)$$

$$x^* = 1 \qquad\qquad x^* = -1$$

$$f(1) = \frac{1}{1 + 1} = \frac{1}{2} \quad f(-1) = \frac{-1}{1 + 1} \quad -\frac{1}{2}$$

The function has critical points at $\left(1, \frac{1}{2}\right)$ and $\left(-1, -\frac{1}{2}\right)$.

Use the first-derivative test with $f'(x) = \dfrac{1 - x^2}{(x^2 + 1)^2}$.

For point $\left(1, \frac{1}{2}\right)$ the curve:

On the left: Let $x = 0$, $f'(0) = \dfrac{1}{(1)^2} = +$ is increasing

At point $\left(1, \frac{1}{2}\right)$:   levels off

On the right: Let $x = 2$, $f'(2) = \dfrac{1 - 4}{(5)^2} = -$ is decreasing

The critical point $\left(1, \frac{1}{2}\right)$ is a relative maximum.

For point $\left(-1, -\frac{1}{2}\right)$ the curve:

On the left: Let $x = -2$, $f'(-2) = \dfrac{1 - 4}{(5)^2} = -$ is decreasing

At point $\left(-1, -\frac{1}{2}\right)$:   levels off

On the right: Let $x = 0$, $f'(0) = \dfrac{1}{(1)^2} = +$ is increasing

The critical point $\left(-1, -\frac{1}{2}\right)$ is a relative minimum.

15. The critical point $(0, 0)$ is a relative maximum, and the critical point $(6, 12)$ is a relative minimum.

C16. a.   \Y₁   (Y=) (((X,T,θ,n) (^) (2) (−) (X,T,θ,n) (+) (1) ( )

         (÷) (((X,T,θ,n) (^) (2) (+) (1) ( ) (ENTER)

     Graph in the standard viewing window:

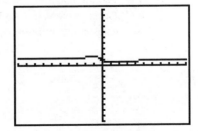

b. Press (TRACE) (ZOOM) (2) (ENTER).

Press (2nd) (CALC) (4) and find the relative maximum at $(-1, 1.5)$.

Press (2nd) (CALC) (3) and find the relative minimum at $(1, 0.5)$.

c. The critical point $(-1, 1.5)$ is an absolute maximum point.

The critical point $(1, 0.5)$ is an absolute minimum point.

d. $f(x) = \dfrac{x^2 - x + 1}{x^2 + 1}$

$$f'(x) = \frac{(x^2 + 1)\dfrac{d}{dy}(x^2 - x + 1) - (x^2 - x + 1)\dfrac{d}{dy}(x^2 + 1)}{(x^2 + 1)^2}$$

$$= \frac{(x^2 + 1)(2x - 1) - (x^2 - x + 1)(2x)}{(x^2 + 1)^2}$$

$$= \frac{2x^3 - x^2 + 2x - 1 - 2x^3 + 2x^2 - 2x}{(x^2 + 1)^2}$$

$$= \frac{x^2 - 1}{(x^2 + 1)^2}$$

Critical points occur where $f'(x) = 0$.

$$0 = \frac{x^2 - 1}{(x^2 + 1)^2}$$

$$= x^2 - 1$$

$$= (x - 1)(x + 1)$$

$$x - 1 = 0 \qquad x + 1 = 0$$

$$x^* = 1 \qquad x^* = -1$$

If $x^* = 1$, $f(1) = \dfrac{1^2 - 1 + 1}{1^2 + 1} = \dfrac{1}{2} = 0.5$ and $(1, 0.5)$ is a critical point.

If $x^* = -1$, $f(-1) = \dfrac{(-1)^2 - (-1) + 1}{(-1)^2 + 1} = \dfrac{1 + 1 + 1}{1 + 1} = \dfrac{3}{2} = 1.5$ and $(-1, 1.5)$ is a second critical point.

The determination that $(-1, 1.5)$ is a relative maximum and, in fact, an absolute maximum can be made from the graph in part a. There is no need to use either of the tests. The same is true for classifying the point $(1, 0.5)$ as a relative (absolute) minimum.

C17. a. Graph in the standard viewing window:

b. There is only one critical point, $(0, 6)$, and it is a relative maximum.

c. The absolute maximum value is $f(x) = 6$, and it occurs when $x = 0$. The function has no absolute minimum.

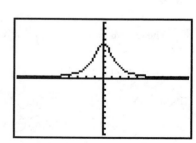

18. a. $f(x) = 2x^3 - 21x^3 + 60x + 10$

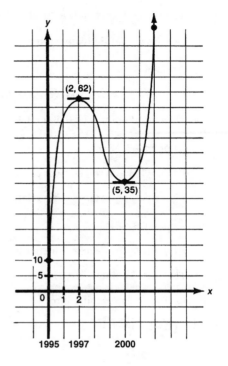

The function is a cubic with $a = 2$.

The y-intercept is at 10.

$f'(x) = 6x^2 - 42x + 60$

$0 = 6x^2 - 42x + 60$

$0 = 6(x^2 - 7x + 10)$

$0 = 6(x - 2)(x - 5)$

$x^* = 2$

$f(2) = 16 - 84 + 120 + 10$

$= 62$

$x^* = 5$

$f(5) = 250 - 525 + 300 + 10$

$= 35$

The function has critical points at (2, 62) and (5, 35).

b. Restricted domain: $0 \le x$

c. $f(6) = 2(216) - 21(36) + 60(6) + 10$
$= 432 - 756 + 360 + 10$
$= 46$

There were 46 paying guests in 2001.

d. The present

e. 2000

f. 35

g. The number of guests at the resort was decreasing.

h. It is estimated that the number of guests will continue to increase. The club's prospects look good.

UNIT 17

1. The critical point (0, –3) is a relative minimum.

2. $f(x) = x^4 - 18x^2$

$f'(x) = 4x^3 - 36x$

$f''(x) = 12x^2 - 36$

Critical points are located where $f'(x) = 0$.

$0 = 4x^3 - 36x$

$0 = 4x(x^2 - 9)$

$0 = 4x(x + 3)(x - 3)$

$x^* = 0 \qquad x^* = -3 \qquad x^* = 3$

$f(0) = 0 \quad f(-3) = -81 \quad f(3) = -81$

There are critical points at $(0, 0)$, $(-3, -81)$, and $(3, -81)$.

Use the second-derivative test with $f''(x) = 12x^2 - 36$.

At point $(0, 0)$,

$f''(0) = 0 - 36 = -$, curve is concave down,

The critical point at $(0, 0)$ is a relative maximum.

At point $(-3, -81)$,

$f''(-3) = 108 - 36 = +$, curve is concave up,

The critical point $(3, -81)$ is a relative minimum.

At point $(3, -81)$,

$f''(3) = 108 - 36 = +$, curve is concave up,

The critical point $(3, -81)$, is a relative minimum.

3. $(0, 0)$ is a relative minimum.

4. $f(x) = 2 - x^5$

   $f'(x) = -5x^4$

   $f''(x) = -20x^3$

   Critical points are located where $f'(x) = 0$.

   $\quad 0 = -5x^4$

   $\quad x^* = 0$

   $f(0) = 2 - 0 = 2$, and $(0, 2)$ is a critical point.

   Use the second-derivative test with $f''(x) = -20x^3$.

   $f''(0) = 0$ and test fails.

   Rework using first-derivative test with $f'(x) = -5x^4$.

For point $(0, 2)$		the curve:
On the left:	Let $x = -1$, $f'(-1) = -5 = -$	is decreasing
At point $(0, 2)$:		levels off
On the right:	Let $x = 1$, $f'(1) = -5 = -$	is decreasing

   The critical point $(0, 2)$ is a stationary inflection point.

5.   There are no critical points.

6. $f(x) = 10x^6 + 24x^5 + 15x^4 + 2$

   $f'(x) = 60x^5 + 120x^4 + 60x^3$

   $f''(x) = 300x^4 + 480x^3 + 180x^2$

   Critical points are located where $f'(x) = 0$.

$0 = 60x^5 + 120x^4 + 60x^3$

$0 = 60x^3(x^2 + 2x + 1)$

$0 = 60x^3(x + 1)^2$

$\qquad x^* = 0 \qquad x^* = -1$

$\qquad f(0) = 2 \quad f(-1) = 10 - 24 + 15 + 2 = 3$

There are critical points at $(0, 2)$ and $(-1, 3)$.

Use the second-derivative test with $f''(x) = 300x^4 + 480x^3 + 180x^2$.

At $(-1, 3)$, $f''(-1) = 300 - 480 + 180 = 0$, test fails

At $(0, 2)$, $f''(0) = 0$, test fails

Use first-derivative test with $f'(x) = 60x^5 + 120x^4 + 60x^3$

$$= 60x^3(x + 1)^2$$

For point $(-1, 3)$		the curve:
On the left:	Let $x = -2$, $f'(-2) = (+)(-)^3(-)^2 = -$	is decreasing
At point $(-1, 3)$:		levels off
On the right:	Let $x = -0.5$, $f'(-0.5) = (+)(-)^3(+)^2 = -$	is decreasing

The critical point $(-1, 3)$ is a stationary inflection point.

For point $(0, 2)$		the curve:
On the left:	Let $x = -0.5$, $f'(-0.5) = -$	is decreasing
At point $(0, 2)$:		levels off
On the right:	Let $x = 1$, $f'(1) = (+)(+)^3(+)^2 = +$	is increasing

The critical point $(0, 2)$ is a relative minimum.

7. The critical point $(1, \frac{1}{4})$ is a relative maximum. The critical points $(0, 0)$ and $(2, 0)$ are relative minima.

8. $f(x) = (x - 5)^3$

$f'(x) = 5(x - 5)^2$

Critical points are located where $f'(x) = 0$.

$\quad 0 = 5(x - 5)^2$

$x^* = 5$

$f(5) = (5 - 5)^3$

$\quad = 0$

There is a critical point at $(5, 0)$. $f(x)$ is a cubic. A cubic with only one critical point means that $(5, 0)$ must be a stationary inflection point.

9.   From Exercise 2,   (0, 0) is a relative maximum,

(3, –81) is a relative minimum,

(–3, –81) is a relative minimum.

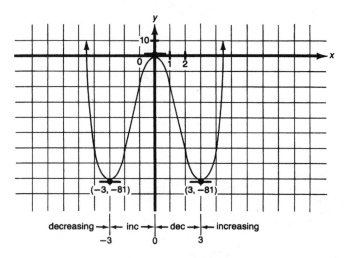

10.   a.   The absolute minimum value of $f(x)$ is –81 and occurs when $x = -3$ or $x = 3$.

b.   The function is increasing over the intervals $-3 < x < 0$ and $3 < x$.

c.   The function is decreasing over the intervals $x < -3$ and $0 < x < 3$.

11.

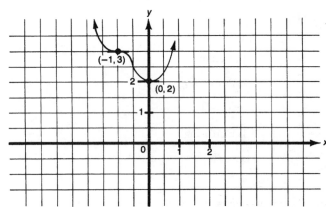

12.   a.   The absolute minimum value of $f(x)$ is 2 and occurs when $x = 0$.

b.   The function is increasing over the interval $0 < x$.

c.   The function is decreasing over the intervals $x < -1$ and $-1 < x < 0$.

13.   There are critical points at $(-1, -2)$ and $(1, 2)$.

14.   Use the original-function test with $f(x) = \dfrac{x^2 + 1}{x}$.

At point $(-1, -2)$

On the left: Let $x = -2$, $f(-2) = \dfrac{(-2)^2 + 1}{-2} = \dfrac{4 + 1}{-2} = \dfrac{5}{-2} = -2.5$

At point $(-1, -2)$: $f(-1)$ $= -2$

On the right: Let $x = -0.5$, $f(-0.5) = \dfrac{(-0.5)^2 + 1}{-0.5} = \dfrac{+0.25 + 1}{-0.5} = -2.5$

The critical point (–1, –2) is a relative maximum.

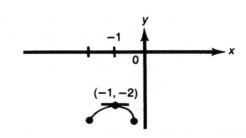

At point (1, 2)

On the left: Let $x = 0.5$, $f(0.5) = \dfrac{(0.5)^2 + 1}{0.5} = \dfrac{1.25}{0.5} = 2.5$

At point (1, 2): $f(1)$ $= 2$

On the right: Let $x = 2$, $f(2) = \dfrac{(2)^2 + 1}{2} = \dfrac{5}{2}$ $= 2.5$

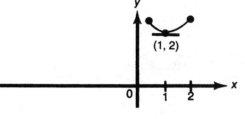

The critical point (1, 2) is a relative minimum.

15. At point (–1, –2):   The curve is increasing, levels off, then is decreasing.

        The critical point (–1, –2) is a relative maximum.

     At point (1, 2):   The curve is decreasing, levels off, then is increasing.

        The critical point (1, 2) is a relative minimum.

16. $f'(x) = \dfrac{x^2 - 1}{x^2}$

$f''(x) = \dfrac{x^2 \dfrac{d}{dx}(x^2 - 1) - (x^2 - 1)\dfrac{d}{dx}(x^2)}{x^4}$

$= \dfrac{x^2(2x) - (x^2 - 1)(2x)}{x^4}$

$= \dfrac{2x^3 - 2x^3 + 2x}{x^4}$

$= \dfrac{2x}{x^4}$

$= \dfrac{2}{x^3}$

Use the second-derivative test with $f''(x) = \dfrac{2}{x^3}$

At point (–1, –2),

$f''(-1) = \dfrac{2}{(-1)^3} = -$, curve is concave down, ⌢

The critical point (–1, –2) is a relative maximum.

At point (1, 2),

$f''(1) = \dfrac{2}{1} = +$, curve is concave up, ⌣

The critical point (1, 2) is a relative minimum.

17. The critical points are (–1, –2), a relative maximum, and (1, 2) a relative minimum, no matter which method was used. Test preference is individual.

18. $f(x) = 2x^3 - 15x^2 + 300$

$f'(x) = 6x^2 - 30x$

$0 = 6x(x - 5)$

$x^* = 0$ $\qquad\qquad$ $x^* = 5$

$f(0) = 0 - 0 + 300$ $\quad$ $f(5) = 2(125) - 15(25) + 300$

$\qquad = 300$ $\qquad\qquad\qquad = 250 - 375 + 300$

$\qquad\qquad\qquad\qquad\qquad\quad = 175$

Critical points are at (0, 300) and (5, 175). The function is a cubic with $a = 2$, opening 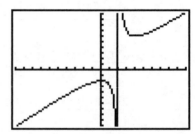.

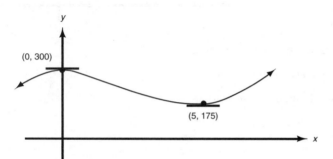

The critical point (0, 300) is a relative maximum, and the critical point (5, 175) is a relative minimum.

C19. Graph in the standard viewing window:

The critical point (0, –2) is a relative maximum, and the critical point (4, 6) is a relative minimum.

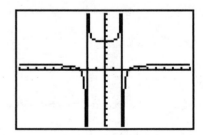

C20. \Y₁ (Y=) ((2 ((X,T,θ,n) ^ 2 – 9)
( ) ÷ ((X,T,θ,n) ^ 2 – 4)
(ENTER)

Graph in the standard viewing window:

a. Press (2nd)(CALC)(3) to locate the relative minimum at point (0, 4.5).

b. There are no absolute extrema.

21. a. There was a loss of $16,000 in the second year of operation.

b. In 1998 the restaurant experienced a loss of $27,000.

c. Last year was the most profitable.

d. The year 1999 was the least profitable.

e. In 1999, the least profitable year, losses totaled $32,000.

f. In 2001, the restaurant broke even, incurring neither a profit nor a loss.

g. Excellent

22. $R(x) = 40x - x^2$

$R'(x) = 40 - 2x$

$0 = 40 - 2x$

$2x = 40$

$x^* = 20$

$R(20) = 40(20) - (20)^2$

$= 800 - 400$

$= 400$

The function is a parabola with $a = -1$, opening $\bigcap$. The critical point (40, 400) is an absolute maximum.

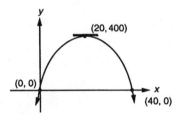

$R(30) = 40(30) - (30)^2$

$= 1200 - 900$

$= 300$

At a price of $30, 10 cases will be sold per day, and total revenue will be only $300.

The best price to charge would be $20 per case. At this price, 20 cases will be sold and the maximum possible revenue of $400 will be achieved.

UNIT 18

1. See Example 4, Unit 15, for the graph and determination of the critical points.

The absolute maximum is -10 and occurs when $x = 1$.

The absolute minimum is -35 and occurs when $x = 6$.

2. $y = f(x) = 5x - 1 \qquad 0 \le x \le 5$

There are no critical points; $f(x)$ is linear.

$f(0) = 0 - 1 = -1$

$f(5) = 25 - 1 = 24$

The absolute maximum is 24 and occurs when $x = 5$.

The absolute minimum is -1 and occurs when $x = 0$.

3. See Example 7, Unit 15, for the graph and determination of critical points.

    The absolute maximum is 1 and occurs when $x = 0$.

    The absolute minimum is -8 and occurs when $x = 3$.

4. $y = f(x) = -x^3 - 3x^2 + 1 \qquad -2 \le x \le 2$

    See Example 8, Unit 15, for the graph and determination of critical points.

    $f(-2) = 8 - 12 + 1 = -3$

    $f(0) = 0 - 0 + 1 = 1$

    $f(2) = -8 - 12 + 1 = -19$

    The absolute maximum is 1 and occurs when $x = 0$.

    The absolute minimum is -19 and occurs when $x = 2$.

5. The absolute maximum is 16 and occurs when $x = 4$.

    The absolute minimum is -4 and occurs when $x = 2$ and $x = -1$.

6. $y = f(x) = x^4 - 8x^3 + 18x^2 - 27 \qquad 1 \le x \le 2$

    See Example 5, Unit 16, for the graph and determination of critical points.

    $f(1) = 1 - 8 + 18 - 27 = -16$

    $f(2) = 16 - 64 + 72 - 27 = -3$

    The absolute maximum is -3 and occurs when $x = 2$.

    The absolute minimum is -16 and occurs when $x = 1$.

7. The absolute maximum is 3 and occurs when $x = 8$.

    The absolute minimum is 0 and occurs when $x = -1$.

8. $y = f(x) = 3x^4 - 16x^3 + 18x^2 \qquad 0 \le x \le 2$

    See Example 4, Unit 17, for the graph and determination of critical points.

    $f(0) = 0 - 0 + 0 = 0$

    $f(1) = 3 - 16 + 18 = 5$

    $f(2) = 48 - 128 + 72 = -8$

    The absolute maximum is 5 and occurs when $x = 1$.

    The absolute minimum is -8 and occurs when $x = 2$.

9. The absolute maximum is 17 and occurs when $x = -1$.

    The absolute minimum is 0 and occurs when $x = 0$.

10. $y = f(x) = 3x^5 - 50x^3 + 135x + 20 \qquad -2 \le x \le 1$

    $f'(x) = 15x^4 - 150x + 135$

    $0 = 15(x^4 - 10x^2 + 9)$

    $0 = 15(x^2 - 9)(x^2 - 1)$

    $0 = 15(x + 3)(x - 3)(x + 1)(x - 1)$

    $\cancel{x = -3} \qquad \cancel{x = 3} \qquad x = -1 \qquad x = 1$

$y = f(-1) = 3(-1)^5 - 50(-1)^3 + 135(-1) + 20 = -3 + 50 - 135 + 20 = -68$

$y = f(1) = 3(1)^5 - 50(1)^3 + 135(1) + 20 = 3 - 50 + 135 + 20 = 108$

$y = f(-2) = 3(-2)^5 - 50(-2)^3 + 135(2) + 20 = 3(-32) - 50(-8) + 135(2) + 20 = 54$

The absolute maximum is 108 and occurs when $x = 1$.

The absolute minimum is -68 and occurs when $x = -1$.

11. The absolute maximum is $\frac{1}{8}$ and occurs when $x = 3$.

The absolute minimum is -1 and occurs when $x = 0$.

12. $y = q(x) = \sqrt{x} + x \qquad 0 \le x \le 4$

$\qquad = x^{1/2} + x$

$q'(x) = \frac{1}{2}x^{-1/2} + 1$

$\qquad = \frac{1}{2\sqrt{x}} + 1$

$0 = \frac{1}{2\sqrt{x}} + 1$

$0 = 1 + 2\sqrt{x}$

$-1 = 2\sqrt{x}$ has no solution

There are no critical points.

$y = q(0) = \sqrt{0} + 0 = 0 + 0 = 0$

$y = q(4) = \sqrt{4} + 4 = 2 + 4 = 6$

The absolute maximum is 6 and occurs when $x = 4$.

The absolute minimum is 0 and occurs when $x = 0$.

C13. There are no critical points on interval $2 \le x \le 4$.

The absolute maximum is 92.646 and occurs when $x = 4$.

The absolute minimum is 18.646 and occurs when $x = 2$.

C14. From the graph it is obvious that neither critical point is in the specified interval; therefore there is no need to find them. The absolute extrema will occur at the endpoints.

Recall that there are several ways to compute them. This is a review of one method.

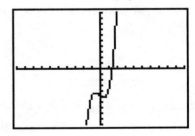

Press (2nd) (CALC) (1) (3) (ENTER), and $Y = 55$ appears at the bottom of the screen.

The absolute minimum is 55 and occurs when $x = 3$.

Press (2nd) (CALC) (1) (5) (ENTER), and $Y = 265$ appears at the bottom of the screen.

The absolute maximum is 265 and occurs when $x = 5$.

UNIT 19

1. To maximize profit 18,500 units should be produced.

2. $\qquad P(x) = 50x - 0.01x^2 - 1,000 \qquad x \geq 0$

$\qquad P'(x) = 50 - 0.02x$

$\qquad \quad 0 = 50 - 0.02x$

$\qquad 0.02x = 50$

$\qquad \quad 2x = 5,000$

$\qquad \quad x = 2,500$

$\qquad P(2,500) = 50(2,500) - 0.01(2,500)^2 - 1,000$

$\qquad \qquad \quad = 125,000 - 62,500 - 1,000$

$\qquad \qquad \quad = 61,500$

The function, $P(x)$, is quadratic, opening down, absolute maximum.

The maximum profit of $61,500 will be achieved when 2,500 units are produced.

3. To maximize profit 24,000 units should be produced, which would lead to a maximum profit of $57,520,000.

4. Revenue = price · quantity

$\qquad R(p) = p(1,152 - 5p)$

$\qquad \quad = 1,152p - 5p^2$

$\qquad R'(p) = 1,152 - 10p$

$\qquad \quad 0 = 1,152 - 10p$

$\qquad 10p = 1,152$

$\qquad \quad p = 115.2$

$\quad$ If $p = \$115.20$ and $q = 1,152 - 5p$,

$\qquad \qquad$ then $q = 1,152 - 5(115.20) = 576$

$R(115.20) = (115.20)(576) = \$66,355.20$

a. A price of $115.20 will result in maximum total revenue.

b. At a price of $115.20, maximum revenue of $66,355.20 can be expected.

c. At a price of $115.20 per unit, revenue will be maximized, and 576 units will be sold.

d. parabola

$\qquad$ x-intercepts

$\qquad 0 = p(1,152 - 5p)$

$\qquad p = 0 \qquad \quad 1,152 - 5p = 0$

$\qquad \qquad \qquad \qquad \quad 5p = 1,152$

$\qquad \qquad \qquad \qquad \quad p = 230.4$

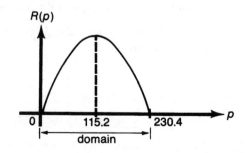

e. Restricted domain: $0 \leq p \leq 230.40$

5. a. $R = (10 - 0.25q)q$

$\quad$ b. 20

    c. $100

    d. 20

    e. $5

6. Profit = total revenue − total cost

$$= \text{(price)(quantity)} - \text{(total cost)}$$

$$P(x) = 50x - (110 - 2.5x^2 + x^3) \qquad x \geq 0$$

$$= 50x - 110 + 2.5x^2 - x^3$$

$$P'(x) = 50 + 5x - 3x^2$$

$$0 = -3x^2 + 5x + 50$$

$$0 = 3x^2 - 5x - 50$$

$$0 = (3x + 10)(x - 5)$$

$$x^* = 5 \qquad 3x + 10 = 0$$

$$\cancel{3x = -10}$$

        Test: $P(x)$ is a cubic,

          $x^* = 5$ is absolute maximum on domain

$$P(5) = 50(5) - 110 + 2.5(25) - (125)$$

$$= 250 - 110 + 62.5 - 125$$

$$= 77.5$$

The company should manufacture 5 units to maximize profit at $77.50.

$R(x) = 50x$; therefore $MR = R'(x) = 50$.

$C(x) = 110 - 2.5x^2 + x^2$; therefore $MC = C'(x) = -5x + 3x^2$.

$R'(5) = 50$

$C'(5) = -5(5) + 3(5)^2$

$$= -25 + 75$$

$$= 50$$

When profit is maximized at $x = 5$, marginal cost equals marginal revenue.

7. $P(q) = 199.5q - q^2 - 5{,}000$

To maximize profit, 99.75 units of the commodity should be produced.

8. $P(x) = R(x) - C(x)$

$$= 20x - (0.01x^2 + 0.2x + 40) \qquad x \geq 0$$

$$= 20x - 0.01x^2 - 0.2x - 40$$

$$= 19.8x - 0.01x^2 - 40$$

$$P'(x) = 19.8 - 0.02x$$

$$0 = 19.8 - 0.02x$$

$$0.02x = 19.8$$

$$x = 990$$

$$R(990) = 20(990)$$

$$= 19{,}800$$

$$C(990) = 0.01(990)^2 + 0.2(990) + 40$$

$$= 0.01(980{,}100) + 0.2(990) + 40$$

$$= 9{,}801 + 198 + 40$$

$$= 10{,}039$$

$$P(990) = R(990) - C(990)$$

$$= 9{,}761$$

a. Profit will be maximized when 990 units are produced and sold.

b. The maximum profit will be $9,761.

c. Total revenue = (selling price)(quantity sold)

We were given $R(x) = 20x$, where x is quantity sold; therefore the selling price is 20, which is constant.

The unit price at which profit is maximized is $20.

d. The total cost of producing 990 units will be $10,039.

e. Average cost = $\dfrac{\text{total cost}}{\text{quanity}}$

$$A(x) = \frac{10{,}039}{990}$$

$$= 10.14$$

The average cost per unit when profit is maximized will be $10.14.

9. a. 100 students

(100, 1,000) is an absolute minimum.

b. The minimum total survey cost will be $1,000.

10. $C(x) = x^2 + 4x + 16 \qquad x > 0$

$$A(x) = \frac{x^2 + 4x + 16}{x}$$

a. $C(10) = 100 + 40 + 16 = 156$

The total cost of producing 10 units is $156.

b. The average cost per unit for producing 10 units is $15.60.

c. $A'(x) = \dfrac{x \dfrac{d}{dx}(x^2 + 4x + 16) - (x^2 + 4x + 16)\dfrac{d}{dx}(x)}{x^2}$

$$= \frac{x(2x + 4) - (x^2 + 4x + 16)(1)}{x^2}$$

$$= \frac{2x^2 + 4x - x^2 - 4x - 16}{x^2}$$

$$= \frac{x^2 - 16}{x^2}$$

$$0 = \frac{x^2 - 16}{x^2}$$

$$0 = x^2 - 16$$

$$0 = (x + 4)(x - 4)$$

$$x^* = 4 \qquad \cancel{x^* = -4} \text{ (not in domain)}$$

$$A(4) = \frac{16 + 16 + 16}{4} = \frac{48}{4} = 12$$

d. The average cost per unit is minimized if 4 units are produced, in which case the average cost per unit is $12.

11. The average cost per unit is minimized if 400 units are produced, in which case the average cost per unit is 100.

12. What is to be maximized in the problem? Profit

The equation for profit was given as:

$$y = P(x) = 25x - x^2 - 100 \qquad 6 \le x \le 14$$

We are being asked to find a maximum on a closed interval.

Recall from Unit 18 that a three-step procedure is used.

1. Find the x-coordinates of critical points within the interval.

$$y = P(x) = 25x - x^2 - 100 \qquad 6 \le x \le 14$$

$$P'(x) = 25 - 2x$$

$$0 = 25 - 2x$$

$$2x = 25$$

$$x = 12.50$$

2. Determine the y-coordinate for each of the critical points within the interval and the lower and upper limits of interval.

$$y = P(12.50) = 25(12.5) - (12.5)^2 - 100 = 312.5 - 156.25 - 100 = 56.25$$

$$y = P(6) = 25(6) - (6)^2 - 100 = 150 - 36 - 100 \qquad\qquad = 14$$

$$y = P(14) = 25(14) - (14)^2 - 100 = 350 - 196 - 100 \qquad\quad = 54$$

3. The absolute maximum is 56.25 and occurs when $x = 12.50$.

Even though the owner would allow Pen to charge as much as $14 for a haircut, Pen should charge only $12.50, thereby maximizing his daily profit at $56.25.

I especially like this problem. So often the tendency is to charge as much as possible, whereas the problem illustrates that Pen will actually achieve a greater profit by charging less.

13. $TR(x) = -20x^2 + 2,000x$

To maximize total revenue, the ABC Company should continue to charge $50 per person. At that price average demand would be 1,000 customers per week, which would maximize revenue at $50,000.

14. $P(x) = 3x - \dfrac{x^3}{1,000,000}$ $0 \le x \le 1,500$

        $P'(x) = 3 - \dfrac{3x^2}{1,000,000}$

        $0 = 3 - \dfrac{3x^2}{1,000,000}$

        $0 = 3,000,000 - 3x^2$

        $0 = 1,000,000 - x^2$

        $0 = (1,000 - x)(1,000 + x)$

        $x^* = 1,000$ $x = \text{—}1{,}000$

        $P(0) = 0 - 0 = 0$

        $P(1,000) = 3(1,000) - \dfrac{(1,000)^3}{1,000,000} = 2,000$

        $P(1,500) = 3(1,500) - \dfrac{(1,500)^3}{1,000,000} = 1,125$

a.   To maximize profit, 1,000 units should be produced and sold.

b.   The maximum profit will be $2,000.

UNIT 20

1.   To maximize profit, the company should spend $4,000 on advertising.

2. $C(x) = 20x + \dfrac{1,280}{x}$

        $= 20x + 1,280x^{-1}$

     $C'(x) = 20 + 1,280(-x^{-1-1})$

        $= 20 - 1,280x^{-2}$

     $0 = 20 - \dfrac{1,280}{x^2}$

     $0 = 20x^2 - 1,280$

     $20x^2 = 1,280$

     $x^2 = 64$

     $x = \pm 8$

        $= 8$

          Test: $C''(x) = -1,280(-2x^{-2-1})$

                      $= 2,560x^{-3}$

                      $= \dfrac{2,560}{x^3}$

          $C''(8) = +$, concave up, minimum

To minimize the cost, the company should operate 8 machines.

3. To minimize the cost, the order size should be 100 units per order.

4. Maximize area.   Constraint:

    Area = length · width $60 = 2x + 2y$

    $\qquad = y \cdot x$ $2y = 60 - 2x$

    $A(x) = (30 - x)x$ $y = 30 - x$

    $\qquad = 30x - x^2$

    $A'(x) = 30 - 2x$

    $\qquad 0 = 30 - 2x$

    $\qquad 2x = 30$

    $\qquad x = 15$

    If $x = 15$,   Test: $A(x)$ is a parabola,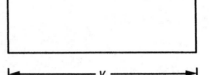

    then $y = 30 - 15$ $x^* = 15$ is an absolute maximum.

    $\qquad = 15$

    To maximize the area, the rectangle should be 15' x 15'.

    It's area will be 225 square feet.

5. To enclose the largest possible area, the length of the side parallel to the river should be 500 feet.

6. Maximize area.   Constraint:

    Area = length · width $800 = 4x + 2y$

    $\qquad = y \cdot x$ $2y = 800 - 4x$

    $A(x) = (400 - 2x)x$ $y = 400 - 2x$

    $\qquad = 400x - 2x^2$

    $A'(x) = 400 - 4x$

    $\qquad 0 = 400 - 4x$

    $\qquad 4x = 400$

    $\qquad x = 100$

    If $x = 100$,

    then $y = 400 - 2x$

    $\qquad = 400 - 2(100)$

    $\qquad = 200$

    Test: $A(x) = $ a quadratic, opening.

    $\qquad$ (100, 200) is an absolute maximum.

    Area $= y \cdot x$

    $\qquad = (200)(100)$

    $\qquad = 20,000$

    With 800 feet of fencing, the largest total area that Richard McHenry can enclose is 20,000 square feet.

7. $x = 100$ and $y = 200$

8. Maximize area.

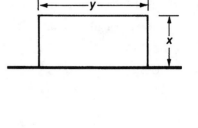

Area = length · width

$= y \cdot x$

$A(x) = (50 - 1.25x)x$ ends at $1.25 per foot

$= 50x - 1.25x^2$ front at $2.00 per foot

$A'(x) = 50 - 2.5x$ $100 for fencing

$0 = 50 - 2.5x$

$2.5x = 50$ Constraint:

$25x = 500$ $100 = 2.0(y) + 1.25(2x)$

$x = 20$ $100 = 2y + 2.5x$

$2y = 100 - 2.5x$

$y = 50 - 1.25x$

If $x = 20$,

then $y = 50 - 1.25(20)$

$= 50 - 25$

$= 25$

To maximize area the rectangular garden should be 20′ × 25′ with the long side as the front. The area would be 500 square feet.

9. $x = \sqrt{\dfrac{2Qs}{c}}$

10. Minimize cost.

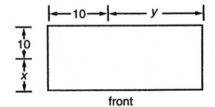

Cost = cost of front + cost of three sides

$= 6(10 + y) + 12(x + 10 + 10 + y + x + 10)$

$= 6(10 + y) + 12(2x + y + 30)$ front $6 per foot

$= 60 + 6y + 24x + 12y + 360$ sides $12 per foot

$= 24x + 18y + 420$

Constraint:

$$\text{area} = 1{,}600$$

$$(x + 10)(y + 10) = 1{,}600$$

$$xy + 10x + 10y + 100 = 1{,}600$$

$$xy + 10x + 10y = 1{,}500$$

$$xy + 10y = 1{,}500 - 10x$$

$$y(x + 10) = 1{,}500 - 10x$$

$$y = \frac{1{,}500 - 10x}{x + 10}$$

$$C(x) = 24x + 18\left(\frac{1,500-(10x)}{x+10}\right) + 420$$

$$C'(x) = 24 + 18\left[\frac{(x+10)(-10)-(1,500-10x)}{(x+10)^2}\right]$$

$$= 24 + 18\left[\frac{-10x-100-1,500+10x}{(x+10)^2}\right]$$

$$= 24 + \frac{18(-1,600)}{(x+10)^2}$$

$$0 = 24 - \frac{28,800}{(x+10)^2}$$

$$0 = 24(x+10)^2 - 28,800$$

$$0 = (x+10)^2 - 1,200$$

$$1,200 = (x+10)^2$$

$$\sqrt{1,200} = x+10$$

$$34.6 = x+10$$

$$x = 24.6$$

Test: $C'(x) = 24 - 28,800(x+10)^{-2}$

$$C''(x) = -28,800(-2(x+10))^{-3}$$

$$= \frac{57,600}{(x+10)^3}$$

$C''(24.6) = +$, concave up, minimum

If $x = 24.6$

then $y = \dfrac{1,500 - 10(24.6)}{24.6 + 10}$

$$= \frac{1,500 - 246}{34.6}$$

$$= \frac{1,254}{34.6}$$

$$= 36.2$$

Thus the side, which was $x + 10$, will be 34.6 and the front, which was $y + 10$, will be 46.2.

To minimize the cost, the rectangle should be 46.2 feet by 34.6 feet with the front the longer side.

11. a. The stone will reach its maximum height in 3.75 seconds.

   b. The maximum height reached by the stone will be 625 feet.

12. a. $h = -16t^2 + 64t + 80$

   $h' = -32t + 64$

   $0 = -32t = 64$

   $32t = 64$

   $t = 2$

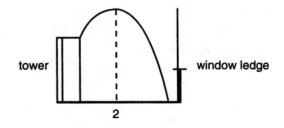

tower   window ledge

The stone will reach maximum height in 2 seconds.

b.   If $t = 2$,

then $h = -16(2)^2 + 64(2) + 80$

$= -16(4) + 128 + 80$

$= 144$

The maximum height reached by the stone will be 144 feet.

c.   The stone will hit the ground when $h = 0$.

Thus $0 = -16t^2 + 64t + 80$

$0 = t^2 - 4t - 5$

$0 = (t - 5)(t + 1)$

$t = 5$ ~~$t = -1$~~

The stone will hit the ground in 5 seconds.

13.   To maximize his rental income, the manager should charge $600 per month.

14.   Maximize profit.

Profit = total revenue − total cost

$\qquad$ = (price)(quantity) − (fixed cost + total variable costs)

$P(x) = (2,000 - 5x)(100 + x) - (125,000 + 500x)$

$= 200,000 + 2,000x - 500x - 5x^2 - 125,000 - 500x$

$= 75,000 + 1,000x - 5x^2$

$P'(x) = 1,000 - 10x$

$0 = 1,000 - 10x$

$10x = 1,000$

$x = 100$, which is not in the domain

Because the limitation on seating for the plane, that is, 150 people, must be considered.

$100 \le$ $\qquad\qquad$ people on tour $\qquad\qquad$ ≤ 150

$100 \le$ $\quad$ $100 + x$ $\qquad\qquad\qquad$ ≤ 150

$\qquad\qquad$ Subtracting through by 100 gives $0 \le x \le 50$.

Since the only critical point is not in the closed interval, we need consider only the two endpoints:

$P(0) = 75,000 + 1,000(0) - 5(0)^2 = 75,000$

$P(50) = 75,000 + 1,000(50) - 5(50)^2$

$= 75,000 + 50,000 - 12,500$

$= 112,500$

The absolute maximum occurs when $x = 50$.

Since price $= 2,000 - 5x$

If $x = 50$, then $5(50) = 250$.

The price should be lowered by $250.

15. The price should be lowered by $500.

16. Maximize output.

    Total output = number of workers $\cdot$ output per worker

    $$TO(x) = (12 + x)(600 - 30x)$$

    $$= 7,200 - 360x + 600x - 30x^2$$

    $$= 7,200 + 240x - 30x^2$$

    $$TO'(x) = 240 - 60x$$

    $$0 = 240 - 60x$$

    $$60x = 240$$

    $$x = 3$$

    If $x = 3$

    then $TO(3) = 7,200 + 240(3) - 30(9)$

    $$= 7,200 + 720 - 270$$

    $$= 7,650$$

    Test: $TO(x)$ is a parabola, opening ⌒.

    The critical point is an absolute maximum.

    a. The firm should use 15 workers on the assembly line to maximize total output.

    b. The maximum number of units that can be produced per week is 7,650.

17. a. The company should price its new stereo at $820 to lead to maximum profit of $150,100.

    b. If an $8 tax is imposed, the company should price its stereo at $824 to lead to maximum profit of $148,544.

    c. The tax increases the price by $4.

    d. Half, or 50%, of the tax is passed on to the consumer.

18. $V(x) = x(20 - 2x)(10 - 2x)$ (See the accompanying drawing)

    $$= x(200 - 60x + 4x^2)$$

    $$= 200x - 60x^2 + 4x^3$$

    $$V'(x) = 200 - 120x + 12x^2$$

    $$0 = 200 - 120x + 12x^2$$

    $$0 = 50 - 30x + 3x^2$$

    $$0 = 3x^2 - 30x + 50$$

    Use quadratic formula.

    $$x = \frac{-b \pm \sqrt{b^2 - 4ac}}{2a}$$

    $$= \frac{-(-30) \pm \sqrt{(-30)^2 - 4(3)(50)}}{2(3)}$$

$$= \frac{30 \pm \sqrt{900 - 600}}{6}$$

$$= \frac{30 \pm \sqrt{300}}{6}$$

$$\approx \frac{30 \pm 17.3}{6}$$

$$x = \frac{30 + 17.3}{6} \qquad x = \frac{30 - 17.3}{6}$$

$$= 7.9 \qquad\qquad = 2.1$$

Test: $V(x)$ is a cubic, $\curvearrowright\curvearrowleft\uparrow$.

$x^* = 2.1$ is the relative maximum.

$$V(2.1) = 2.1(20 - 2(2.1))(10 - 2(2.1))$$

$$= 2.1(20 - 4.2)(10 - 4.2)$$

$$= 2.1(15.8)(5.8)$$

$$= 192.4$$

a.   The volume of the largest box that can be formed
     is 192.4 cubic inches.

b.   To maximize the volume, the squares cut from each corner must be 2.1" × 2.1".

19. The squares should be 2.35" × 2.35".

20. Minimize cost.

Cost = cost of refrigerated area + cost of warehouse

= \$6 (area of 4 interior refrigerated walls) + \$2 (area of 4 interior warehouse walls)

$$= 6(8x + 8x + 8y + 8y) + 2(8x + 8x + 8y + 8y)$$

$$= 6(16x + 16y) + 2(16x + 16y)$$

$$= 96x + 96y + 32x + 32y$$

$$= 128x + 128y$$

Constraint:

floor space = 3,200 square feet

$$2xy = 3,200$$

$$y = \frac{1,600}{x}$$

$$C(x) = 128x + 128\frac{1,600}{x}$$

$$= 128x + 204,800x^{-1}$$

$$C'(x) = 128 - 204,800x^{-2}$$

$$0 = 128 - \frac{204,800}{x^2}$$

$0 = 128x^2 - 204{,}800$

$0 = x^2 - 1{,}600$

$x = 40$ $x = -40$ (not in domain)

If $x = 40$, then $y = \dfrac{1{,}600}{40} = 40$.

To minimize costs, the refrigerated area should be $40' \times 40'$.

UNIT 21

1. $\dfrac{dy}{dx} = -21x^2$

2. $\qquad x^2 + 5y = 10$

$\dfrac{d}{dx}(x^2) + \dfrac{d}{dx}(5y) = \dfrac{d}{dx}(10)$

$2x + 5\dfrac{dy}{dx} = 0$

$5\dfrac{dy}{dx} = -2x$

$\dfrac{dy}{dx} = \dfrac{-2x}{5}$

3. $\dfrac{dy}{dx} = 6x + 4$

4. $\qquad y^2 = x^3$

$\dfrac{d}{dx}(y^2) = \dfrac{d}{dx}(x^3)$

$2y\dfrac{dy}{dx} = 3x^2$

$\dfrac{dy}{dx} = \dfrac{3x^2}{2y}$

5. $\dfrac{dy}{dx} = \dfrac{2x - 1}{3y^2}$

6. $\qquad x^2 + y^2 = 9$

$\dfrac{d}{dx}(x^2) + \dfrac{d}{dx}(y^2) = \dfrac{d}{dx}(9)$

$2x + 2y\dfrac{dy}{dx} = 0$

$2y\dfrac{dy}{dx} = -2x$

$\dfrac{dy}{dx} = \dfrac{-2x}{2y}$

$= \dfrac{-x}{y}$

7. $\dfrac{dy}{dx} = \dfrac{-3x^2}{2y}$

8. $\qquad x^2 - y^2 + 3y = 7$

$\dfrac{d}{dx}(x^2) - \dfrac{d}{dx}(y^2) + \dfrac{d}{dx}(3y) = \dfrac{d}{dx}(7)$

$2x - 2y\dfrac{dy}{dx} + 3\dfrac{dy}{dx} = 0$

$3\dfrac{dy}{dx} - 2y\dfrac{dy}{dx} = -2x$

$\dfrac{dy}{dx}(3 - 2y) = -2x$

$\dfrac{dy}{dx} = \dfrac{-2x}{3 - 2y}$

9. $\dfrac{dy}{dx} = \dfrac{-3}{8y - 1}$

10. $\qquad x^3 y = 2$

$\dfrac{d}{dx}(x^3 y) = \dfrac{d}{dx}(2)$

$x^3\dfrac{d}{dx}(y) + y\dfrac{d}{dx}(x^3) = 0$

$x^3\dfrac{dy}{dx} + y(3x^2) = 0$

$x^3\dfrac{dy}{dx} = -3x^2 y$

$\dfrac{dy}{dx} = \dfrac{-3x^2 y}{x^3}$

$= \dfrac{-3y}{x}$

11. $\dfrac{dy}{dx} = \dfrac{-4x^3 - y^3}{3xy^2 + 1}$

12. $\qquad x^2 - y^2 - x = 1$ at $(2, 1)$

$\dfrac{d}{dx}(x^2) - \dfrac{d}{dx}(y^2) - \dfrac{d}{dx}(x) = \dfrac{d}{dx}(1)$

$$2x - 2y \frac{dy}{dx} - 1 = 0$$

$$-2y \frac{dy}{dx} = 1 - 2x$$

$$\frac{dy}{dx} = \frac{1 - 2x}{-2y}$$

at (2, 1): $$\frac{dy}{dx} = \frac{1 - 2(2)}{-2(1)}$$

$$= \frac{1 - 4}{-2}$$

$$= \frac{3}{2}$$

13. $\frac{4}{7}$

14. $xy = 15$ at (3, 5)

$$\frac{d}{dx}(xy) = \frac{d}{dx}(15)$$

$$x \frac{d}{dx}(y) + y \frac{d}{dx}(x) = 0$$

$$x \frac{dy}{dx} + y = 0$$

$$x \frac{dy}{dx} = -y$$

$$\frac{dy}{dx} = \frac{-y}{x}$$

at (3, 5): $$\frac{dy}{dx} = \frac{-5}{3}$$

15. a. $\dfrac{dy}{dx} = \dfrac{2y - 3x^2}{2y - 2x}$

b. $\dfrac{3}{4}$

c. $3x - 4y = -9$

16. $y = \sqrt{x}$

$$y^2 = x$$

$$\frac{d}{dx}(y^2) = \frac{d}{dx}(x)$$

$$2y \frac{dy}{dx} = 1$$

$$\frac{dy}{dx} = \frac{1}{2y}$$

17. $\dfrac{dy}{dx} = \dfrac{6x + 7}{2y}$

18. $q = p^2 - 40p + 2{,}150$ $0 \le p \le 20$

a. $$\frac{d}{dq}(q) = \frac{d}{dq}(p^2) - \frac{d}{dq}(40p) + \frac{d}{dq}(2{,}150)$$

$$1 = 2p \frac{dp}{dq} - 40 \frac{dp}{dq} + 0$$

$$1 = \frac{dp}{dq}(2p - 40)$$

$$\frac{dp}{dq}(2p - 40) = 1$$

$$\frac{dp}{dq} = \frac{1}{2p - 40}$$

b. $q = 100 - 400 + 2{,}150 = 1{,}850$

c. At $p = 10$, $\dfrac{dp}{dq} = \dfrac{1}{20 - 40} = \dfrac{1}{-20} = -0.05$

UNIT 22

1. There are no critical points.

2. $f(x) = e^{x^2}$

$$f'(x) = \frac{d}{dx}(x^2) \cdot e^{x^2}$$

$$= 2x e^{x^2}$$

Critical points occur where $f'(x) = 0$.

$$0 = 2x e^{x^2}$$

$2x = 0$ $e^{x^2} = 0$ has no solution.

$x^* = 0$

If $x^* = 0$, $f(0) = e^0 = 1$;

thus $(0, 1)$ is the only critical point.

Classify critical point.   First-derivative test.

At $(0, 1)$ with $f'(x) = 2xe^{x^2}$ the curve:

On the left: Let $x = -1$, $f'(-1) = (+)(-)(+) = -$ is decreasing

At point $(0, 1)$:   levels off

On the right: Let $x = 1$, $f'(1) = (+)(+)(+) = +$ is increasing

The critical point $(0, 1)$ is a relative minimum.

The minimum value for $f(x)$ is 1 and it occurs when $x = 0$.

3.   The critical point $(0, e)$ is a relative minimum.

4. $f(x) = \dfrac{e^x}{x^2}$

$$f'(x) = \frac{x^2 \dfrac{d}{dx}(e^x) - e^x \dfrac{d}{dx}(x^2)}{(x^2)^2} \qquad \text{(Quotient Rule)}$$

$$= \frac{x^2 e^x - e^x\,(2x)}{x^4}$$

$$= \frac{xe^x\,(x-2)}{x^4}$$

$$= \frac{e^x\,(x-2)}{x^3}$$

Find critical points.

$$0 = \frac{e^x\,(x-2)}{x^3}$$

$$0 = e^x(x-2)$$

$x^* = 2$ $e^x = 0$ has no solution.

If $x^* = 2$, $f(2) = \dfrac{e^2}{4}$ and $\left(2, \dfrac{e^2}{4}\right)$ is a critical point.

Classify critical point.   First-derivative test

For point $\left(2, \dfrac{e^2}{4}\right)$ with $f'(x) = \dfrac{e^x\,(x-2)}{x^3}$ the curve:

On the left: Let $x = 1$, $f'(1) = \dfrac{(+)(-)}{(+)} = -$ is decreasing

At point $\left(2, \dfrac{e^2}{4}\right)$:   levels off

On the right: Let $x = 3$, $f'(3) = \dfrac{(+)(+)}{(+)} = +$ is increasing

The critical point $\left(2, \dfrac{e^2}{4}\right)$ is a relative minimum.

5.   a. $AC = q(x) = \dfrac{e^x}{x}$

     b.   The total cost of producing two units would be $7.389.

     c.   The average cost per unit of producing three units would be $6,695.

     d.   To minimize the average cost per unit, only one unit should be produced. Refer to Example 14.

6.   a. $f(x) = x^2 e^x$

$$f'(x) = x^2 \frac{d}{dx}(e^x) + (e^x)\frac{d}{dx}(x^2) \qquad \text{(Product Rule)}$$

$$= x^2 e^x + e^x(2x)$$

$$= x^2 e^x + 2xe^x$$

$$= xe^x(x + 2)$$

Find all critical points.

$$0 = xe^x(x + 2)$$

$$x^* = 0 \qquad x + 2 = 0 \qquad e^x = 0 \text{ has no solution.}$$

$$x^* = -2$$

If $x^* = 0$, $f(0) = 0e^0 = 0$,

thus $(0, 0)$ is a critical point.

If $x^* = -2$, $f(-2) = (-2)^2 e^{-2} = 4e^{-2}$,

thus $(-2, 4e^{-2})$ or $\left(-2, \dfrac{4}{e^2}\right)$ is a critical point.

Classify the critical points.

Use the first-derivative test with $f'(x) = xe^x(x + 2)$ the curve:

$(0, 0)\begin{cases}\text{On the left:} & \text{Let } x = -1, f'(-1) = (-)(+)(+) = - & \text{is decreasing} \\ \text{At point } (0, 0): & & \text{levels off} \\ \text{On the right:} & \text{Let } x = 1, f'(1) = (+)(+)(+) = + & \text{is increasing}\end{cases}$

$(-2, 4e^{-2})\begin{cases}\text{On the left:} & \text{Let } x = -3, f'(-3) = (-)(+)(-) = + & \text{is increasing} \\ \text{At point } (-2, 4e^{-2}): & & \text{levels off} \\ \text{On the right:} & \text{Let } x = -1, f'(-1) = (-)(+)(+) = - & \text{is decreasing}\end{cases}$

The critical point $(0, 0)$ is a relative minimum.

The critical point $(-2, 4e^{-2})$ is a relative maximum.

     b.   Graph in the standard viewing window:

The absolute minimum is 0 and occurs when $x = 0$.

There is no absolute maximum.

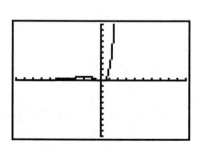

7. a. $M'(t) = 80e^{0.08t}$

    b. After 5 years the total amount will be $1,491.80.

    c. The interest that will be earned in the sixth year is estimated to be $119.34.

8. $p = 5e^{-x}$ $0 \le x \le 2$

    where p = price, in dollars,

    and x = quantity, in thousands

    Revenue is to be maximized on restricted domain.

    Revenue = price $\cdot$ quantity

    $$= 5e^{-x} \cdot (x)$$

    $$R(x) = 5xe^{-x}$$

    $$R'(x) = 5x \frac{d}{dx}(e^{-x}) + (e^{-x}) \frac{d}{dx}(5x)$$

    $$= 5x(-e^{-x}) + (e^{-x})(5)$$

    $$= 5e^{-x}(-x + 1)$$

    $$-x + 1 = 0 \qquad 5e^{-x} = 0 \text{ has no solution.}$$

    $$x = 1$$

    Looking for maximum value on restricted domain

    critical point   If $x = 1$, $R(1) = 5e^{-1} = 5(0.36788) = 1.8394$

    lower limit   If $x = 0$, $R(0) = 5(0)e^{0} = 5 \cdot 0 \cdot 1 = 0.000$

    upper limit   If $x = 2$, $R(2) = 5(2)e^{-2} = 10(0.13534) = 1.3534$

    Maximum value is 1.8394 and occurs when $x = 1$.

    When $x = 1$, $p = 5e^{-1} = 5(0.36788) = 1.8394 \approx 1.84$.

    The weekly revenue will be maximized at a price of $1.84 per box of popcorn.

9. a. Currently monthly sales are $2,000.

    b. It is estimated that, 2 months after advertising is stopped, monthly sales will be $602.38.

    c. $y'(t) = -1,200e^{-0.60t}$

    d. In the third month after the advertising has been stopped, the monthly sales will drop by about $361.428.

C10. a. When the car is purchased, $t = 0$.

    $$V(t) = 19,000e^{-0.225(0)}$$

    $$= 19,000e^{0}$$

    $$= 19,000(1)$$

    $$= 19,000$$

    The original purchase price for the car was $19,000.

b. The rate of change of V with respect to t is the derivative, $V'(t)$.

$$V(t) = 19{,}000e^{-0.225t}$$

$$V'(t) = \frac{d}{dt}(-0.225t) \cdot 19{,}000e^{-0.225t}$$

$$= -0.225(19{,}000e^{-0.225t})$$

$$= -4{,}275e^{-0.225t}$$

$$V'(0) = -4{,}275e^{-0.225(0)}$$

$$= -4{,}275e^{0}$$

$$= -4.275(1)$$

$$= -4.275$$

The trade-in value the first year is expected to decrease by about $4,275. Or, the car will depreciate in value by about $4,275 the first year.

$$V(5) = 19{,}000e^{-0.225(5)}$$

$$= 19{,}000e^{-1.125}$$

$$= 19{,}000(0.3247)$$

$$= 6{,}169.30$$

$$V'(5) = -4{,}275e^{-0.225(5)}$$

$$= -4{,}275(0.3247)$$

$$= -1{,}388.09$$

The trade-in value of the car after 5 years will be about $6,169.30 with depreciation for the sixth year estimated to be about $1,388.09.

c. $$V(10) = 19{,}000e^{-0.225(10)}$$

$$= 19{,}000e^{-2.25}$$

$$= 19{,}000(0.1054)$$

$$= 2{,}002.60$$

After 10 years the trade-in value for the car will be $2,003.

C11. a. At 8 A.M. the temperature of the coffee was 117°F.

b. At 8:30 A.M., or $t = 0.5$, the coffee had cooled to 88.6°F.

By 9 A.M., or $t = 1$, the coffee's temperature was 78.1°F.

And by noon, or $t = 4$, the coffee had cooled to 72°F.

c. The viewing window's dimensions are [0, 12] by [0, 200].

d. The temperature of the room was 72°F.

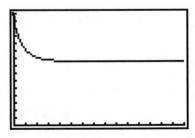

UNIT 23

1. $f'(x) = \dfrac{7}{7x-1}$

2. $f(x) = \ln 3x$

$f'(x) = \dfrac{\frac{d(3x)}{dx}}{3x}$

$= \dfrac{3}{3x}$

$= \dfrac{1}{x}$

3. $f'(x) = \dfrac{50}{x+5}$

4. $f(x) = \ln (x^2 - 5)$

$f'(x) = \dfrac{\frac{d(x^2-5)}{dx}}{(x^2-5)}$

$= \dfrac{2x}{x^2-5}$

5. $f'(x) = 2 + 2\ln x$

6. $f(x) = \ln x^2 - 5$

$f'(x) = \dfrac{\frac{d(x^2)}{dx}}{x^2} - 0$

$= \dfrac{2x}{x^2}$

$= \dfrac{2}{x}$

7. a. If \$5,000 is spent on advertising, the company's sale will be \$904,720.

b. $S'(x) = \dfrac{500}{x}$

c. If the advertising budget is increased from \$5,000 to \$6,000, sales will increase by about \$100,000.

8. $f(x) = \dfrac{2x}{\ln x}$

$f'(x) = \dfrac{(\ln x) \frac{d}{dx}(2x) - (2x) \frac{d}{dx}(\ln x)}{(\ln x)^2}$

$= \dfrac{(\ln x)(2) - 2x\left(\frac{1}{x}\right)}{(\ln x)^2}$

$= \dfrac{2\ln x - 2}{(\ln x)^2}$

$= \dfrac{2(\ln x - 1)}{(\ln x)^2}$,

if you prefer the answer factored

9. $y' = 2 + \ln x^2$

10. $f(x) = e^x \ln x$

$f'(x) = e^x \dfrac{d}{dx}(\ln x) + (\ln x)\dfrac{d}{dx}(e^x)$

$= e^x \left(\dfrac{1}{x}\right) + (\ln x)(e^x)$

$= e^x \left(\dfrac{1}{x}\right) + e^x \ln x$

$= e^x \left(\dfrac{1}{x} + \ln x\right)$, if factored

11. a. The total cost of producing three units will be \$5,016.09.

b. $TC'(x) = \dfrac{10}{x+2}$

c. The marginal cost of the fourth unit will be \$2.00.

C12. a. $f(x) = \ln (x^2 - 6x + 25)$

$f'(x) = \dfrac{\frac{d}{dx}(x^2 - 6x + 25)}{(x^2 - 6x + 25)}$

$= \dfrac{2x - 6}{x^2 - 6x + 25}$

Critical points occur where $f'(x) = 0$.

$0 = \dfrac{2x - 6}{x^2 - 6x + 25}$

$0 = 2x - 6$ (cleared of fractions)

$x^* = 3$

If $x^* = 3$ and $f(x) = \ln (x^2 - 6x + 25)$,

then $f(3) = \ln (9 - 18 + 25)$

$= \ln 16$

The function has one critical point at $(3, \ln 16)$.

b. Graph on the standard viewing window:

c. The function's one critical point (3, ln 16) is a relative
minimum. It is also an absolute minimum. In other
words, the absolute minimum value of the function is
ln 16 or 2.77, which occurs when $x = 3$.

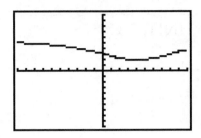

13. a. On the third day after a training period, a worker should be able to produce almost 365 items.

b. On the fourth day, the worker should be able to increase his or her output by about 300 items.

14. $f(x) = 4 \ln x - 2x$

$$f'(x) = 4 \left(\frac{1}{x} \right) - 2$$

$$= \frac{4}{x} - 2$$

Find all critical points.

$$0 = \frac{4}{x} - 2$$

$$0 = 4 - 2x \quad \text{(cleared of fractions)}$$

$$2x = 4$$

$$x^* = 2$$

If $x^* = 2$, then $f(2) = 4 \ln 2 - 4$ and $(2, 4 \ln 2 - 4)$ is the critical point.

If you wish, $4 \ln 2 - 4$ can be rewritten as a decimal approximation of $4(0.69315) - 4 \approx -1.2274$.

Classify critical point. Use first-derivative test.

For point $(2, 4 \ln 2 - 4)$ with $f'(x) = \dfrac{4}{x} - 2$ the curve:

On the left: Let $x = 1$, then $f'(1) = 4 - 2 = +$ is increasing

At point $(2, 4 \ln 2 - 4)$:   levels off

On the right: Let $x = 3$, then $f'(3) = -\dfrac{4}{3} - 2 = -$ is decreasing

The critical point $(2, 4 \ln 2 - 4)$ is a relative maximum. Or, the maximum value of $f(x)$ is $4 \ln 2 - 4$ and occurs when $x = 2$.

UNIT 24

1. $\dfrac{2x^5}{5} + \dfrac{x^3}{3} + C$

2. $\displaystyle \int \left(x + \frac{1}{x} \right) dx = \int x \, dx + \int \frac{1}{x} \, dx$

$$= \frac{x^2}{2} + \ln |x| + C$$

3. $\dfrac{x^8}{8} + C$

4. $\displaystyle\int x^{1/3}\, dx = \dfrac{x^{(1/3)+1}}{\frac{1}{3}+1} + C$

$\quad\quad\quad = \dfrac{x^{4/3}}{\frac{4}{3}} + C$

$\quad\quad\quad = \dfrac{3x^{4/3}}{4} + C$

5. $5\ln|x| + C$

6. $\displaystyle\int 2x^{-5}\,dx = 2\int x^{-5}\,dx$

$\quad\quad\quad = 2\dfrac{x^{-5+1}}{-5+1} + C$

$\quad\quad\quad = \dfrac{2x^{-4}}{-4} + C$

$\quad\quad\quad = \dfrac{2}{-4x^4} + C$

$\quad\quad\quad = \dfrac{-1}{2x^4} + C$

7. $3x + C$

8. $\displaystyle\int (15x^2 - \sqrt{2})\, dx = 15\int x^2 dx - \sqrt{2}\int dx$

$\quad\quad\quad\quad = 15\dfrac{x^3}{3} - \sqrt{2}x + C$

$\quad\quad\quad\quad = 5x^3 - \sqrt{2}x + C$

9. $3e^x + C$

10. $\displaystyle\int \ln 3\; dx = \ln 3 \int dx$

$\quad\quad\quad = (\ln 3)x + C$

Remember that ln 3 is a constant.

11. $2x\ln x - 2x + C$

12. $\displaystyle\int 5ax^4 q\; dq = 5ax^4 \int q\; dq$

$\quad\quad\quad = \dfrac{5ax^4 q^2}{2} + C$

13. $\dfrac{3\sqrt[3]{x^4}}{4} + C$

14. $\displaystyle\int\left(x^3-\frac{2}{x^2}+2\right)dx = \int x^3\,dx - 2\int x^{-2}\,dx + 2\int dx$

$$= \frac{x^4}{4} - 2\frac{x^{-2+1}}{-2+1} + 2x + C$$

$$= \frac{x^4}{4} - \frac{2x^{-1}}{-1} + 2x + C$$

$$= \frac{x^4}{4} + \frac{2}{x} + 2x + C$$

or, if you prefer, $\displaystyle = \frac{1}{4}x^4 + 2x^{-1} + 2x + C$

15. $2x - x\ln x + C$

16. $\displaystyle\int\frac{e^x-2x}{2}\,dx = \frac{1}{2}\int(e^x-2x)dx$

$$= \frac{1}{2}\left(\int e^x\,dx - 2\int x\,dx\right)$$

$$= \frac{1}{2}\left(e^x - 2\frac{x^2}{2}\right) + C$$

$$= \frac{1}{2}(e^x - x^2) + C$$

or $\displaystyle = \frac{e^x-x^2}{2} + C$

17. $50\ln|x| + C$

18. $\displaystyle\int(3x^{-2}+x^{-1})dx = 3\int x^{-2}dx + \int x^{-1}dx$

$$= \frac{3x^{-2+1}}{-2+1} + \ln|x| + C$$

$$= -3x^{-1} + \ln|x| + C$$

$$= \frac{-3}{x} + \ln|x| + C$$

19. $\dfrac{x^8}{4} + 2x^3 - x + C$

20. $\displaystyle\int\frac{\sqrt{x}}{2}\,dx = \frac{1}{2}\int\sqrt{x}\,dx$

$$= \frac{1}{2}\int x^{1/2}\,dx$$

$$= \frac{1}{2}\cdot\frac{x^{(1/2)+1}}{\frac{1}{2}+1} + C$$

$$= \frac{1}{2} \cdot \frac{x^{3/2}}{\frac{3}{2}} + C$$

$$= \frac{1}{2} \cdot \frac{2}{3} x^{3/2} + C$$

$$= \frac{x^{3/2}}{3} + C$$

$$= \frac{\sqrt{x^3}}{3} + C$$

or $\qquad = \frac{x\sqrt{x}}{3} + C$

21. $\frac{x^4}{4} - 4x^3 + x^2 + C$

22. $\displaystyle\int x^3 \left(\frac{2}{x} + 4x \right) dx = \int (2x^2 + 4x^4) \, dx$

$$= 2 \int x^2 dx + 4 \int x^4 dx$$

$$= \frac{2x^3}{3} + \frac{4x^5}{5} + C$$

23. $\frac{x^3}{3} - 2x^2 - 5x + C$

24. $\displaystyle\int \frac{1+x}{x^2} \, dx = \int \left(\frac{1}{x^2} + \frac{x}{x^2} \right) dx$

$$= \int \left(x^{-2} + \frac{1}{x} \right) dx$$

$$= \frac{x^{-2+1}}{-2+1} + \ln |x| + C$$

$$= \frac{x^{-1}}{-1} + \ln |x| + C$$

$$= \frac{-1}{x} + \ln |x| + C$$

25. $2x^4 + 4 \ln |x| + C$

26. $\quad$ If $f'(x) = 2x^3 - x + 5,$

$\quad$ then $\quad f(x) = \displaystyle\int (2x^3 - x + 5) dx$

$$= 2 \int x^3 \, dx - \int x \, dx + 5 \int dx$$

$$= \frac{2x^4}{4} - \frac{x^2}{2} + 5x + C$$

Point (3, 38) satisfies the function;

$$f(x) = \frac{1}{2}x^4 - \frac{1}{2}x^2 + 5x + C$$

$$38 = \frac{1}{2}(3)^4 - \frac{1}{2}(3)^2 + 5(3) + C$$

$$38 = 40.5 - 4.5 + 15 + C$$

$$C = -13$$

The equation of the curve is

$$f(x) = \frac{1}{2}x^4 - \frac{1}{2}x^2 + 5x - 13.$$

27. The function is $f(x) = x^2 - 5x + 4$.

28. $MC = 6x + 1$

$$TC = \int (6x + 1)dx$$

$$= \frac{6x^2}{2} + x + C$$

$$= 3x^2 + x + C \quad \text{if fixed cost is \$50}$$

$$TC(x) = 3x^2 + x + 50$$

$$TC(10) = 3(100) + 10 + 50$$

$$= 360$$

The total cost of producing 10 units is \$360.

29. $xy^3 + x^2y^2 + 7y + C$

30. $MP = (100 - 2x)$

$$TP = \int (100 - 2x)dx$$

$$= 100x - \frac{2x^2}{2} + C$$

$$= 100x - x^2 + C$$

Given that, when 10 units are produced and sold, total profit = \$150; then

Profit $= 100x - x^2 + C$

$$150 = 100(10) - (10)^2 + C$$

$$150 = 1,000 - 100 + C$$

$$C = -750$$

The total profit function is $P(x) = 100x - x^2 - 750$.

UNIT 25

1. $\dfrac{(x-2)^6}{6} + C$

2.

$$\int e^{3-5x}\, dx = \int e^u\, \dfrac{du}{-5}$$

   Let $u = 3 - 5x$. $= -\dfrac{1}{5}\int e^u\, du$

   $\dfrac{du}{dx} = -5$ $= -\dfrac{1}{5}e^u + C$

   $du = -5\, dx$ $= -\dfrac{1}{5}e^{3-5x} + C$

   $\dfrac{du}{-5} = dx$

3. $\dfrac{(x^4 + 2x)^4}{4} + C$

4.

$$\int \dfrac{x}{(5+x^2)^4}\, dx = \int \dfrac{x}{u^4}\, \dfrac{du}{2x}$$

   Let $u = 5 + x^2$. $= \dfrac{1}{2}\int \dfrac{du}{u^4}$

   $\dfrac{du}{dx} = +2x$ $= \dfrac{1}{2}\int u^{-4}\, du$

   $du = 2x\, dx$ $= \dfrac{1}{2} \cdot \dfrac{u^{-3}}{-3} + C$

   $\dfrac{du}{2x} = dx$ $= \dfrac{u^{-3}}{-6} + C$

   $= \dfrac{1}{-6u^3} + C$

   $= \dfrac{1}{-6(5+x^2)^3} + C$

5. $\dfrac{(2x^5 + 1)^3}{3} + C$

6.

$$\int \dfrac{2x^3}{x^4 + 1}\, dx = \int \dfrac{2x^3}{u}\, \dfrac{du}{4x^3}$$

   Let $u = x^4 + 1$. $= \dfrac{2}{4}\int \dfrac{du}{u}$

   $\dfrac{du}{dx} = 4x^3$ $= \dfrac{1}{2}\ \ln |u| + C$

   $du = 4x^3\, dx$ $= \dfrac{1}{2}\ \ln |x^4 + 1| + C$

$$\frac{du}{4x^3} = dx \qquad\qquad\qquad = \frac{1}{2}\ln(x^4+1)+C$$

Note: The absolute value sign is unnecessary since x^4+1 is a positive number.

7. $\dfrac{(x^3-5)^8}{24}+C$

8.
$$\int(2x-2)e^{x^2-2x+1}\,dx \;=\; \int (2x-2)e^u\,\frac{du}{(2x-2)}$$

Let $u = x^2 - 2x + 1$. $\qquad\qquad = \displaystyle\int e^u\,du$

$$\frac{du}{dx} = 2x-2 \qquad\qquad = e^u + C$$

$$du = (2x-2)\,dx \qquad\qquad = e^{x^2-2x+1} + C$$

$$\frac{du}{2x-2} = dx$$

9. $\dfrac{e^{x^2+1}}{2}+C$

10.
$$\int 3x^2\ln(x^3-2)\,dx \;=\; \int 3x^2 \ln u\,\frac{du}{3x^2}$$

Let $u = x^3 - 2$. $\qquad\qquad = \displaystyle\int \ln u\,du$

$$\frac{du}{dx} = 3x^2 \qquad\qquad = u\ln u - u + C$$

$$du = 3x^2\,dx \qquad\qquad = (x^3-2)\ln(x^3-2)-(x^3-2)+C$$

$$\frac{du}{3x^2} = dx \qquad\qquad = (x^3-2)\ln(x^3-2)-x^3+2+C$$

$$\qquad\qquad = (x^3-2)\ln(x^3-2)-x^3+C$$

11. $\dfrac{(\ln x)^3}{3}+C$

12.
$$\int \frac{dx}{x+1} \;=\; \int\frac{du}{u}$$

Let $u = x + 1$. $\qquad\qquad = \displaystyle\int\frac{1}{u}\,du$

$$\frac{du}{dx} = 1 \qquad\qquad = \ln|u| + C$$

$$du = dx \qquad\qquad = \ln|x+1| + C$$

13. $2\ln|x^3+2|+C$

14.
$$\int 36x^2 \sqrt{6x^3 + 1}\ dx = \int 36x^2\ (6x^3 + 1)^{1/2}\ dx$$

Let $u = 6x^3 + 1$.
$$= \int 36x^2 u^{1/2}\ \frac{du}{18x^2}$$

$$\frac{du}{dx} = 18x^2 \qquad\qquad = \frac{36}{18} \int u^{1/2}\ du$$

$$du = 18x^2\ dx \qquad\qquad = 2\frac{u^{3/2}}{\frac{3}{2}} + C$$

$$\frac{du}{18x^2} = dx \qquad\qquad = \frac{4}{3} u^{3/2} + C$$

$$= \frac{4}{3}(6x^3 + 1)^{3/2} + C$$

15. $\dfrac{(4x^2 - 2)^{3/2}}{12} + C$

16.
$$\int 2x\ (5x^3 - 2x)^7\ dx = \int 2x\ u^7 \frac{du}{15x^2 - 2}$$

Let $u = 5x^3 - 2x$.   This problem cannot be integrated using substitution.

$$\frac{du}{dx} = 15x^2 - 2$$

$$du = (15x^2 - 2)\ dx$$

$$\frac{du}{15x^2 - 2} = dx$$

17. $\ln|7 + e^x| + C$

18.
$$\int \frac{4x^2}{\sqrt{x^3 - 7}}\ dx = \int \frac{4x^2}{(x^3 - 7)^{1/2}}\ dx$$

Let $u = x^3 - 7$.
$$= \int \frac{4x^2}{u^{1/2}} \frac{du}{3x^2}$$

$$\frac{du}{dx} = 3x^2 \qquad\qquad = \frac{4}{3} \int u^{-1/2}\ du$$

$$du = 3x^2\ dx \qquad\qquad = \frac{4}{3} \cdot \frac{u^{1/2}}{\frac{1}{2}} + C$$

$$\frac{du}{3x^2} = dx \qquad\qquad = \frac{8}{3} u^{1/2} + C$$

$$= \frac{8}{3}(x^3 - 7)^{1/2} + C$$

$$= \frac{8}{3}\sqrt{x^3 - 7} + C$$

UNIT 26

1. $\ln \left| x + \sqrt{x^2 - 7} \right| + C$

2. $\dfrac{e^{2x}}{2} + C$

3. $(x - 1)e^x + C$

4. $\displaystyle\int \dfrac{dx}{x^2 - a^2} = \dfrac{1}{2a} \ln \left| \dfrac{x - a}{x + a} \right|$

   $\displaystyle\int \dfrac{dx}{x^2 - 1} = \dfrac{1}{2} \ln \left| \dfrac{x - 1}{x + 1} \right| + C$ Rule 11 with $a = 1$

5. $\ln \left(x + \sqrt{x^2 + 81} \right) + C$

6. $\displaystyle\int (ax + b)^n \, dx = \dfrac{(ax + b)^{n+1}}{(n + 1)a}$

   $\displaystyle\int (2x - 3)^3 \, dx = \dfrac{(2x - 3)^{3+1}}{(3 + 1)(2)} + C$ Rule 25 with $a = 2$, $b = -3$, and $n = 3$

   $\qquad\qquad = \dfrac{(2x - 3)^4}{8} + C$

7. $\displaystyle\int \dfrac{dx}{x \ln x} = \ln \left| \ln x \right| + C$ Rule 8

8. $\displaystyle\int \dfrac{dx}{\sqrt{x^2 - a^2}} = \ln \left| x + \sqrt{x^2 - a^2} \right|$

   $\displaystyle\int \dfrac{dx}{\sqrt{x^2 - 100}} = \ln \left| x + \sqrt{x^2 - 100} \right| + C$ Rule 12 with $a^2 = 100$

9. $\dfrac{1}{3} \ln \left| \dfrac{x}{10x + 3} \right| + C$

10. $\displaystyle\int x^n \ln x \, dx = x^{n+1} \left[\dfrac{\ln x}{n + 1} - \dfrac{1}{(n + 1)^2} \right]$

    $\displaystyle\int x^5 \ln x \, dx = x^6 \left(\dfrac{\ln x}{6} - \dfrac{1}{36} \right) + C$ Rule 16 with $n = 5$

11. $7 \ln \left| \dfrac{x}{2 + 3x} \right| + C$

12. $\displaystyle\int x^n e^x \, dx = x^n e^x - n \int x^{n-1} e^x \, dx$

    $\displaystyle\int x^2 e^x \, dx = x^2 e^x - 2 \int x^{2-1} e^x \, dx$ Rule 20 with $n = 2$

    $\qquad\qquad = x^2 e^x - 2 \int x e^x \, dx$

$$= x^2 e^x - 2[(x-1)e^x] + C \qquad \text{Rule 9}$$

$$= x^2 e^x - 2[xe^x - e^x] + C$$

$$= x^2 e^x - 2xe^x + 2e^x + C$$

$$= e^x(x^2 - 2x + 2) + C$$

13. $\dfrac{\sqrt{x^2 - 9}}{9x} + C$

14. $\displaystyle\int x(ax+b)^{m/2}\, dx = \dfrac{2(ax+b)^{(m+4)/2}}{a^2(m+4)} - \dfrac{2b(ax+b)^{(m+2)/2}}{a^2(m+2)}$

$\displaystyle\int x(x-6)^{1/2}\, dx = \dfrac{2(x-6)^{(1+4)/2}}{(1)(1+4)} - \dfrac{2(-6)\,(x-6)^{3/2}}{(1)\,(3)} + C \qquad \text{Rule 26 with } a = 1, b = -6, m = 1$

$\qquad\qquad = \dfrac{2(x-6)^{5/2}}{5} - \dfrac{-12(x-6)^{3/2}}{3} + C$

15. $\dfrac{1}{2}\ln|\ln x| + C$

16. $\displaystyle\int x^n (\ln x)^m\, dx = \dfrac{x^{n+1}(\ln x)^m}{n+1} - \dfrac{m}{n+1}\int x^n(\ln x)^{m-1}\, dx$

$\displaystyle\int x^3 (\ln x)^2\, dx = \dfrac{x^4(\ln x)^2}{4} - \dfrac{2}{4}\int x^3(\ln x)\, dx \qquad \text{Rule 22 with } n = 3 \text{ and } m = 2$

$\qquad\qquad = \dfrac{x^4(\ln x)^2}{4} - \dfrac{2}{4}\left[x^4\left(\dfrac{\ln x}{4} - \dfrac{1}{(4)^2}\right)\right] + C \qquad \text{Rule 16 with } n = 3$

$\qquad\qquad = \dfrac{x^4(\ln x)^2}{4} - \dfrac{1}{2}\left[\dfrac{x^4(\ln x)}{4} - \dfrac{x^4}{16}\right] + C$

$\qquad\qquad = \dfrac{x^4(\ln x)^2}{4} - \dfrac{x^4(\ln x)}{8} + \dfrac{x^4}{32} + C$

17. $\dfrac{x^2 e^{8x}}{8} - \dfrac{xe^{8x}}{32} + \dfrac{e^{8x}}{256} + C$

18. $\displaystyle\int x^n e^{ax}\, dx = \dfrac{x^n e^{ax}}{a} - \dfrac{n}{a}\int x^{n-1} e^{ax}\, dx$

$\displaystyle\int x^2 e^{3x}\, dx = \dfrac{x^2 e^{3x}}{3} - \dfrac{2}{3}\int x^{2-1} e^{3x}\, dx \qquad \text{Rule 24 with } n = 2, a = 3$

$\qquad\qquad = \dfrac{x^2 e^{3x}}{3} - \dfrac{2}{3}\int xe^{3x}\, dx$

$\qquad\qquad = \dfrac{x^2 e^{3x}}{3} - \dfrac{2}{3}\left(\dfrac{xe^{3x}}{3} - \dfrac{1}{3}\int x^{1-1} e^{3x}\, dx\right) \qquad \text{Rule 24 with } n = 1, a = 3$

$\qquad\qquad = \dfrac{x^2 e^{3x}}{3} - \dfrac{2}{3}\left(\dfrac{xe^{3x}}{3} - \dfrac{1}{3}\int x^0 e^{3x}\, dx\right)$

$\qquad\qquad = \dfrac{x^2 e^{3x}}{3} - \dfrac{2xe^{3x}}{9} + \dfrac{2}{9}\int e^{3x}\, dx$

Rule 23 with $a = 23$

$$= \frac{x^2 e^{3x}}{3} - \frac{2x e^{3x}}{9} + \frac{2}{9} \left(\frac{e^{3x}}{3} \right) + C$$

$$= \frac{x^2 e^{3x}}{3} - \frac{2x e^{3x}}{9} + \frac{2 e^{3x}}{27} + C$$

UNIT 27

1. 36

2. $\displaystyle\int_{1}^{2} x^2 \, dx = \frac{x^3}{x} \Big|_{1}^{2}$

$$= \frac{8}{3} - \frac{1}{3}$$

$$= \frac{7}{3}$$

3. $e - 1$

4. $\displaystyle\int_{2}^{10} (-x + 10) \, dx = \frac{-x^2}{2} + 10x \Big|_{2}^{10}$

$$= \left(\frac{-100}{2} + 100 \right) - \left(\frac{-4}{2} + 20 \right)$$

$$= -50 + 100 + 2 - 20$$

$$= 32$$

5. $\ln 5 - \ln 2 = 0.916$

6. $\displaystyle\int_{0}^{1} (x - x^2) \, dx = \frac{x^2}{2} - \frac{x^3}{3} \Big|_{0}^{1}$

$$= \left(\frac{1}{2} - \frac{1}{3} \right) - 0$$

$$= \frac{3 - 2}{6}$$

$$= \frac{1}{6}$$

7. $\dfrac{21}{2}$

8. $\displaystyle\int_{1}^{e} \ln x \, dx = x \ln x - x \Big|_{1}^{e}$

$$= (e \ln e - e) - (\ln 1 - 1)$$

$$= e \ln e - e - \ln 1 + 1$$

$$= e - e - 0 + 1 \qquad\qquad \text{Reminder:}$$

$$= 1 \qquad\qquad\qquad\qquad \ln e = 1$$

$$\qquad\qquad\qquad\qquad\qquad\quad \ln 1 = 0$$

9. $\dfrac{255}{4}$

10. $\displaystyle\int_{1}^{5} (x^2 - 6x + 5)\, dx = \dfrac{x^3}{3} - 3x^2 + 5x \Big|_{1}^{5}$

$$= \left(\dfrac{125}{3} - 75 + 25\right) - \left(\dfrac{1}{3} - 3 + 5\right)$$

$$= \dfrac{125}{3} - 75 + 25 - \dfrac{1}{3} + 3 - 5$$

$$= \dfrac{124}{3} - 52$$

$$= -10\dfrac{2}{3}$$

11. $e^8 - 1$

12. Area $= \displaystyle\int_{-1}^{4} (x^3 + 8)\, dx$

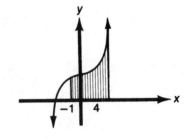

$$= \dfrac{x^4}{4} + 8x \Big|_{-1}^{4}$$

$$= \left(\dfrac{256}{4} + 32\right) - \left(\dfrac{1}{4} - 8\right)$$

$$= 64 + 32 - 0.25 + 8$$

$$= 103.75 \text{ square units}$$

13. 32 square units

14. Area $= \left| \displaystyle\int_{0}^{4} (x^2 - 4x)\, dx \right|$

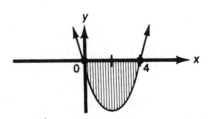

$$= \left| \dfrac{x^3}{3} - \dfrac{4x^2}{2} \Big|_{0}^{4} \right|$$

$$= \left| \dfrac{x^3}{3} - 2x^2 \Big|_{0}^{4} \right|$$

$$= \left| \left(\dfrac{64}{3} - 32\right) - 0 \right|$$

$$= \left| 21\dfrac{1}{3} - 32 \right|$$

$$= \left| -10\dfrac{2}{3} \right|$$

$$= 10\dfrac{2}{3} \text{ square units}$$

15. $7\dfrac{2}{3}$ square units

16. Graph $f(x) = x^3 - 4x$.

Cubic with $a = 1$

Curve opens up:

y-intercept is 0.

x-intercepts are 0, 2, –2

because $0 = x^3 - 4x$

$= x(x^2 - 4)$

$= x(x + 2)(x - 2).$

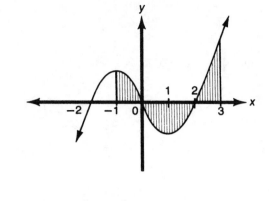

$$\text{Area} = \int_{-1}^{0} (x^3 - 4x)\, dx - \int_{0}^{2} (x^3 - 4x)\, dx + \int_{2}^{3} (x^3 - 4x)\, dx$$

$$= \left[\frac{x^4}{4} - 2x^2\bigg|_{-1}^{0}\right] - \left[\frac{x^4}{4} - 2x^2\bigg|_{0}^{2}\right] + \left[\frac{x^4}{4} - 2x^2\bigg|_{2}^{3}\right]$$

$$= \left[(0 - 0) - \left(\frac{1}{4} - 2\right)\right] - \left[\left(\frac{16}{4} - 8\right) - (0)\right] + \left[\left(\frac{81}{4} - 18\right) - \left(\frac{16}{4} - 8\right)\right]$$

$$= \left(-\frac{1}{4} + 2\right) - (4 - 8) + \left(\frac{81}{4} - 18 - 4 + 8\right)$$

$$= -\frac{1}{4} + 2 - 4 + 8 + 20\frac{1}{4} - 18 - 4 + 8$$

$$= 12 \text{ square units}$$

17. 8 square units

18. $f(x) = 4$

$g(x) = x^2$

$$\text{Area} = \int_{-2}^{2} [\, f(x) - g(x)]\, dx$$

$$= \int_{-2}^{2} (4 - x^2)\, dx$$

$$= 4x - \frac{x^3}{3}\bigg|_{-2}^{2}$$

$$= \left(8 - \frac{8}{3}\right) - \left(-8 - \frac{-8}{3}\right)$$

$$= 8 - \frac{8}{3} + 8 - \frac{8}{3}$$

$$= 8 - 2.67 + 8 - 2.67$$

$$= 10.67 \text{ square units}$$

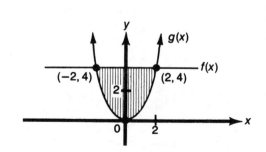

19. Use the graph from Example 14. Notice that from $x = -2$ to $x = 0$, the top function is $g(x) = x^2 - 2x + 1$.

$$\text{Area} = \int_{-2}^{0} [\, g(x) - f(x)\,]\ dx$$

$$= \int_{-2}^{0} [(x^2 - 2x + 1) - (x + 1)]\ dx$$

$$= \int_{-2}^{0} (x^2 - 2x + 1 - x - 1)\ dx$$

$$= \int_{-2}^{0} (x^2 - 3x)\ dx$$

$$= \frac{x^3}{3} - \frac{3x^2}{2} \Big|_{-2}^{0}$$

$$= 0 - \left(\frac{-8}{3} - \frac{12}{2}\right)$$

$$= 2.67 + 6$$

$$= 8.67 \text{ square units}$$

20. $M'(x) = 9x^2 + 50$

Total maintenance from 5 to 10 years after the house was built:

$$M(x) = \int_{5}^{10} M'(x)\ dx$$

$$= \int_{5}^{10} (9x^2 + 50)\ dx$$

$$= \frac{9x^3}{3} + 50x \Big|_{5}^{10}$$

$$= 3x^3 + 50x \Big|_{5}^{10}$$

$$= [3(10)^3 + 50(10)] - [3(5)^3 + 50(5)]$$

$$= 3{,}000 + 500 - 375 - 250$$

$$= \$2{,}875$$

It is estimated that the total maintenance costs for a small house for the period from 5 to 10 years after the house was built would be $2,875.

C21. 1.75 square units

C22. $\backslash Y_1$ (Y=) (2) (2nd) (e^x) (X,T,θ,n) ( )) (+) ((6) (÷) (X,T,θ,n) ( )) (−) (7) (X,T,θ,n) (^) 2) (ENTER)

Graph on the standard viewing window:

The desired area is nonnegative on the interval; thus it can be determined by the definite integral. To evaluate the definite integral, press (2nd) (CALC) (7).

At the <u>Lower Limit?</u> prompt, press (0.5) (ENTER).

At the <u>Upper Limit?</u> prompt, press (1) (ENTER).

The answer appears at the bottom:

$$\int f(x)\ dx = 4.2563375$$

The desired area is 4.256 square units.

UNIT 28

1. a. $6x$

   b. $20y^4$

   c. 0

   d. 6

2. $f(x, y) = (3x - 7y)^5$

   a. $f_x = 5(3x - 7y)^4(3)$

   $$= 15(3x - 7y)^4$$

   b. $f_y = 5(3x - 7y)^4(-7)$

   $$= -35(3x - 7y)^4$$

3. a. $24x^5 - 24x^2 - 7 + 6y + 3x^2y^5$

   b. $6x + 8 + 5x^3y^4$

4. $f(w, x, y) = 2xy + 3x^2yw + w^2 + 10$

   a. $f_w = \dfrac{\partial(2xy)}{\partial w} + 3x^2y\,\dfrac{\partial(w)}{\partial w} + \dfrac{\partial(w^2)}{\partial w} + \dfrac{\partial(10)}{\partial w}$

   $$= 0 + 3x^2y(1) + 2w + 0$$

   $$= 3x^2y + 2w$$

b. $f_x = 2y \dfrac{\partial(x)}{\partial x} + 3yw \dfrac{\partial(x^2)}{\partial x} + \dfrac{\partial(w^2)}{\partial x} + \dfrac{\partial(10)}{\partial x}$

$= 2y(1) + 3yw(2x) + 0 + 0$

$= 2y + 6wxy$

c. $f_y = 2x \dfrac{\partial(y)}{\partial y} + 3x^2w \dfrac{\partial(y)}{\partial y} + \dfrac{\partial(w^2)}{\partial y} + \dfrac{\partial(10)}{\partial y}$

$= 2x(1) + 3x^2w(1) + 0 + 0$

$= 2x + 3wx^2$

5.   a. $3x^2 - 8y + 6x^2y^2$

b. $2y - 8x + 4x^3y$

c. $6x + 12xy^2$

d. $2 + 12x^2y$

e. $-8 + 12x^2y$

f. $-8 + 12x^2y$

6. $f(x, y) = 3x^4y^3 + 2xy$

$f_x = 3y^3 \dfrac{\partial(x^4)}{\partial x} + 2y \dfrac{\partial(x)}{\partial x}$

$= 3y^3(4x^3) + 2y(1)$

$= 12x^3y^3 + 2y$

$f_y = 3x^4 \dfrac{\partial(y^3)}{\partial y} + 2x \dfrac{\partial(y)}{\partial y}$

$= 3x^4(3y^2) + 2x(1)$

$= 9x^4y^2 + 2x$

$f_{xx} = 12y^3 \dfrac{\partial(x^3)}{\partial x} + \dfrac{\partial(2y)}{\partial x}$

$= 12y^3(3x^2) + 0$

$= 36x^2y^3$

$f_{yy} = 9y^4 \dfrac{\partial(y^2)}{\partial y} + \dfrac{\partial(2x)}{\partial y}$

$= 9x^4(2y) + 0$

$= 18x^4y$

$f_{xy} = 12y^3 \dfrac{\partial(x^3)}{\partial x} + \dfrac{\partial(2y)}{\partial x}$

$= 12x^3(3y^2) + 2(1)$

$= 36x^3y^2 + 2$

$f_{yx} = 9y^2 \dfrac{\partial(x^4)}{\partial x} + 2 \dfrac{\partial(x)}{\partial x}$

$= 9y^2(4x^3) + 2(1)$

$= 36x^3y^3 + 2$

7.   The critical point (4, 3, –58) is a relative minimum.

8. $f(x, y) = 3 - x^2 - y^2$

$f_x = -2x$

$f_y = -2y$

Critical points occur where $f_x = 0$ and $f_y = 0$.

$-2x = 0 \qquad -2y = 0$

$x = 0 \qquad\quad y = 0$

Critical point is located at (0, 0, 3).

To classify, use second-derivative test with

$f_{xx} = -2$

$f_{yy} = -2$

$f_{xy} = 0$

Let $D = f_{xx} \cdot f_{yy} - (f_{xy})^2$

$= (-2)(-2) - 0$

$= 4$

Since D is positive, the critical point is either a relative maximum or a relative minimum. Look at f_{xx}. Since f_{xx} is negative, the critical point (0, 0, 3) is a relative maximum.

9.   a.   If Professor Poplin spends $10 a week on advertising and farms 5 acres, his annual profit will be $10,000.

b. $30x^2 - 10y$

c. $40y - 10x$

d. $2,950

e. $100

f. $60x$

g.   600

h.   Profit would increase by about $2,950.

i.   Profit would increase by about $100.

j.   Profit would increase by $3,370.

10. $f(x, y) = 4x^2 + 100y^2 + 2x + 5y + 5{,}000$

a. $f_y(x, y) = 200y + 5$

$f_y(100, 10) = 200(10) + 5$

$= 2{,}000 + 5$

$= 2{,}005$

Currently there are 10 rooms and a weekly advertising budget of $100. If one additional room is constructed, profits can be expected to increase by $2,005.

b. $f_x(x, y) = 8x + 2$

$f_x(100, 10) = 8(100) + 2$

$= 800 + 2$

$= 802$

If an additional dollar per week is spent on advertising, annual profits can be expected to increase by $802.

Index

MAXIMIZE YOUR MATH SKILLS!

BARRON'S EASY WAY SERIES

Specially structured to maximize learning with a minimum of time and effort, these books promote fast skill building through lively cartoons and other fun features.

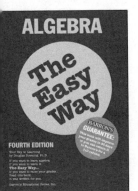

ALGEBRA THE EASY WAY
Fourth Edition
Douglas Downing, Ph.D.
In this one-of-a-kind algebra text, all the fundamentals are covered in a delightfully illustrated adventure story. Equations, exponents, polynomials, and more are explained. 320 pp. (0-7641-1972-9) $14.95, Can. $21.95

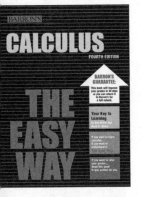

CALCULUS THE EASY WAY
Fourth Edition
Douglas Downing, Ph.D.
Here, a journey through a fantasy land leads to calculus mastery. All principles are taught in an easy-to-follow adventure tale. Included are numerous exercises, diagrams, and cartoons which aid comprehension. 228 pp. (0-7641-2920-1) $14.99, Can. $21.99

GEOMETRY THE EASY WAY
Third Edition
Lawrence Leff
While other geometry books simply summarize basic principles, this book focuses on the "why" of geometry: why you should approach a problem a certain way, and why the method works. Each chapter concludes with review exercises. 288 pp. (0-7641-0110-2) $14.95, Can. $21.00

TRIGONOMETRY THE EASY WAY
Third Edition
Douglas Downing, Ph.D.
In this adventure story, the inhabitants of a faraway kingdom use trigonometry to solve their problems. Covered is all the material studied in high school or first-year college classes. Practice exercises, explained answers, and illustrations enhance understanding. 288 pp. (0-7641-1360-7) $14.95, Can. $21.00

BARRON'S MATHEMATICS STUDY DICTIONARY
Frank Tapson, with consulting author, Robert A. Atkins
A valuable homework helper and classroom supplement for middle school and high school students. Presented here are more than 1,000 math definitions, along with worked-out examples. Illustrative charts and diagrams cover everything from geometry to statistics to charting vectors. 128 pp. (0-7641-0303-2) $12.99, Not Available in Canada

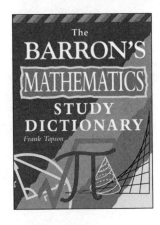

ESSENTIAL ATLAS OF MATHEMATICS
Parramón Studios
This easy-to-comprehend volume explains principles of mathematics, the universal language. It makes imaginative use of charts and diagrams to explain mathematical reasoning. There are color photos and illustrations on every page. 96 pp. (0-7641-2712-8) $10.95, Can. $15.95

Barron's Educational Series, Inc.   (#44) R 10/05
250 Wireless Boulevard
Hauppauge, New York 11788
In Canada: Georgetown Book Warehouse
34 Armstrong Avenue • Georgetown, Ontario L7G 4R9
Visit our web site at: www.barronseduc.com